THE SHAPE OF SPEED

BRUCE FARR
RUSSELL BOWLER

THE
SHAPE
OF
SPEED

John Bevan-Smith

REED

Published by Reed Books, a division of Reed Publishing (NZ) Ltd, 39 Rawene Rd, Birkenhead, Auckland. Associated companies, branches and representatives throughout the world.

www.reed.co.nz

ISBN 0 7900 0691 X

First published 1999

Printed in Singapore

Cover designed by Sunny H. Yang
Text design by Sharon Whitaker

for Debbie

and

for Graham and Jenny

In memory of

Alan Farr

and

Keith Chapman

The trawling net fills, then the biographer hauls it in,
sorts, throws back, stores, fillets and sells. Yet consider what
he doesn't catch: there is always far more of that.

Julian Barnes, *Flaubert's Parrot*

God if you quote me I'm dead. I'll be caught for libel and killed ...

Michael Ondaatje, *Running in the Family*

Contents

PART III

PART I

Chapter 1
Diaspora

Anchor me
In the middle of your deep blue sea,
Anchor me, anchor me, anchor me.
                    Don McGlashan, New Zealand singer-songwriter with
                                                The Mutton Birds[1]

For a population of three-plus million, New Zealand's per
capita rate of success in sailing is ungodly high. It's a total
blowout compared with any other country.
                                    Paul Cayard, American yachtsman[2]

Wind

Stretched and squeezed between 33 and 53 degrees south and 166 and 180 degrees east, Aotearoa New Zealand lies in what is arguably the windiest corner of the Pacific. Shifting masses of air swirl all about its shores, confusing its weather vanes and playing delinquent games with those who forecast its weather.

Its most populous city, Auckland, built on an isthmus of extinct volcanoes, sits in the path of powerful westerlies which sweep across the Tasman Sea. With the power of bursting boilers these roar down its western ranges to rattle the slender structure of its busy harbour bridge.

Rumour has it that number eight wire is used to keep hats and coats in contact with their owners and rod rigging houses with their plots when vicious winds fan out across Cook Strait to wage war on the country's capital, Wellington.

And in its major South Island cities, Christchurch and Dunedin, winds from neighbouring Antarctica have been known to blow so hard and cold that they have done unspeakable things to the statues of the city fathers.

On the seaboards of the USA, regattas can be cancelled when winds reach 20 knots. In Aotearoa New Zealand, with winds rising 30, Kiwi yachties are known to laugh maniacally and immediately put to sea.

Sea

The nation's 268,867 square kilometres of land is surrounded by a vast amount of water. A huge barrier to cross, it forms a massive moat of protection. No point of land is far from the sea. Sprinkled with numerous beaches, estuaries and offshore islands, its long indented coastline, of nearly six thousand kilometres, provides a playground of great beauty for its sea-loving inhabitants.

A geological child, the country clung to the bottom of the world a patient five million years before its first human visitors arrived. These were the Polynesian sailors who charged down foaming paths of ocean to beach their great double hulls on its shrouded shores. The impetus for their migrations is unclear. But what seems likely is that these remarkable Stone Age navigators, using the stars as guides, were surfing the great Pacific while William was crossing the Channel to take a pot shot at Harold.

Much later, the Spanish, Dutch and French, in their lumbering square-riggers, would cross the same expanse of ocean to the very same shores. Most soon left. Those who didn't were eaten, buried or simply vanished along with their dreams. It was only after an Englishman, Captain James Cook, arrived in 1769 that its coastline was circumnavigated and its inverted fish-hook shape revealed to the world.

Land

Initially part of the land mass comprising South America, Africa, India, Australia and Antarctica, Aotearoa fell off the edge of Gondwanaland 80 million years ago. It proceeded to rise and fall as two gigantic plates did battle on the ocean floor. About 55 million years later, its present shape began surging towards the surface.

A country of mountains, hills, plains and rivers, 75 percent of its primeval land was covered in dense evergreen forests. Among the ancient animals in residence prior to the initial land mass separation were several species of flightless birds. With only one scarce predator these land birds flourished. But when the Polynesians began patrolling the plains and burning off the lowlands they not only wiped out the giant moa, they reduced the kiwi to near extinction. That, however, was little more than a Sunday School picnic compared with the violent impact the Europeans were to have on the land. They stripped it of its native forests, turning all but the most problematic areas into a giant farm.[3]

People

Where was the renowned city of Wellington? Did those mud hovels scattered along the beach, or those wooden huts which appeared every here and there, did these represent the City of Wellington? Yes, this was the City of Wellington! Where then is the fine fertile land which shall produce such outstanding crops? Surely not these steep and wooded mountains? Oh no! These mountains are part of the City of Wellington. Where then, was the cry of our passengers, are our hundreds of acres? At Wanquanin, was the answer. And where in the World is Wanquanin? Rather more than a hundred miles off!

Dr Featherston aboard the *Olympus*, Wellington Harbour, 1841.[4]

The good doctor and his fellow passengers weren't the only ones sucked in by the erudite conman Edward Gibbon Wakefield. Lauded as the founder of an innovative system of colonisation, Gibb was also gifted in the art of seduction. Or perhaps he just got lucky when the first under-age female he abducted fell in love with him, gave him two children and an income when she died. But when, as a 30-year-old widower, he chanced his arm again and whisked fifteen-year-old Ellen Turner off to gay Paris, her wealthy silk-manufacturing father would have none of it. Convicted of "'diverse subtle stratagems", forgery, flagrant lies and intimidation "for sake of lucre and gain"'[5], Gibbon and his brother-accomplice William both did time.

Three years in Margate Prison (1827–30) might have been enough to break a weaker-spirited person. Not Edward Gibbon. His two forays into the field of abduction had taught him an important lesson: present the powers-that-be with a fait accompli and you've got a fifty-fifty chance of pulling it off. His parliamentary ambitions may have been thwarted but prison gave him the opportunity to hone his writing skills. This he did in his colourful *Letters from Sydney* published under a pseudonym before his release. It would have climbed to the top of a bestsellers list had there been one. Nevertheless, it soon brought him to the attention of a readership seduced by the sparkle of his prose. His ideas not only appealed to the poor, the homeless and the growing numbers of unemployed. Politicians read votes in their subtexts. And businessmen, like London's leading shipowner Joseph Somes, saw pounds sterling flying from their yardarms. Gibb and Will were under way. And so was the 'systematic' colonisation of New Zealand.

The plan was as simple as it was ingenious: sell the land before you buy it and use the gross receipts to finance the entire operation. So successful was the speculation-driven lottery that it took only three months to sell, at one pound per acre, the first 99,999 acres put up for sale. And that was before Colonel Will had arrived at Wellington with just five thousand and forty pounds two shillings and twopence (plus assorted muskets, blankets and trinkets) to purchase the said acreage. True, part of the massive profits were to be used for surveying the sites and paying the outward passage of the poor 'colonists'. But margins — no doubt improved on by the resounding salute fired by Somes' old *Tory* before Will disembarked to do business with Maori — would never come better than that.

It was all there at the beginning: greed, boats, guns and money. So it wasn't surprising that the Wakefields' race to beat the British Government to settle the central districts of New Zealand sent them on a collision course with successive governors-general, caused bloody wars and brought the country to the point of bankruptcy. Short-term profits may have been high. But they carried unimaginable long-term costs. Although sanitised by the involvement of British politicians and big business, the complex actions of the New Zealand Company remained the biggest case of fraud and theft in the country's history. Until, that is, corporate New Zealand pumped enough cash through its offshore laundries in the 1980s and '90s to have sent the Wakefields' schemes into spin-dry mode.

It was a bizarre Victorian melodrama. On the one hand it starred the gentlemen thugs. On the other the Colonial Office: the beleaguered boys of Downing Street badgered by businessmen and missionaries and their own James Busby to claim the fragile frontier for the greater good of all. Yet in spite of this, the Wakefields and a reluctant British Government did manage to send a fascinating assortment of settlers on the four- to five-month journey to the outskirts of civilisation. Mainly of humble British, Scottish and Irish stock, these early colonists possessed a wide range of skills.[6] They would need every one in order to tame the land, learn to live with Maori, and to turn their fledgling districts into functioning parts of a cohesive whole.

Within their clusters of communities, the colonists quickly turned themselves into resilient, self-reliant folk who could turn their hands to anything. They could fell a forest, fix a pump. They could build a boat, a house. Erect a town. Carve a road through a forest. Heave a railway track across a gorge. They could turn grasslands into sheep runs to catch the rising price of wool. Or raise cattle in burnt-off bush to supply their growing towns with milk. Others grew grain and wheat. Still others, fruit and vegetables. Some of the more reckless, by way of smuggling spirits and tobacco, not only deprived the administration of revenue but unwittingly created a future national pastime. School and church committees and the paying of calls formed the palette of social life. Various sports, including racing horses on the beach and the very English game of cricket, provided daubs of brighter colour. The colony's first concerts were performed in its first decade and, by its end, the first Grand Ball had been held in Government House. Forever the sage and seer, Edward Gibbon Wakefield propounded that 'a shortage of young women was the greatest evil of all'.[7] It was a nice piece of projection from Monsieur le Ravisseur. Fortunately, more grounded folk, like Mary Marshall of Wellington, saw it rather differently: 'Fancy ... a woman ... [with] a wooden leg, a son of twenty-two, and six children has just been married again. No one need despair after that, I think!'[8]

What shaped the country was its landscape. What gave it definition was its seascape. What gave it spirit was its stock. The motivation was its isolation. To survive that far from the rest of the world you had to adapt. To adapt you had to rely on each other. Only then could you begin to mint the currency of character and create opportunity. And much sooner than expected they began to see a change. They saw that they had become a people rugged like their environment, determined and resourceful. Yet more than that. The sea, land, wind and the back-breaking process of colonisation had turned them into a nation of lateral thinkers.

Some showed up early. Like Kate Sheppard, the pioneer-feminist whose leadership and courage were instrumental in her country extending the vote to women before any other in the world.[9] And Ernest Rutherford, the fourth son of a New Zealand Company wheelwright, who divided the indivisible and ushered in the atomic age.[10] And Richard Pearse, the farmer's reclusive son, who flew a heavier-than-air machine before the American Wrights.[11] But that was just the beginning.

***

Matthew Thomas Clayton first saw Wellington Harbour in 1846, just six years after the first New Zealand Company settlers had landed with their pianos, pots and pans on the beach at Pito-One. He was a fifteen-year-old apprentice aboard the *London*. He would become an officer on the frigate-built *Kent* in 1856 and within a few years assume its command. During the Victorian gold rush he carried both passengers and gold bullion between Melbourne and London. In 1862 he had the distinction of beating home to London four China clippers the *Kent* had encountered just north of the equator. Within hours of arrival that success had made him the talk of the London Exchange and the toast of the Blackwall Line. It also secured his reputation as a brilliant seaman.

After settling in Auckland in 1864, Captain Clayton exchanged his sextant for an easel, raised eight children with his wife Ellen and became one of the country's finest seascape painters. His 1905 painting of Cook's *Endeavour* beating into Poverty Bay in 1769 decorated the back of the New Zealand one pound note. And his canvas depicting Governor Hobson's landing at Waitangi in 1840 was for many years displayed in the Treaty House. His paintings were hung in galleries all around the world. And his love of the sea, his skill on the helm and his eye for line and design would one day find their way into a sumptuous gene pool inherited by two young lads called Farr growing up in Leigh, north of Auckland.

Edmund William Duke Clayton, Captain Clayton's grandson, was a yachtsman, tenor and part-time physical education instructor with the Auckland YMCA. He spent most summer weekends crewing on the *Ladye Wilma*, a Logan-designed 43-foot cutter. He was training to be an architect when he joined the New Zealand Medical Corps as an orderly. Like others drafted into World War I, he would complete his final examinations overseas. In Duke's case he would do so in stately Edinburgh while there on furlough. But he nearly didn't make it.

Sent to Cairo for further training, he left for Gallipoli on 29 September 1915, aboard the Red Cross steamer *Marquette*. Salonika was off the port quarter and Duke talking with chums in a hold when a torpedo burst through the plating. There wasn't time to call out let alone scream. Duke's friends bought it just feet from where he stood, blown out of his life mid-sentence. Suddenly all around him were the sounds of a dying ship. The mingled cries of panic and command. Steam pipes hissing. The sound of water rushing through the fatal wound. Duke, his head covered in blood, managed to scramble to the listing upper deck and throw himself over the side. He was the last to be plucked from the sea alive, eight hours after the sinking. He would later say it was the steel upright he'd been leaning against that had saved his life.

Captain Clayton knew his grandson had spent from October 1915 to January 1916 in Salonika recovering from a shrapnel wound to the head. He knew that on 12 June he had sailed from Egypt bound for England, then France and the cauldron of the Front. But for years no one really knew the extent of Duke's injuries or that he would carry the scars long after the war was over. His daughter Ileene recalls:

> Like many returned servicemen my father would talk little about the war. But when I was twelve I noticed he was having a lot of trouble with his hands. It was only then he told me the reason for the scars all over his face and body. It was while he was doing his rounds in a makeshift wooden hospital that the spirit stove went up. He knew at once that both the structure and the patients were in danger. And that's how it happened. He raced to the blazing stove, picked it up, and carried it outside to safety.
>
> I think he was shell-shocked when he returned home. So many of the young men were. He taught architecture at Auckland Technical Institute. But still not fully recovered, he decided to go farming and moved north to Mata, just south of Whangarei. It took him a few years to come right. Which was when he returned to Auckland and established a successful business building high-quality homes.[12]

Duke Clayton was one of the 'lucky' casualties of Winston Churchill's plan to knock away a Central Powers' prop by attacking Constantinople. Duke, of course, had no idea that Churchill's alternative to trench warfare in Flanders and France was about to blood his country into nationhood. He might have had an inkling, though, had he heard the Commander-in-Chief selling the plan to his troops: 'You will hardly fade away until the sun fades out of the sky and the earth sinks into the universal blackness. For already you form part of that great tradition of the Dardenelles which began with Hector and Achilles. In another few thousand years the two stories will have blended into one ...'[13]

Not surprisingly, Sir Ian Hamilton's preparation was as flawed as his speech. Not only would his love of classical antiquity and lack of detailed maps of the area prove a lethal cocktail. The poet-general's failure to have the peninsula reconnoitred had missed the Turkish build-up — a threefold reinforcement. As Aubrey Herbert wrote before leaving Lemnos for Gallipoli: 'The whole thing is so incredibly unreal. We are like ghosts called upon to make a pageant on the sea. Every way one turns, from the African savage on shore to the Etonian fop, to the wooden Horse of Troy or to the wily Greek of this place, it is all a dream.'[14] That dream would become a nightmare as the waters of Anzac Cove turned red with blood. 'Dig-dig-dig until you are safe!' Hamilton commanded the ANZACs as Turkish shells exploded in their faces. They were Diggers after that.[15]

**

Fourteen months before, as the Austro-Hungarian guns opened fire across the Danube, Odile Jacquart had been in the home of a neighbour learning dressmaking. Possessed of a handsome beauty, the eighteen-year-old lived quietly with her parents, four brothers and a sister in a French border town affectionately known as 'Petit Paris'. All their married lives Alfred and Sidoni Jacquart had woven fabric by hand in their Wattrelos home for a factory in nearby Roubaix. Odile, their third-youngest child, had for the last five years walked across the road to a convent school in Belgium. Like the other inhabitants of Wattrelos she had no idea that in less than three weeks invading German troops would be marching down that very same road. Nor that just a few miles away, on the edge of the marshy plain of Flanders, some of the bitterest fighting in the history of mankind was about to take place. For it was here, in and around the medieval city of Ypres, with its towering pinnacles and elaborate monuments, that both sides were about to learn that neither possessed the technology to defeat the other. Nor the imagination.

We saw the first Germans come into France. They all went past our way. It was a complete occupation of everything. They had the most marvellous equipment, beautiful horses. You could see why they were winning. They would come into our house and go right through everything. They would be looking for food, soap, anything to take with them to the Front. They would lie on the floor to sleep at night then get up in the morning and go off towards Ypres. Every day we heard the guns. And at night they would light up the sky.[16]

For the duration of the war, history placed Wattrelos on the German side of the Western Front. Reports of the fighting were sketchy, especially for civilians in an occupied town. They could hear the guns but had little idea of the carnage: 380,000 dead in the first three weeks of fighting. Nor that during the third and final battle for Ypres nearly half a million men would die from gas and bursting shells, from rifle and machine-gun fire and from ignominious drowning in the bog beside the duckboards. Odile did not see the slow procession underground, the eviscerated bodies, or the countless limbs in odd detachment. What she saw was a bedraggled, war-weary bunch, mud-covered, with heads bandaged, arms in slings, no longer singing their marching songs as they retreated back through Wattrelos four years later.

We were fortunate. We had a big garden out the back and grew a lot of food so we were never hungry. But the curfew was a problem. We were shut in from two o'clock in the afternoon and for four years had no social life. I think that is why so many young people went away as soon as the war was over.

I met Harry Farr at the end of the war, 1918. He was an artillery observer. He had to calculate how far away the enemy was and to pass that information to the gunners. He was very good at mathematics — he did it for a hobby — which is why he had that job. He was among the last to leave, part of a group billeted at our home while they waited to be demobilised. They were all very nice, these boys, and we would play games together when they were off duty. I remember the day he left. He looked very smart in his uniform. But with his full kit on he couldn't get out our door! So he had to take everything off, put his kit back on when he got outside, then say goodbye to my family all over again. Perhaps he was nervous!

After everyone had gone I didn't think too much about things. You just got on with life. So I was shocked when I received a letter from Scotland saying he was coming back to marry me. But I knew he was serious — there was an engagement ring in the envelope! I think I was in a bit of a muddle, really. But when I arrived in Paisley the Farr family made me very welcome. And my mother came across and stayed for three months when Jim was born. Then we went back to Wattrelos to visit when Jim was six weeks old. My younger sister, Marie, she came and stayed with us for a year. I was never lonely, although I sometimes got confused — speaking in English but thinking in French, sometimes the other way round. So it was always very helpful that my husband was fluent in French.

We didn't stay long in Scotland. Harry went to Halifax to find a job. He was a master printer and always found work. Then we went to London for more work, then Bristol where we lived for three years. A close friend of Harry's there had a relation in Christchurch, New Zealand, and he found him a job at Whitcombe and Tombs. So we came here in 1927 when Jim was seven years old. Christchurch is a pretty place but I found it cold. So we moved to Auckland the following year. [17]

History chose the same place in Auckland for the Farr and Clayton families to meet as Bishop Selwyn had selected for his Melanesian Mission. As a 20-foot mullety glided into Mission Bay on a midsummer's day in 1946, a 20-year-old nurse was coasting down the hill towards the same fashionable bathing spot. Ileene Clayton and her girlfriend had ridden their bikes from the nearby suburb of Grafton. Jim Farr, having rowed his young brother Bob ashore, had just taken a walk to the shops. It was Ileene's smile that stopped him in his tracks, there on the beach, as he made his way back to the *Vagrant's* tender. But that wasn't the only distraction for the normally shy young man. The piping hot sand was hurting his feet like billyoh.

'Like to go for a sail?'

Not the most original in pick-up lines. But for a 20-year-old from a distinguished line of sailors it was one she could hardly refuse. Hotfooting it to the water's edge, the yachtie was a happy man, the pretty nurse all smiles. It was an invitation that would change the lives of Ileene May Clayton and James Alfred Jacquart Farr — and a lot more besides.

Chapter 2
Revelation

Russ and I locked horns a few times in our younger days both on and off the water. And yes, he sold me one of those home-made aluminium masts of his — sent it down to Wellington — and it didn't work for crap!

Geoff Stagg, President, Farr International Inc, USA[1]

'That's a Geoffreyism! The masts worked like a dream. Admittedly they were an acquired taste. But they were keenly sought after. For a while Don — my brother-in-law and forward hand — and I had quite a business going. We would make these things by shoving one piece of aluminium tubing into another. Then we'd press the sleeves together by backing them into a pohutukawa tree with my red MG! In fact we were using one of our home-made masts when we won our second Cherub National title in 1968.'

'Was that the championship sailed on Wellington Harbour the weekend the *Wahine* sank?'[2]

'Yes.'

'When Bob and Peter Walker came second to you and Don?'

'That's right. It was a scary weekend, but memorable. With four wins in four races, we only had to start in the last race to win the title. Which was fortunate because we'd already sold *Mecca* to a Napier buyer. I say fortunate because by then the waves had become so powerful they were tossing boats in the air like matchsticks. And if the waves didn't get you, the side force of the wind would knock you flat. There was no way we could have stayed upright for the whole race. We were first around the first mark, then we retired. Even the sailing instructions were frightening. They included what to do if you came across a dead body on the course. In one race, as we made our way towards Muritai, we could see the remains of a 50-foot launch being pounded against the cliffs. It was a very visible reminder of just how serious things had become. Winds had hit 160 knots that weekend.'

'Didn't Barbara Kendon fly down to join Peter and you?'

'Yes. Apparently the guy next to her in the plane asked what she was doing going to Wellington on such a lousy weekend. She told him to watch a regatta. "You're nuts", he said. "The only reason I'm going is to identify a corpse!" '

***

A Tauranga Cup winner, twice national Cherub champion, Interdominion twelve-footer champion, Australian Cherub champion, inaugural world Cherub champion and national eighteen-footer champion, Russell Bowler was one of the most gifted of the many talented centreboard helmsmen who flew their craft around the country's harbours in the 1960s and '70s.[3] He was a pioneer in the use of exotic materials. He produced the first round-bilged foam sandwich hull in New Zealand. He introduced the pocket-luff mains to the twelve-footer class and, with Bruce Farr, twin trapezing. He adapted the space-frame and brought it to small boat construction. His ideas were plagiarised on both sides of the Tasman. He was the most original small-boat designer and builder of his day, his designs the most radical.

***

'Go back to bed, Johnnie, go back to bed. It's all right. Mum's fine. Everything's going to be okay. It's a boy.'[4]

If ever a child was adored it was Geoffrey Russell Bowler. Home-delivered at 67 Epsom Avenue, Auckland, on 9 February 1946, he had, nevertheless, arrived somewhat unexpectedly. Born nine years after Jeanette and five after Katharine, he was the boy twice hoped for by Ces and Jean. It was not coincidental that Jeanette's nickname was Johnnie.

But it was Jeanette to whom Russell was closest in those first years of his life. She was the one who would mother him, read *Little Toot* to him at bedtime. It was she who would take him on trips to the zoo, for rides on trams, to the movies, on visits to Aunties Con and Olive, their father's older sisters.

Being the middle child and of a more competitive disposition, Katharine saw things rather differently. Especially when the new arrival grew into a shadow that followed her when she went to visit her friends. Being displaced at home was one thing, having her girlfriends stolen quite another. If K sensed the shadow closing in, she wasn't beyond reminding her blue-eyed blond-haired brother of the only blemish she could think of.

'With those big ears of yours, if you stand in the wind you'll probably fly away.'

Ironically it would be K, with her Zeddy and Frostbite yachts, who would spark her brother's interest in sailing and provide him with the best darn crew he ever had.

Cecil, Russell's father, was born in Hastings in 1900, the youngest of Harry and Jessie Bowler's three children. On his father's side he was the grandson of a Lancashire spindle-maker who had migrated to New Zealand in the early days of the colony and had become a South Island farmer. On the Longstaff side he was the nephew of Emma Beaufoy, the Poverty Bay personality who had named the district where she held her two thousand acres with the Maori equivalent of her maiden name: Rakauroa.[5]

At the age of 20 and with little formal education, Cecil arrived in Auckland with his family. He put himself through night school while working as a clerk and gained a Bachelor of Laws degree. It was quite a feat for a country boy. He also gained a soulmate for life when he married Jean Victoria Dick in 1933. Cecil spent his first thirty years as a lawyer representing the Auckland Transport Board. And on retiring at age sixty he promptly bought into a conveyancing practice and worked another twenty years. At Spencer & Spencer he represented a cross-section of Auckland businesses as well as a clutch of personalities, including writer Barry Crump. His gentlemanly charm and dry sense of humour earned him a following from a number of female clients who, increasingly as they aged, would do nothing legal unless it was through their 'Mr Bowler'. Ces would be well rewarded for his skill and faithfulness. Two of his wealthy clients, bereft of heirs, left him their entire estates. Ces adored Jean and the kids, wrote a mean lyric and enjoyed nothing better than a singalong. Large chunks of his spare time were given to the local presbytery of Greyfriars and the Kohimarama Yacht Club. But it was the law that turned his clock. It was both his passion and his pastime.

Jean Dick came from a family of Scottish ancestry. Like the Bowlers and the Longstaffs, it had strong filial ties and a long Christian tradition. Her father, Russell James Dick, had come to Wellington from Melbourne as a teenager, and worked for Whitcombe and Tombs before becoming a successful Auckland printer. The eldest of Jean's brothers, after whom Russell is named, was Surveyor-General of New Zealand and awarded the Queen's Service Medal in 1954. Jean, who excelled in mathematics, trained as a teacher. Illness prevented her from continuing in her chosen profession and, at age 21, she was sent to Indianapolis by her father. There, with Auckland girlfriend Connie Creamer, she worked at the Church of Christ headquarters. It was a time when the United States was pushing the frontiers of technology, churning out inventions and rewarding the efficiency and effectiveness of both production and labour. It was an unusual thing in the 1920s for a New Zealand woman to do — go off and work overseas. And this, along with the Dick family environment, would leave Jean noticeably outward and progressive in her thinking. She brought the echoes of American culture back home with her. The hard copy followed. For years the *Saturday Evening Post*, with its Norman Rockwell covers, ads for Chev Bel Air cars and jingoistic articles arrived in the Bowler letterbox. Slowly they shuffled their way into family lore and, over time, began to stir the imagination of her youngest.

> Russ had such a solid childhood, the best of both worlds. He not only had his doting older sisters when he was small. He also had their boyfriends as mates when he was growing up. And by the time he hit his teens, with his sisters both married, his parents were able to devote even more time to him. Ces and Jean took a natural joy and pride in his sailing success. They were just so positive for him. And such characters themselves. Jean the bridge-playing lover of water and Ces the charmer who was not beyond playing the clown. Much to the amusement of everyone, he would take that battered bugle everywhere. He even used it to communicate the birth of Jeanette and Bob's first while Russ was sailing for the Tauranga Cup. They were older parents, though you'd hardly have noticed. They were always in the thick of things, working on the yacht club committee and right into the social activities with the other parents who would have been ten or more years their junior. So in this rather lovely way Russell's success provided them with an opportunity they otherwise might not have had. And even in her old age, after she had lost Ces, Jean was still so positive, still so contemporary, still so full of fun. Talking to her was like talking to someone my age.[6]

For fourteen years the Bowlers' bach at Lake Rotoma was their home away from home. The bach had begun life humbly enough. Jean's brother Jim had purchased a house with a stand-alone garage on Maori leasehold land and talked Ces into joining him. Set in native bush on the edge of the lake, the dwelling needed to be gerried into shape before it could be occupied. With the help of neighbours and an assortment of demolition material they got to work that first summer. A wall and two windows were thrown up to replace the original doors. To the delight of the night-sliding possums, a sloping corrugated iron roof was nailed into place. A long-drop was dug out the back and a weatherboard surround erected. With time and money running out it was decided that the bunk-room extension would have to wait another year. There was neither washing machine nor shower. Just buckets and a basin. And when the water tanks ran dry they washed in the lake and drank from it with never a thought of giardia.

This was a holiday community typical of the times. It mirrored the egalitarian society from which its members came. Judges drank with plumbers. Carpenters mixed with businessmen, including cannery king James Wattie. Shopkeepers swapped yarns with lawyers. The old laughed with the young. And all were friendly with the local farmer. A secluded stretch of water, Rotoma had everything. Native timber to the shoreline. The bell calls of the tui. Fantails flitting along bush tracks. Tributaries of crystalline water trickling to its edge. Not far away there was even a cluster of hot mineral pools where the Korean war could be refought with Uncle Jim and couples could slide off into dark corners.

Even at an early age, Russell showed traits inherited from his parents: the calm determination from his father, the love of water from his mother. It was there when he first borrowed a sheet from his sister's bed, attached it to a slender ti-tree trunk and stuck an oar out the back of their clinker dinghy. Horrified to discover that their five-year-old had almost reached the other side of the lake, Ces and Jean urgently despatched Bob Holley to bring back their budding Chris Columbus.

Six to ten years older than me, the boys dating my sisters were my heroes. Being invited on an expedition with them was like heaven to me. They not only sailed yachts. They were into power boats as well. People at the lake were putting plywood boxes together, throwing Ford Ten engines in them and going aquaplaning. Watching the whole scene was a powerful influence on me. I was fascinated with the making of waves, with the speed of boats, and just trying to figure the whole thing out.[7]

It was the Christmas the Bowlers and their Rotoma neighbours purchased 1.5 hp British Anzani outboards that Russell had his first revelation. Only seven years old, he had already figured out that their lighter nine-foot dinghy would be faster than the neighbour's heavier twelve-foot one. To watch the longer boat creeping ahead of theirs, with both Anzanis on full throttle, left him both bewildered and shattered. It was his sisters' boyfriends who took him aside when he came ashore.

'Look, young fella ... no need to be so upset ... it's a bit complicated mathematically, but it goes something like this: the longer you are on the waterline the faster it is you go.'

It was the following summer that Russ was allowed to go solo in the family P Class *Charles*. After his first embarrassing capsize, he spent count-less hours sailing up and down that beautiful vale of water learning all about the wind. A small lake with high hills all around, gusts would come squirrelling down. It taught him how the wind was stronger here and not there, and that if you wanted to get clear air you had to go out into the mid-dle. In another summer of learning he marvelled at how Don McGlashan[8] and his mates could build these boats called Cherubs out of eighth-inch ply. And how a guy called Mickey Orchard could knock the frames out of them and they'd still stay together.

At ten Russell had two more revelations: you could find second-hand boats for sale in Saturday morning's *New Zealand Herald*, and if you spoke nicely to your father he might buy you one. Which was how he acquired his first yacht, a Junior Moth, and a year later the funds to build his first boat, a canoe. But it wasn't until 1960, when the Bowlers moved to the harbourside suburb of Kohimarama, that Ces dug deep and parted with sixty-five quid for Russell's first P Class yacht: *Marika Jnr.*

***

Harry Highet, a Ministry of Works construction supervisor, couldn't swim. So when he scaled down his fourteen-foot design to seven feet it had to be unsinkable. He sailed his prototype for the first time in the Onerahi New Year's Day Regatta, 1921. After increasing its beam to 3 ft 6 in, and after receiving an approach from boat-builder Brian Carter, ten modified versions were built in time for the summer of 1924. That was the year the P (for primer) Class officially began. It was also the year Harry Highet, now transferred to Tauranga, gave copyright ownership and the right to administer the class to the Tauranga Yacht & Power Boat Club. By 1985 it was estimated that over 3000 had been built from 1100 lines plans sold — a statistic that spoke volumes about the way a country valued its artists. But knowing he had created a boat in which children could learn to sail was reward enough for Harry. 'My heart is too full for words,' he said as he watched one hundred and fifty-three of his progeny race on Auckland's Waitemata Harbour before the Queen and the Duke of Edinburgh in 1963.

The list of those who have won either or both national titles keenly contested in this class — the Tanner Cup (the inter-provincial competition) and the Tauranga Cup (the inter-club competition) — reads like a *Who's Who* of New Zealand yachting: Mander, Gilpin, Townson, Paterson, Bowler, Moyes, Thom, Gilberd, Barnes, Dickson, Coutts, Greenwood, Egnot, Monk, Bilger ...

The first time I met Russell was the day Dad and I rescued him. It was his first sail at the Kohimarama Yacht Club. It was blowing the usual gale for the October shake-down race. My boat wasn't ready so Dad suggested he and I go out in our ten-foot clinker. An experienced sailor, he knew there'd be trouble with all these brand new P Class skippers. And sure enough, within minutes there were capsized craft all along the waterfront. It was blowing so hard many of them couldn't get their boats upright. The patrol boat was having trouble coping with the mayhem. So there I was, a little Grace Darling, racing around with my father rescuing as many of these kids as we could. Russell had become so exhausted from getting his boat up then having it knocked down that he just couldn't do it again. Poor Jean and Ces were becoming frantic on the beach, watching their son drifting farther and farther away. We finally got to him somewhere off St Heliers. I can still see this skinny kid sitting beside me in the back of the dinghy, shaking himself to death. The moment we got him ashore Jean and Ces thanked us profusely and raced him off home.[9]

The highlight of the long summer months of competition was the end-of-season weekend of Lidgard hospitality on Bon Accord Harbour. At 0700 hours on Saturday morning, the P Class yachts from the Kohimarama and Point Chevalier clubs were loaded onto Jim Lawler's Fairmile *Ngaroma* (later on Claude Millar's scow *Alert* when they became too numerous) and shipped the thirty nautical miles to Kawau Island. There, in between feasts from the 40-foot barbecue and bedlam after lights-out, the four-race regatta was held.

Like a litany of comic footnotes to the Kohi club's activities were Gaffer Gooseneck's commentaries.

Dinner Menu:
Roast octopussises flavoured with Billy goat sauce; bingaroo sausages; bludjeoned bacon specially selected and sliced from wild boar; Vegies — Kawau 'Murphies' boiled to blazes and garnished with Mrs Lidgard's cockroach sauce; special requests — boiled cabbage seasoned with seaweed and oyster shells.
Breakfast:
Leftover cockroach sauce with sour milk; tough Kornies; cold hard hot-dogs garnished with sand.
Lunch:
Fried Weta-roos; barbecued jelly fish on sticks; boiled cabbage on toast; remains of Kave man barbecued last night and left in underground storage.[10]

Gaffer's description of the young skippers and their boats was no less picaresque. His mind, as in this final postscript to the Kawau Island weekend of 1961, seemed able to dial up ideas from some previously unknown literary universe.

Any skipper who thinks he can get Ces Bowler's Kave man to stow away in the ladies' Powder room hasn't got a show. He's gone dragging our fattest Yacker Yacker away with her husband's gold wrist watch around his neck![11]

***

He was probably that little bit smarter than most of us.
He was always a cunning yachtsman.
Peter Shaw, owner/skipper, P Class *Viti Lailai*[12]

At thirteen and still small for his age, Russell Bowler had begun competitive sailing later than most. Soon to turn fourteen and with an age limit of fifteen for the class, he had to catch up fast. He sailed twice every weekend — Saturdays at Tamaki, Sundays at Kohimarama, and at nearby Glendowie Yacht Club when the other two weren't holding races. His progress was impressive. By the end of his first half season of competitive racing he had become a scratch sailor at Tamaki Yacht Club and had been promoted from Kohi's Third Division to its First. That was enough to persuade Ces to purchase a Terylene mainsail to replace the original cotton one. Support and performance were, by now, sailing cheek by jowl. Based on finishing times expressed as percentages, handicaps were everything. Mark Paterson, the current holder of the Tanner and Tauranga Cups, sailed off the lowest handicap (scratch) and was the person to beat. Russ had raced to the next lowest, two percent, but Peter Shaw and Bruce Legg, each on four percent, were breathing down his neck.

Momentum continued to build for the rising star. So did his family's support. The winter before his last summer season they went looking for a newer, lighter boat. They tried to buy a Mark Paterson trial boat. When that failed they turned to Claude Millar and bought P35, a well-performed second-hand yacht. The budding designer-builder took the deck off *Tranquil* and removed everything he thought unnecessary. After replacing the deck he radiussed more corners. He reduced the size of the splashboards to get the windage down. He acquired a new mast. He also received two new mainsails from Leo Bouzaid's Sails & Covers. Russ Bowler was now ready to take on the best young sailors in the country.

> That last season Mark [Paterson] and I had some tremendous battles. He was a very disciplined sailor. He virtually lived on the water. He was only three when his father built him a dinghy and pushed him out off Kohi beach. And Bill Paterson was into two-boat testing long before the idea reached the America's Cup fraternity. As well, Bill was buddies with prominent yachties like Ralph Roberts who would come and watch Mark and give him advice. They'd work on the little things: how he sat, the way he hunched over to reduce windage. Mark was a major influence on my racing career. He showed me that discipline worked. It was an intense period of learning for me. If you set the boat up with one thing wrong you got punished. If you didn't cover or didn't allow yourself room to move on a shift you lost. And if you allowed Mark to get more than a boat's length ahead it was all over.[13]

Russell may have finished second in the Auckland Tanner Cup trials that year. But in a storming run to the finish of his P Class career he would win the Freshwater Championship on Lake Rotorua, the Auckland Anniversary Regatta's P Class event, and become the only person to break Mark Paterson's remarkable three-year domination of the class.

January 20–21, 1962. With Westhaven as the venue for the Tauranga Cup, the 36 competitors not only had each other to contend with. They also had the tide which ebbed and flooded beneath the harbour bridge at a rate only slightly less than the theoretical speed of their tiny craft. It was like playing billiards on a sloping billiard table; getting across the current in the right position was everything. From the first gun the battle was intense. By the second, as the yachting correspondent in the *8 O'Clock* reported,[14] it had become red hot.

> The old, old story of yacht racing, two boats covering each other so closely that a third could sneak by and win the race ... Paterson (Kohimarama) and Bowler (Torbay) were so engrossed in their battle for first place that they appeared to be oblivious of the threat of two others, J. Faire (Hamilton) and G. Adams (Wanganui). The pair had perceived, which Paterson and Bowler apparently had not, that the tide was now coming in and not out. A straight line had therefore become the shortest distance between the two points, and it was along this line that Faire and Adams were rapidly travelling ...[15]

It was a contest in which Russell would learn two valuable lessons: the importance of consistency and the power of a protest. Having won the first three races, Mark Paterson's form fell away dramatically in the resailed second and last two races. With both boats on starboard in the second race, Russell had tacked onto port, but the protest committee disqualified Mark for not avoiding the collision. The shock decision scuttled the champ. And after slipping that lesson under his kapok life jacket, Russell, the Torbay Boat Club representative for this tournament, went on to win the series without winning a race. His three seconds and two sixths had given him the prized national title by a slim nine points.

That night Bill Paterson duly paid the loss of his wager to Ces Bowler in the form of six miniature (and highly prized) bottles of whisky. The next day he took all thirty-six competitors on his 28-foot keeler to the far side of Waiheke Island. Arriving back at dusk to find the waterfront lined with parents, Bill decided the day's fun wasn't over. Like a crusty Peter Pan he offered his lost children king-sized bottles of Coke if they could put *Kismet's* spreaders in the water. The already anxious parents watched in horror as their offspring began climbing up the mast and out along the already swinging boom. There began a slow back-and-forth rocking of the keel-boat. The children's whoops of delight increased in volume as the swaying gained momentum. Fathers on the shore called out in helpless protest; mothers sank fainting to their knees. But there was no real cause for alarm. Bill had built his boat from scratch and knew its righting moment better than his wife's. After being laid on her beam-end with her spreaders well and truly dunked, *Kismet* slowly but surely hauled herself into a position more appropriate (and demure) for a lady of the sea at anchor.

***

Although the sixties was a decade of great cultural change, it was still a time when reality was not yet virtual and television and the corporates were neither the makers of myth nor the determiners of social mores. Indeed, it was an era when performance not perception was still the benchmark of success.

Soon after his Tauranga Cup success, Russell was to sail against two young helmsmen who were destined to play significant but very different roles in his future career. For winning the P Class race in Auckland's 1962 Anniversary Regatta, his last in the class, Russell received the Noel Cole Cup. In that same race, sailing a boat called *Pee Bee*, was a boy called Peter Blake. Already big for his age, he had a handicap to match: eight percent. In March of that year, Russell and Ron Blakey won the Auckland Secondary School Championships for Auckland Grammar. Sailing with Brian Wade into second place for Takapuna Grammar was a teenager with silken helming skills. His name was Peter Walker. Russ and Peter were to become close friends. And not many years hence, after crossing Australia in a clapped-out Ford, the pair would join forces to contest an inaugural world championship on the Swan River in Perth.

Chapter 3
Resolution

Good morning everybody! Good morning every-
body ...! It's getting very nice now. It'll soon be all
right. Just a few showers perhaps today. But it's
nice and it's warm and comfortable and quite all
right and everybody can go shopping ...
Aunt Daisy, The Commercial Broadcasting Service,
Radio New Zealand, Wellington[1]

Like a firecracker on speed, Maud Ruby Basham (Aunt Daisy) left breathless even the most voluble politicians of her day. She could talk her adoring public into buying anything from greeting cards to blinds, meringue whip to Weet-Bix, recorder-grams to fly spray. But although 17 May 1949 had dawned beautifully fine in Auckland, Ileene May Farr wasn't about to do Aunt Daisy's bidding. For starters she'd been unwittingly put to sleep while the nursing staff dealt with a log-jam of deliveries. And when she came to, there were far more pressing things to think about. Like the relaxation techniques that had proved so helpful last time she had given birth, and which she now invoked. Sharp at one o'clock she went into labour, and was delivered of a 6 lb 6 oz baby boy a few minutes later.

***

Jim and Ileene had begun married life near Odile and Harry Farr's home in St Heliers, then moved to One Tree Hill just before their second son, Bruce Kenneth, was born. Two months later they moved into a state house in Orakei. It was here, at Orakei Primary, that Bruce and his fifteen-month-older brother Alan received their first years of schooling. And, for a time, Sunday School instruction at the feet of a talented helmsman, Bob Stewart.

Orakei was also where Bruce began his boat-building education. Jim had not only built a Sabot as tender for their 20-foot keeler *Jeanette*. He had also managed to fit a 24-foot Airborne onto the front lawn of 17 Ngaio

Street. The brainchild of English yacht designer Uffa Fox, it was one of only two imported into New Zealand for training purposes during World War II. The father of the planing dinghy had taken his idea and, with Lord Brabazon's blessing, turned it into a streamlined lifeboat to be slung beneath aircraft and dropped to ditched crews in the English Channel.

Watching his dad and boat-builder pal Bernie Lovegrove modify the Airborne would become one of Bruce's earliest memories. Over the winter the pair stripped it down, added a spade rudder, a keel with a 500 lb bulb to replace the centreboard, and extra freeboard by raising the gunwale. They gave it a new deck and added a cabin. Little did the four-year-old know, as he watched *Caper* being taken to Okahu Bay, that it would be one of the first fin keel-spade rudder yachts to be launched in Auckland since the Logan brothers' *Sunbeam* in 1900. Unconventional, perhaps. Unusual, definitely. But it was the shape of things to come. The shape of speed.

Boats were Jim Farr's passion. From his boyhood dinghies, to *Vagrant* which carried the honeymooning couple around the Bay of Islands, to their small keelers, Jim had been messing about in boats since his family had moved from Christchurch to Auckland's marine suburb of Devonport. Now his and Ileene's free weekends and summer holidays were spent enjoying the same pleasures as the rest of Auckland's waterborne community — cruising, *en famille*, the islands and bays of the Hauraki Gulf.

It was while on one of these holidays that Bruce had his first sailing lesson. They had not long been anchored in Te Kouma Harbour, Coromandel, when Jim rigged the Sabot. He fitted his seven-year-old with a life jacket and helped him over the side. He'd just pushed him away from Jeanette when a williwaw rushed down the hills. Like an Exocet it raced across the harbour. The wind may have had the last laugh as it spun Bruce in a gybe. But the determined look on his face as it bobbed up past the gunwale told his parents he wouldn't be upended quite so easily next time around.

A painfully shy student from his first day in class, Bruce ended his schooling in Auckland with a prophetic comment alongside his '1' in Arithmetic: 'Top. Has produced some outstanding work.' The year was 1957.

***

In 1857 the Matheson brothers, Duncan and Angus, and a number of Nova Scotian families set sail on six vessels bound for New Zealand. Most would settle in Waipu. Those from the Mathesons' brigantine *Spray* would choose Omaha, a district 65 kilometres north of Auckland. Originally under the ownership of Chief Te Kiri, it was a hilly bush-clad region of rivers, lakes and islands. It also included the harbours of Whangateau and Leigh.

Angus, already married with one child, and Duncan carried 97 passengers plus crew when they set sail for Auckland on 10 January 1857. It was an arduous six-month journey. Two years later the brothers chose a picturesque spot just south of Leigh in which to settle permanently. It was at either end of this bay, surrounded by pohutukawa and puriri trees, that the brothers built their rough-sawn homes. It was also the perfect setting for building their trading schooners. They had an ample supply of suitable native timber. There was an area between the headlands flat as a factory floor, a sloping sandy beach, deep anchorage, and a reef and small offshore island providing them with shelter. It was here that *Coquette*, *Rangatira*, *Three Cheers* and *Rhino*, among others, were knocked from their chocks and sent scudding into the waters of the bay.

As in other areas of the colony, corn, wheat and vegetables were grown and family orchards established. The growing community also lived off the district's greatest asset, the sea. The fish they caught, if not eaten or sold, was smoked and held in sheds for the winter. Fortnightly the collective crayfish catch was put on board the *Kawau* which, having just dropped off supplies to the families of the area, would take its precious cargo back to Auckland for sale. It wasn't until the end of World War II that Leigh was hooked up to the national grid. It was another ten years before power reached Goat Island, where members of a new generation of the Matheson family had begun a dairy farm. Before electricity, homework was done by candlelight, milking by an old Anderson one-stroke. And it could take all day to boil, rinse and crank-handle the family washing through the wringer.[2]

***

Although Leigh was isolated, it was becoming a flourishing fishing centre by the time Jim Farr nudged his eighteen-foot keeler *Santorin* into Whangateau Harbour. Progress up the channel was slow. It became even slower as the tide began ebbing. Caught in the lee of the land without a motor, the skipper, his pregnant wife and baby boy were going nowhere fast. Fortunately, neither was Ernie Torkington. It wasn't the first time he'd observed the slapstick humour of the tide as it funnelled in and out of the narrow channel. Having enjoyed the joke just long enough, he ran the short distance from his farmhouse to the Ti Point wharf. There the Matheson relative by marriage put out in his little blue runabout.

'Ernie's the name!'

'Jim Farr, Ernie.'

'Here, grab this line, Jim. Now, how are you off for milk? Need any fresh water? And what about those nappies, Mrs Farr?'

Ernie's welcome personified Leigh. After making their boat secure, the Farrs went ashore to meet up with friends holidaying from Auckland. They ended up making more. As they spun out idyllic days within the little community, it wasn't long before a thought began to grow in Jim Farr's mind: one day he would quit his printing job and go commercial fishing.

It took nine years for the dream to become a reality. But when the time was right he packed in his job as production manager at greeting card manufacturer Tanner Couch and became manager of Leigh Fisheries, the local co-operative. It was a good way to learn the trade and meet the locals. A year later Jim would realise his ambition and become the proud owner of his first fishing boat.

As with many who defy convention to follow their hearts, moving to Leigh set Jim Farr free. Riding the ocean swell as he baited his longlines, as the diesel clattered beneath his feet and Great Barrier rose and fell on the horizon, was a dream come true. To line up three points on land and be able to support his family from the ensuing snapper catch was everything he'd ever hoped for.

> The Leigh community was great. I'm sure it had a big influence on Alan and Bruce. It was a community that relied on itself. They built their own houses and boats. They did things for each other. And if you didn't know something, somebody would teach you. I wasn't a handyman when I went there but I was when I left. There was a chap, Bob Brewster, who'd sit on the end of the wharf. One day he called out: 'Hey Jim, isn't it about time you gave that motor a decarb and a valve grind?' I'd never pulled an engine apart in my life. 'Then I'd better show you,' he said. That's the sort of people they were. And between them they could do anything. There was Joe Joyce, from a family in Puhoi, who could fix anything mechanical — qualified mechanics would come to him when they were stumped — and Len Knaggs who could throw together a beautiful clinker dinghy in three days flat.
>
> That first year in Leigh we rented a house in Matheson Bay. That's where I built the Flying Ant — on the verandah — with help from Alan and Bruce. The second year we bought a section just up the road. And on the days I couldn't go fishing because of the weather I worked on the house with a carpenter friend. There was no power to the site so we did it all by hand, with Ileene and the boys pitching in. And it was under this house, as an eleven-year-old, that Bruce designed and built his very first yacht.[3]

Resolution, this 'nifty little boat which worked pretty well',[4] was launched on New Year's Day 1961 from the very same spot where the last of the Matheson ships had slipped into Matheson Bay. And like *Rangatira* and *Coquette* before it, the first Farr design would one day win its class in Auckland's Anniversary Regatta.

***

As much as we change the landscape, the landscape changes us. If that is true of a young country, it was particularly so of the youngsters who sailed their odd assortment of craft on the Whangateau in the late 1950s and 60s. A tidal harbour, it is full of moving sandbanks and snaking channels. You can push out from Ti Point at high tide and barely get your feet wet. Come back after a race and you can have a hundred metres of mudflat to traverse. Winds funnel down the northern hills to crack off massive macrocarpas or race across its shallow face from the landlocked south and west to the ocean-facing east. Its physical characteristics would have a profound effect on the brothers Farr and their friends. Just as Alan would one day write programs for the country's first computers, so this unique environment would write a mass of data as reveal codes on the minds of these young sailors.

Formed in June 1963, the Whangateau Boating Club soon became a focal point for a number of local families. Eleven fathers, including councillor Rex Collings who had mooted the idea, Jim Farr as secretary-treasurer, Ernie Torkington and Gordon Lovegrove, made up its first committee. Alan Farr and Anthony Bullock were its first junior members. The rest of the gang soon followed. Bruce Farr:

> We ended up with quite a fleet, fifteen or twenty boats of various types. There were Idle Alongs and Flying Ants, Starlings and Herons, Arrows, Cherubs, Zephyrs and Moths. The start-finish line was off Omaha Wharf, the course heading down the channel. One of the neat things about the Whangateau is that it's tidal. This meant you quickly learnt about tides and puffs. You learnt how to get into eddies and how not to break your centreboard by recognising where the shoals were. That was hugely valuable in terms of learning how to sail, something you might not have got off Torbay or Kohimarama. So while we weren't into high-class racing, we did get very good at boat handling, at being smart about currents and changeable winds. And when we weren't racing, we'd go sailing, often all day, all over the harbour — Lindsay Lovegrove, the Torkington boys and their sister Marie, Des Matheson, Alan and I. We'd do that all summer.
>
> Looking back, the whole of the Leigh area was a really beautiful environment in which to grow up. The sandy beach and crystal-clear waters of Matheson Bay; the Whangateau, a safe and interesting harbour on which to learn how to sail. Just a great way to live. In one sense we didn't have a lot. In another, we had everything.[5]

When it came to boats, Bruce was his father's shadow. If Jim was at the local boat-builder's having a cup of tea, Bruce was right there with him. When Jim was building the Spencer Flying Ant for his boys, Bruce was carting the sheets of ply and Alan adding the immaculate finish. Soon, however, one of those roles would be reversed; it would be Bruce doing the building and Jim the assisting.

It all came about because Bruce was small for his age and the £150 for a Zephyr was beyond his personal budget. Gordon Lovegrove had heard of the Farrs' move north through his brother Bernie and also thought Leigh might be a good place to settle. Along with their household possessions they brought their Jack Logan father-and-son knockabout. But death was knocking at the Arrow Class door, and Gordon and his son Lindsay were soon looking for a replacement. Tanner Cup winner Des Townson had produced a pretty round-bilged boat which was proving popular in Auckland. Having two of these single-crew Zephyrs on the Whangateau, Lindsay and Bruce decided, would be the way to hone their racing skills. So when Bruce announced his decision to build a smaller boat to his own design, Lindsay was disappointed. He thought he'd lost his competition. He hadn't. On its first outing — a Whangateau Boating Club Open Day — the first Farr design defeated everything of equivalent length and sail area and proved competitive with Lindsay's home-finished Zephyr.

Resolution was an amalgam of ideas. In part it was Bruce translating his Zephyr dream into a boat he could keep upright. But he gave it a hard chine, which was easier and cheaper to build and, like their Flying Ant, aft buoyancy tanks. But he did away with the Ant's straight stem. There would be no wayward behavioural features with the first Farr design; a sloping stem and fuller forward sections would see to that. As in the trend-setting Zephyr, Bruce fitted a traveller amidship and a reefing claw on the boom. He also designed the rig and sail and, with the help of his father's carpenter friend, built its mast and boom. The Farr one-stop shop had arrived.

Bruce was thirteen when we took it down to Auckland for the Anniversary Regatta. He'd entered it in the Pennant Class which had everything from fourteen feet to *Resolution*'s ten foot six. There was a strong westerly blowing straight down the Waitemata and there was a big following sea. Bruce was already up on a plane when he hit the start line off Orakei Wharf. He shot straight into the lead. I wasn't sailing that day. I was standing up by the Savage Memorial and had a great view. I can still see Bruce hanging off the transom, spray flying everywhere, just tearing away from the rest of the fleet.

He went from success to success after that. Soon he was designing and building Restricted Moths. Although they had the same waterline length as Zephyrs, his and Alan's boats walked all over ours.[6]

Ask Bruce Farr how he got into yacht design and he'll tell you he never got out of it. Raise an eyebrow and he'll tell you he was sailing before he was born. That he was still in nappies when he and Alan were bundled into *Santorin*'s forward cabin to ride out storms. He'll tell you that from the day he could walk he was standing beside his father watching him build and rebuild an assortment of boats from dinghies to keel-boats. That he was using hammers and chisels before he was using crayons and pencils. But that explains only part of the story.

What was happening beneath the Matheson Bay home during the winter of 1960 was barely discernible: the blending in a young boy's soul of the gifts inherited from a sumptuous gene pool. And the maternal DNA (aglow with beaux arts) and the paternal DNA (bristling with mathematics) were getting along just fine. A unique bonding of subject and object was in progress; the tying of a Gordian knot between a young designer and the now visible product of his imagination. For an eleven-year-old, with a shyness so debilitating he might not answer the family phone, here was a valid means of self-expression. Already it had earned him parental support. Successfully executed, peer approval and many hours of sailing pleasure should follow. It would not only work. So powerful would this bonding become over the next few years that it would give complete shape and meaning to his life. There would be hardship, failure, tragedy and personal loss. But after that winter neither meaning nor direction would ever desert him.

**

The International (or Europa) Moth was already popular in the United Kingdom and Europe when Hal Wagstaff introduced the Australian version into New Zealand in 1963. He called it the Restricted Moth Class.[7] Sailed under rules based on the Australian Moth Class, its name belied its essential characteristic. It had, in fact, few restrictions save its eleven-foot length and 80 square feet of mainsail. Herein lay its chief attraction. For a comparatively small outlay an amateur could experiment with design and measure performance against others from all around the country. Here was a home for those thoughts that had been growing like a forest in the young designer's head — a properly constituted class of affordable yacht in which his theories could be tested and refined. But it was Alan, who had produced the Whangateau Boating Club's handicapping system, who conducted the correspondence with Hal Wagstaff as the brothers' interest in the Restricted Moth Class grew.

> Alan had a fine intellect. He also had a very great mathematical ability. Years after our first meeting we became firm friends when he moved to Wellington and worked for the DSIR. Both were excellent sailors. Perhaps it was the inner drive to create his own designs that gave Bruce the edge on the water. Just as Bruce McLaren's driving ability was the reason he came to international prominence as a designer of racing cars, so Bruce's helming skills and feeling for a yacht's movement through water were the critical factors in the excellent hull shapes he produced from the start.[8]

Meeting Hal Wagstaff and seeing the fruit of his labour — the double-chine Puriri design — encouraged Bruce to think about his own Moth design. Better still, if he could learn how to draw, maybe he could land a few design-and-build commissions and create a summer job for himself. A long weekend with his engineer uncle, Colin Clayton, in the drawing office of the Whangarei Harbour Board, taught him the basics. *Behemoth*, designed and built for Alan over the 1963–64 Christmas holidays, became his first Moth project. Vern Smith, a Torkington cousin, and David Hook liked what they saw and each ordered a Farr Mark I for delivery that summer. Still only 110 lb and accustomed to capsizing a lot, Bruce added a false floor to facilitate self-draining when he built *Mammoth* for himself the following winter. He also fitted a retractable skeg just forward of the rudder to counter lateral slippage when sailing downwind.

It all came together in 1965, on Lake Rotorua. Weighed down by layers of wet woollen jumpers and with the feeling lost from his feet, Bruce won his first North Island Restricted Moth Championship in windy, ice-cold conditions. For a fifteen-year-old from Leigh it was a taste of the big time. To win in this company was an experience that would linger in his memory for a long time to come.

The issue of weight reduction was by now assuming more and more prominence in the Farr design equation. Later that year, Bruce designed and built his first Mark II Moth. For Alan's *Behemoth II*, Bruce selected the lightest plywood available and set it over more stringers and plywood frames than he had for his previous design. An immaculate working of wood and attention to construction detail were now observable Farr features. Launched just two days before the 1966 national championship, *Behemoth II* was an impressive 30 lb lighter than *Mammoth*.[9]

March 1966 was the month the Farr name was first heard in yacht clubs around the country. For designers of Restricted Moths it was also the month alarm bells began ringing in their ears. Bruce and Alan Farr, David Hook, Vern Smith and Alan Torkington shocked the Restricted Moth Association by winning the inter-club NZRMCA Challenge Flag for the little-known Whangateau Boating Club. Sailed off Auckland's Browns Bay over five races using Olympic courses, Bruce ended the regatta with a 1/1/2/1/2 series. It gave him his first national title: the Lynn Trophy. He also completed a hat-trick of wins by taking the junior championship and the Claymore Designer's Trophy for the boat with most points skippered by its designer. *Sea Spray*'s correspondent described the Farr designs as 'rather basic and wholesome hard-chine hulls with stayed masts, remarkable only for their fine finish and tuning.'[10] The Farrs' performances spoke rather more eloquently than that. Seven of the 32 entrants who had paid ten pounds to the holder of Box 51 Leigh for one of his designs thought so too.[11] You didn't have to ask brother Alan. His third place in the championship was proof enough. Even more so was his winning that year of the New Zealand Universities Summer Tournament title and his University Blue.

Someone who didn't have a Farr design was Rob Blackburn:

> I'd made the decision to get into Restricted Moths and asked Merv Elliott to build me one. I arrived at Browns Bay this morning and remember seeing these weird-looking things belonging to two hairy-legged schoolboys from Leigh. I quickly decided they weren't going to give me and my nice Robert Brooke design any trouble. But they blew us away that day, beating us by ten to fifteen minutes. I thought it was a fluke. But the next weekend they cleaned our clocks again. Soon his designs were catching on fast. In fact, so much so that if you wanted to be in with a chance in the Restricted Moth Class you had to have a Farr.[12]

As sweet as it was, success was also having its drawbacks. Although Bruce won his second North Island title in 1966 with another five straight wins, with more and more of his designs on the water his and Alan's competitive advantage was slowly slipping away. This was particularly noticeable during Bruce's Auckland championship win that year when he was pushed hard by the fiercely competitive Rob Blackburn.

> Right from the word go he seemed to have had an innate understanding of the principles needed to make a hull travel sweetly through water. I remember spending time at their Matheson Bay home. He'd have books by Herreshoff spread around his bedroom and be saying things like, 'Look at that underwater shape, Rob, the hull form's all wrong. With that much under the water he'd have been dragging half the ocean along with him!' Even at that age he was intense. And philosophical. At times like that I thought he was talking another language.[13]

As this North Shore group of sailors grew, so did Rob and Bruce's friendship. Like the P Class fraternity, the parents of the Moth sailors would take their youngsters to regattas all around the country. Their cars would be chock-a-block with bodies and tents, their home-made trailers piled high with hulls, sails and rigging.

By the time he'd won his fourth successive North Island Restricted Moth title in 1968, this time in his Mark III *Dynamoth*, Bruce had established 'a record to be emulated'.[14] But it was now time for bigger and better things, like teaming up with Rob Blackburn to race in open twelve-footers. But the friendships forged in these early days would linger long past this new and intense period of sailing against the likes of Lidgard and Bouzaid, Bowler and McGlashan. Not many years hence, Rob and Bruce, together with their mates, would sail a new style of keel-boat into yachting history.

Chapter 4
Making Waves

Conquest is not heroic. What's heroic is making love last.
Gerard Depardieu, French actor-winemaker[1]

The three Bowler siblings would all marry those they fell in love with in their teens.

Jeanette Bowler was a fifteen-year-old when she first saw Bob Holley on an Epsom tennis court. A Cupid without a bow, Bob was already poised to bat a ball in her direction when she began trading blows across the net with her girlfriend Pamela. That Jeanette was struck inches from her heart didn't seem to matter. From the far side of the court Bob's had been a meritorious shot. Jeanette was now in love.

Like an elegant bird of prey, Mickey Orchard's Cherub glided in ever-diminishing circles around the heavy Z Class yacht. Its crew, a fifteen-year-old and her unflappable mother, looked across the surface of Lake Rotoma for a tell-tale sign of wind. Anything from anywhere, they muttered, to get us away from these Whakatane larrikins. Then a strange thing happened. The closer the two boats drew together the more the Cherub's forward hand became enamoured of the pony-tailed blonde seated at the Zeddie's helm. Soon her feisty gaze was frying his brain, his testosterone levels rocketing out of control. Panicking at this turn of events he chose bravado over tact.

'Hey Doris! Like to go for a sail in a real yacht?'

Such was the power of a Kiwi male's mating call in 1955 that K Bowler would soon be in love. Don McGlashan had not only landed a date on the lake but a future wife as well.

For the youngest Bowler sibling it began one late summer's day in 1966. In the manner of a *deus ex machina*, he made three momentous decisions when he got out of bed that morning:

1. He would not go sailing.
2. He would call Philip Bull.
3. He would wear his favourite pair of pink chequered swimming trunks.

But the ten-inch waves which greeted Russ and Philip at Orewa Beach, 40 kilometres north of Auckland, indicated it would be a day more profitably spent out of the water than in it. Adorned with rack and nine foot boards, they drove the car the length of the beach. In a not inconspicuous spot they parked the MG facing the sea and themselves on the sand beside it. With the size and shape of wave no longer of interest, they could now turn their attention to the size and cut of bikini. It could not have been a more perfect Sunday for two nineteen-year-olds from Auckland. Or could it?

Just around the corner, at Hatfield's Beach, a sixteen-year-old was disengaging herself from the company of her parents. Having agreed to rendezvous with them at the end of the day, she and her friend Jennifer Taylor walked around the rocks in search of better things. They had not long reached Orewa Beach when they heard a shout. They attached it to a torso and knew they'd struck gold. Philip Bull had been ogled many times while on lifeguard duty at their local beach, Browns Bay. Flicking back their hair, they made straight for the boards and bodies. Russ slipped his shades to the top of his head. He blinked, and blinked again. He could hardly believe his eyes. She was not just a picture of perfection, all limbs and breezy smile. The little the willowy blonde was wearing was exactly the same colour and pattern as the little he had on himself.

It was an altogether pleasant afternoon spent in and out of the water. The laughter came in giddy rolls, their chatter increasingly animated as they dried themselves in the sun. There was no denying the tug of attraction. More frequently than he considered discreet, the champion helmsman found his attention slipping from the distinctive wrap-around shades to those strips of pink chequered fabric. Matches are made in heaven, he kept thinking to himself. Can they also be made on a cutting room table? Even as they were saying goodbye the coincidence was still too much and he forgot to get her number. But he didn't forget the name that had come up several times during their conversation. At morning tea on Monday the Ministry of Works cadet made straight for Peter Dorflinger.

'Lynda who?'

'That's where I'm stumped. Lives down the road from you.'

'Blonde?'

'Blonde.'

'Legs?'

'Up to here. And those bumps out the front.'

'Right. Lovely lady. Name's Lynda Smith.'

At first she thought it was her father, his voice notched down a tone or two, calling from a party. She giggled when he told her who it was. Squealed when they hit upon their plan. For a first date Fred and Marge thought that going to the local dance hall with a trainee engineer sounded harmless enough. They weren't to know that the red MG-TD, with its distinctive throaty roar, would shoot right past Milford's Surfside and onto the harbour bridge. It was the first of many rides that Lynda Smith, her model's hair flying in the slipstream, would take along Tamaki Drive to number 227. On this occasion to a beer and banjo party organised in Ces and Jean's absence.

***

Within forty years of its official founding, the inhabitants of the colony were designing and building the first in a long line of centreboard yachts. These were the boats used for netting mullet which abounded in the tidal reaches of the Hauraki Gulf. Fashioned by commercial demand out of local conditions and sporting enormous rigs, they were fast and fun to sail. Soon they were providing bellicose racing on the Waitemata Harbour. They were given letters for their mainsails — H, I, L and N — to denote their various lengths. And when they had reached their use-by date they became the preserve of the buccaneer young. Exchanging pig-iron ballast for the bottled variety, mullet boat crews would celebrate the last days of bachelorhood in red-blooded cruising to Kawau and other islands of the Gulf.

From the Baileys and the Logans and other designers came additions to this alphabetical list of centreboarders. Like the mulleties, their construction and shape were determined by the rigours of locale and cost. Glad Bailey's X Class, designed in 1918, was a fourteen-foot refinement of those early racers. Alf Harvey's Idle Along was designed to counter the blustery conditions of Wellington's harbour. Bob Brown's Z Class gave the Takapuna Boating Club's teenagers the means by which to tame the chop of Auckland's Shoal Bay. Twelve years into retirement, Arch Logan produced his legendary M Class, the first dinghy in the world to use a trapeze.[2] There was the twelve foot nine inch R Class. And the S, T and Y classes, all heavy boats around fourteen feet providing close racing within their fleets. Then there were the unrestricted twelve- and eighteen-footers, Q and V respectively, the most spectacular classes of all.

Uffa Fox held that weight was only important in steamrollers. But it would take another holocaust before that principle found wide acceptance in yacht design. Out of the swirling clouds of World War II came the de Havilland Mosquito, a lightweight bomber made of wood laminates over balsa. From there it was only a small step to the commercial production of plywood. Manufactured under factory conditions using industrial rollers, these multi-directional veneers were held together by synthetic glues derived from benzine. Phenolic resins were not just impervious to water and worm, they were reliable and long-lasting. Together with treated plywood they gave the amateur the means with which to build boats at home or with buddies down at the local yacht club. With materials tailor-made for the lateral thinker, an environment second to none, and New Zealand's exports booming, here was a revolution waiting to happen. All that was needed to turn Auckland into the Florence of the yachting world's Renaissance was a burst of inspiration.

It came from an unexpected source — a shy, retiring architect of a whimsical disposition and a probing mind that was never short of opinion. Born in Australia but raised in New Zealand, John Spencer lit the torch in 1951 when he designed and built a dinghy for Ray Early. *Cherub* proved a great success in the open Pennant Class. It also caught the eye. It seemed the perfect vehicle for the youth of New Zealand yachting who, increasingly as the world economy fed back into their own, were coming into small disposable incomes with which to indulge their passion. A set of rules was drawn up and the Cherub Class was born. And like the Emmy before it, its rules allowed for design development within minimum and maximum measurements. Prominent industrialist and sailor Tom Clark:

> That was his great design feature — making straight sheets of plywood accept a compound curve. That meant it could carry enormous loads provided the curves could be kept in place. And that's where the new glues came in. As a designer I think he was at genius level. But he wasn't interested in money or becoming famous. He didn't want to go there. He loved designing boats for the do-it-yourselfer and talking to his mates in the pub or over a flagon of wine at home. He just wanted to be a sage and pick the fur out of his belly button. That was Spence. A real thinker. In my opinion he got the whole thing rolling.[3]

Faster than the Business Roundtable can say SOE,[4] the country embraced the hard-chine flyer. Its lines plans sold like *8 O'Clocks*. Overnight it seemed the Cherub had become the darling of the dinghy world, making obsolete the clinker and carvel-built hulls which preceded it. Spencer's became a household name. Soon he was adding the Flying Ant, Javelin and a junior version of the Cherub to his stable of designs. By 1968 nearly two thousand Cherubs had been registered. Seven marks over two decades brought many refinements to its hull and its designer the title of the Plywood King. John Spencer had opened the floodgates. But others were having a go too. Des Townson produced the Zephyr, Mistral and Dart. Add to this the growing number of Restricted Moths, twelve- and eighteen-footers — most designed and built by amateurs — and by the mid to late 1960s New Zealand had produced a platform of dinghy designs second to none in the world.

Then there were the yacht clubs, which sprang up the length of the country, supported by an army of enthusiastic parents. A body to oversee the sport — the New Zealand Yachting Federation — was established in 1952. The infrastructure blossomed to rival that of other major sports. There had always been the coastline and the wind. Now there was a new technology, and with it, the added dimension of affordable choice.

***

Before building a sixteen-foot runabout for his family, Russ Bowler built an OK Dinghy, *Neiuffe*, launching it at the end of 1962. The following February he won the Royal Regatta's OK Dinghy event. He won further races in Auckland but knew he lacked the bodyweight for the all-weather rigours of the class. He sold *Neiuffe* to build *Mecca* after crewing with Peter Willcox on his Cherub. But still not weaned from his years of single-handed sailing he borrowed *Neiuffe* from its new owner and took it to Lake Rotorua. There, in April 1963, he won the New Zealand OK Dinghy Freshwater Championship.

Not only did Cherubs provide exciting racing within their class. They could also be sailed as unrestricted twelve-footers and, if selected for the national team, compete annually against Australia for the Silasec Trophy. Built during the winter of 1963, *Mecca* was ready for launching at the end of that year. And as was the fashion, Russell made sure it had a second, bigger rig in order to make the most of its dual-class status.

> There was a lot of Cherub building going on in Auckland at the time. There were some very clever craftsmen putting them together. A lot were exquisite pieces of workmanship. There was one called *Bondi Tram* which was just beautiful. There was *Oliver Cool*. There was *Interlude*, all trimmed around the edges. *Mecca* was the first boat Don and I built together. At the time Don was nearing the end of his training as a civil engineer. He was particularly good in the workshop, and between us we figured out how to make it light yet strong.
>
> *Mecca* was conventional. It was based on what was called the modified Mark VII, the shape everyone was using at the time. You'd get Spencer's design and tweak it around until you felt you had a better boat. It was in this way that I learned some of the basic lessons of design. How you could straighten a boat out and it would go faster in a breeze but slower in the light. How others with more shape and straighter runs aft went better on the reaches but had trouble in seaways. So all the important lessons were there to be learnt in the Cherub. Being short and generally sailed in rough water with tide against wind, the small difference in hull shapes and how they turned into handling characteristics became important lessons to take on board.[5]

Russ and Don McGlashan (now married to Russ's sister Katharine) finished a creditable tenth in their first outing in *Mecca*. That was in the national twelve-footer trials sailed in Auckland in January 1964. The first twelve boats were automatically selected to represent their country, but Russ and Don felt they weren't competitive enough to warrant the cost of the trip to Sydney. Nor did they want to lose the use of their boat for over a month while it was being shipped across the Tasman and back. Instead they stayed at home and practised.

It paid off. In April of that year they took part in their first Cherub nationals at Nelson. The competition was fierce. Ray Stagg and his son Geoffrey, from the Muritai Yacht Club in Wellington, won the invitation race in *Whispers*. Russell Botterill and Wayne Dillon, from the Kohimarama Yacht Club, were the form crew in *Even Stevens*. And Max and Peter Walker, from the Wakatere Boating Club on Auckland's North Shore, were showing great form in *Tantalus*. No one was taking much notice of the lightweight seventeen-year-old in *Mecca*, even if he had won the prestigious Tauranga Cup the previous year and had an OK Dinghy national title under his belt. It occurred to no one that his P Class victory beneath the Auckland harbour bridge had taught him plenty about the vagaries of a swift-running tide.

It was only after every boat had capsized in 50-knot winds that the decision was made to abandon Sunday's race. Monday dawned little improved, the sky sullen, the wind blowing 40 knots. To prevent total mayhem in Tasman Bay the committee decided to resail the fourth race within the confines of the harbour. Rip tides with a rise and fall of twelve feet were a feature of the area. But no one had thought to consider the effects of the recent Alaskan earthquake on the harbour's run of current. *Mecca* was in the middle of the pack passing Nelson wharf when Russ noticed waves smashing against the piles. He yelled above the wind's roar as they rounded the last buoy.

'Don ... something's weird ... the waves on the other side of the harbour were running the other way!'

The moment it twigged he threw. There would be no following the pack over to the Boulder Bank to escape the incoming tide. Instead he pointed *Mecca* back towards the wharf where the tide was rushing out. It was like opening the throttle on his runabout. They rode the waves like a speeding bullet to the finish. Their first gun in a national Cherub championship was like music to their ears.

Two years later *Mecca* would dominate the Cherub Nationals with three wins and two thirds from five starts. Likewise the trials at the end of the year to select a national team for the first Cherub Interdominion contest with Australia. Russ and Don would finish a disappointing third behind Bob and Peter Walker's *Serendipity* in the series sailed on Sydney's Middle Harbour. Being holed at the start of Heat Five and finishing with a buoyancy tank full of water hadn't helped their cause. They were, however, the only boat to win two heats. It was sign of things to come. Don McGlashan:

I would have been twelve and a half stone in those days, which was heavy for a Cherub forward hand. Russ was a paperweight. All up our weight would have been line-ball with other crews. But ours was more balanced. It meant we had grunt in the heavy. And with Russ's skill in the light — he was quite brilliant at picking shifts — we had all conditions covered. And once we had mastered the boat it gave us great consistency.[6]

It had taken two seasons to get *Mecca* really going. Yet by April 1968, when they parted company with their red-hulled flyer at the end of the infamous '*Wahine* weekend', Russ and Don had sailed it to two North Shore, two Auckland and two national titles. They had risen to the very top of the most numerous and competitive class in the country.

They were debating where to go next when they heard of a Q Class owned by Don Lidgard. With its upthrust bow it wasn't exactly a looker. But *Query* was made of a revolutionary new material called polyurethane foam and was said to be extremely light. To the engineering brothers-in-law those two facts screamed at them louder than a 707 on takeoff.

Chapter 5
Designer Full-time

... had Bruce Farr chosen to be a champion sailor he would have been in possession now of Olympic gold medals, world championship titles, and sailing America's Cup yachts instead of designing them.

John Spencer (deceased), yacht designer,
Okaito, Bay of Islands, New Zealand[1]

He was obviously a thinker. He was a top-class yachtsman with a hands-on understanding of what made a boat tick. He was a boat-builder which enlarged and diversified his thinking. He had vision. Plus he was extremely focused. Combine all that with what I regard as a near-genius mathematical ability, and there you have it. Any number of people have got this and not that. Bruce had the lot.

Jim Young, yacht and launch designer,
Auckland, New Zealand[2]

The sixties was a decade when the summers seemed longer than before, the easterlies warmer and more gentle as they slid around Cape Colville to warm Auckland's sandy beaches. It was a decade awash with popular culture. A time of concept albums and complex vocal harmonies, when the young went off to coffee bars and dances and debutante balls and everybody partied. It was also a decade of casting off, when the children of the post-War years refused the hand-me-downs of conformity and constraint. They could afford to. They had known neither Depression nor world war. They were free to wave their banners and march. Say no to Vietnam. Yes to women's rights. Release a piglet in a lecture hall when the Finance Minister came to speak. They wrote what they spoke — *Bullshit and Jelly Beans*[3] — and had plenty to smile about while doing it.

But they were ignorant of Empire.[4] In childhood they had seen a mountaineer become a knight, had visits from the Queen. They had studied potted histories of its virtues but they hadn't read its price tag. Whether you paid at Gallipoli or Ardennes or for the recent wool-price slump, it cost. And as Mother England kissed the kids goodbye and headed for Brother Europe it was going to cost again. Ahead lay years of uncertainty. Of politicians pleading. Of producer boards panicking. To come were oil shocks and borrowing.[5] Think Tanks and Think Bigs. Experiments and failures — all before the economy would become diversified enough to get back on its feet. Everywhere the waters were turbulent. But instead of battening down the hatches the yachting community flung theirs open. Seventy years before, Premier 'King Dick' Seddon had declared that rugby and war would be the chief means by which New Zealand would establish itself on the world stage. Surrounded by so much water you wonder how he missed the sea.

***

Sixty-six kilometres north of Auckland, Warkworth sits pretty as a placemat on the southern bank of the Mahurangi River. History seems to rise above its rocky race and drift around the wooden gables from its settler past. It was bustling with commerce the morning I went in search of Percy Street. Nowhere is far in Warkworth so it wasn't hard to find. But I was thrown when it suddenly ran out of road. Was my Wises Directory not living up to its name? Or had a substantial piece of Percy been thrown out the town planner's window before it had even been laid? I lost count of the number of times I drove around the block before I found number 56. A dog barked when I knocked on the door. Then the call barrelling up the concrete path beside the pale blue bungalow:

'*Down here! I've been waiting for you!*'

I placed a tentative boot on the shed's top step, checked for the dog, then hauled myself inside. The owner of the voice was sitting in a pool of filtered light, an Olivetti at his elbow.

'Lost my hand-held compass in the divorce settlement. Been struggling for direction ever since.'

He laughed, and gestured towards a stool.

'It's not the first time and it won't be the last. Percy Street has a reputation. Here, have a seat.'

Then, without further ado, Harry Bioletti, author and retired high school teacher, pulled out his notes and began reading.

'He was a gifted boy. He was gifted academically and was recognised as such by his fellow pupils. He was also very self-contained. Living in a world of his own, is how one of his fellow pupils described him. That's not to say he was withdrawn. Far from it. In his sixth-form year he was made a prefect, captained the hockey first eleven and represented the school at athletics. Only a waif of a boy, not the rugby-playing type, he was able to stand up to the combined furore of the academics of the school when he decided to chuck it in. When I look back on the young Bruce Farr the central thing for me is this: he must have been tremendously single-minded in knowing what he wanted to do from a very early age.

'I taught him Technical Drawing in the third and fifth forms. His drawings and printing were always beautifully done. He was outstanding in this subject. He got the top mark in the school in School Certificate: 93 percent. But he also excelled at science and mathematics, regularly topping his class with marks in the eighties and nineties. Mind you, it can't have been easy. He had a hard act to follow. His brother Alan had been outstanding. He had gained Mahurangi College's highest-ever School Certificate mark. In his last year he was joint dux and awarded the David Wilson Memorial Prize in mathematics, chemistry and physics. With these results Alan won a scholarship to Auckland Grammar School for his seventh form year.

'Cliff Brooking, the headmaster, was a dogmatic man. He hadn't been there long and naturally wanted his new school to do well. But for it to do well academically he needed his brightest pupils to go on, especially to scholarship level. "You're what?" he said to Bruce when he told him he was going to leave. "You're going to ... to mess around in boats?" It was unheard of for a pupil of Bruce's ability to leave partway through the year. Brooking was livid. He got Bruce's parents in and had them on the mat. To be fair, he wasn't just doing it for the school. He genuinely believed Bruce was throwing his life away. It was at a time, remember, when no one earned a full-time living designing boats. And there were only a few in the country making a name for themselves. A chap called Spencer was one — the person, it was rumoured, who had given Bruce this advice ...'[6]

It had been a shock. But by the time Cliff Brooking came to say goodbye to his star pupil he had regained his equanimity. His letter of testimony was both studied and gracious.

> Bruce Farr ... left the College at the end of the first term 1966 to set up on his own as a boat designer and boat-builder.
>
> Bruce has an excellent record at this College. His undoubted academic and sporting abilities were never displayed ostentatiously ... He was most impressive in any situation demanding intelligence and skilled application of effort. In short he can be recommended without reserve as a most pleasant and intelligent young man who should grace any profession that he chooses to enter.[7]

Convinced that doing what you wanted in life was what mattered most, Jim and Ileene supported their youngest son's decision. Although made just days before he left, it was made with absolute conviction. Still, it wasn't straightforward. Naval architecture was virtually unknown in New Zealand. Nowhere was it taught as a discipline. Nor was there much in the way of reading material available, although Olin Stephens had replied to a letter from Bruce with a list of books he might read on the subject. Then there were the persuasive arguments from vocational guidance teachers. Professions like architecture and engineering had a lot going for them. There was also the vague feeling of unease within the family with Alan's independence, his mixing with the wealthy when he went to Auckland Grammar. His university career, already brilliant, and a serious injury sustained on a forestry holiday job, seemed to make him even more distant. There was talk of his becoming involved with long-haired radicals, and of a growing association with their most colourful spokesman: Tim Shadbolt. Still painfully shy, Bruce even wondered if he'd be able to handle the campus social life. But even as this swirl of decision-making went on around him, sitting there like a faithful friend was the one thing that gave him direction. This, it seemed, was why he breathed. No matter the soundness of conventional wisdom, the dream just wouldn't go away. It was why he had taken to technical drawing 'like a rat up a drain pipe'.[8] Even the commissions kept coming.

In truth the seed of his decision had been planted the previous year when he had visited John Spencer at his Browns Bay yard. Nervous for the first 30 seconds, Bruce thought it a bit like meeting God. Clad in corduroys and a rib-knit sweater, the bearded one crawled out from under the 62-foot plywood racer he was building for industrialist Tom Clark. Spencer would not comment on when Bruce should leave school. But when it came to drawing yachts his advice was unequivocal.

'If you want to be a yacht designer, you'd better learn how to build them.'

Perhaps more attributable to Noah than God, to a fifteen-year-old they still sounded like words from on high. It is advice Bruce still passes on to this day. Drawing boats, he will tell you, is about organising communication for the boat-building process. A designer is just part of that process. If you don't know how to build a boat then it's hard to understand when something goes wrong or when a drawing doesn't communicate itself adequately to the builder.

In reality, the step to a final decision had been less of a formality than a matter of course. On a May day in 1966, at the end of seven kilometres of unsealed road, Bruce Farr stepped off the school bus for the very last time. His formal education was over. At sixteen years of age he was now a full-time yacht designer and builder. Until he could figure out something better, his workshop would be the place where he had built *Resolution* five and a half years earlier: the basement of the family home in Matheson Bay.

Bruce had not only designed and built boats for himself and Alan. He had been designing and building boats for paying customers since 1964, as well as selling his lines plans. There was even a repeat order: *Thermoth* for New Zealand Post Office engineer Vern Smith, who offered to tutor Bruce in engineering and mathematics after he left school. That led to his first commission that wasn't a Moth. In experimental mood, Merv Elliott ordered a single-handed Q Class with a sliding seat. 'A real handful but with moments of genuine speed',[9] it would become the forerunner of the revolutionary Farr 3.7. Then came his only multihull commission, an eighteen-foot A Class catamaran called *Rascal* for Douglas Haigh. Soon after, Gary Banks, an Elliott relative by marriage, stepped forward with the first of the seventeen-year-old's glamour commissions, the eighteen-footer *Cool Leopard*.[10]

It was over the next nine months, as he worked on these design-and-build commissions, that Bruce began to come to grips with some fundamental issues. He was learning a lot about his limitations. Pushing hard against the boundaries of that knowledge, he could hear John Spencer's words ringing in his ears. His life stretched out before him and being a backyard builder of boats wasn't part of the composition. He was also self-employed, a man of independent means, and that didn't seem to sit too well with living at home with his parents.

The only solution, it seemed, was to move to where the work was: Auckland. If he could get a job building boats, if he were able to pay his dues, then maybe, one day, he would be able to set up his own design office and perhaps a boat-building yard in the country's yachting capital. As an interim measure, Jim and Ileene rented a flat in Nile Road, Milford, on Auckland's North Shore. It would do for Alan and Bruce. It would also give them a base as they shuttled back and forth tidying up their affairs in Leigh while they looked for a home in Auckland.

But this was 1967. The industry was quiet. Still dominated by the amateur home builder, it was also shackled by regulation. Boat-builders weren't supposed to use native timber. Yet their clients only wanted boats built of kauri. Even the native timber which could be bought had to be stored for six months before use. This environment would only change as synthetic glues drove home their advantage and the use of exotics became accepted in construction. But if that weren't enough, for a designer of yachts the numbers weren't healthy either. Seventy percent of the market was made up of boats driven by power not wind.

Undeterred, the eighteen-year-old laced up his boots and went labouring for house-builder and friend Graham Crooks. That would keep the money coming in while he went door-knocking. He was even prepared to kiss goodbye to four years of freedom by becoming a boat-builder's apprentice, if that's what it took to realise his dream. But he didn't know before leaving Leigh that there was an industry quota for apprentices. Nor that most boat-builders in Auckland were already up to their limit.

When he wasn't on a building site Bruce was pounding the pavements. His first port of call was the designer who had been his inspiration. John Spencer wished he had an opening. Sadly, he didn't. Bruce knocked on the door of the person widely regarded as the best builder of wooden boats in the country. But Brin Wilson had just apprenticed sons Robert and Richard and had no room either. Bruce tried Max Carter. When that door closed he took a bus to Birkenhead and walked down Hinemoa Street. He knew the ferry service to Auckland no longer ran from the wharf and that Jim Young had taken over the corrugated iron building. The setting looked promising. A stone's throw away a small armada of yachts and launches swung from their Harbour Board moorings, the adjacent park a picture under its panoply of leaves. And Young, he knew, had a reputation to match his bustling business.

He was just another young chap off the street, although I had heard of his success in the Restricted Moths and knew he'd designed and built an A Class catamaran. We had two apprentices at the time and weren't allowed any more. But I needed additional labour. So the way around that was to hire Bruce as a fully fledged boat-builder. It was good for me. And as we were designers as well as boat-builders, that suited Bruce.

Half his time was spent on the shop floor and the other half drawing. It must have been rather boring work. But he did whatever was required: labouring, building, drawing, lofting. He worked hard and was very much part of the team. He'd get into the usual scuffles with the lads and end up getting thrown in the tide. He appreciated my giving him a corner of the shed in which to build *Miss Beazley Homes*. He did that with Rob Blackburn's help in the evenings and on the weekends.

In 1968 he entered the Junior Offshore Group (JOG) design competition run by the Royal Akarana Yacht Club. It was his first keel-boat design. I remember it well because he'd drawn bunks on both sides of the boat so the crew could sleep to windward. Nobody had done that sort of thing before. I thought it probably the best in the competition but knew it wouldn't win. The most outstanding are often a bit unusual. Which, of course, is why a committee never votes for them.

He came into his own when he helped me with a 40-foot rating yacht I had designed for a client. In those days few rating yachts were being built in New Zealand. That was partly because we only had mechanical and electric calculators with which to do the long and complicated sets of calculations. Using long-handed mathematics — well, you could go for five days before you knew whether or not you had made a mistake. So his ability to think mathematically was invaluable. He and I worked on this thing knowing that the

Royal Ocean Racing Club (RORC) Rule had come to an end and the International Offshore Rule (IOR) was going to take over. Problem was, the new owner didn't want to wait until the new rule was out. Which meant Bruce and I had to second-guess it and make my design fit a rule we hadn't seen. Not long after leaving me he designed *Titus Canby*. It was not only a fast boat. With his mathematical ability he was able to adjust his design to get a satisfactory rating within the new IOR. The whole era of yacht design we have now was just getting going then. And Bruce was there at the beginning.[11]

Still very close, Alan and Bruce moved to the bach in Castor Bay their parents had bought for their use. Here the different routes their lives would take were further defined. Bruce remained wrapped up in boats; Alan sailed less. Alan's degree in mathematics and his part-time evening work were involving him more and more with computers. At the time it was a world away from boats, although Bruce did wonder if he could write a program for him converting the measurements of the RORC to the IOR. For the most part Alan's friends were young intellectuals. His world was an esoteric one, understood by few. Nevertheless, he remained the more outgoing of the two brothers. He was the university's Student Association Public Relation's officer. Dubbed 'my Public Apologizer'[12] by Tim Shadbolt, it wasn't surprising. As a capping stunt, Shadbolt and his team of pranksters had just stolen a truck from the Army's maintenance yard at Waiouru where it was in for a service. The drive overnight up the middle of the island to Rudman Gardens in the Auckland Student Quad was embarrassing enough. But what got the Army really going was that the students had made off with machine gun ammo as well.

Bruce, still cripplingly shy, confined his social life to yachting circles. Alan, on the other hand, had recently met and fallen in love with an undergraduate. Two years later, in 1970, he and Pamela Duncan would marry and move to Wellington. There, while working for the DSIR, Alan would complete an honours degree in retrieval systems, and Pam her science degree. In 1973 Alan joined an elite group at Burroughs, a world-wide think tank working on the next generation of computers.

In the meantime, Jim and Ileene bought a pretty wooden bungalow in Devonport: 11A Tudor Street. Jim lowered the basement floor to give it more headroom. If Bruce needed a place to build more boats, that should do the trick. Jim would later add a granny flat, then a room with a view from where they could watch the yachts racing on the harbour.

It was late in the 1969 season, just before the national eighteen-footer championships, that Bruce got a call from Warwick Goldsworthy. Frank Blackburn, the helmsman of his eighteen-footer, *Guinness Lady*, had been tragically killed in a car crash. Would Bruce like to steer? That was like asking a bird if it wanted to fly. Being one of the first three-man eighteens, and with Bruce helming from a trapeze, they proved more than competitive. In a three-week period they won the New Zealand eighteen-footer championships and came close to taking the world title in Brisbane. A third in that world championship cemented Bruce's relationship with the class. And when he arrived home to a new commission, it turned into a love affair. With what was to follow, the timing could not have been better.

1970 was Bruce's third year with Jim Young. By then he had become a full-time draughtsman, churning out stock plans of power boats and New Zealand's biggest-selling launch range: the Vindex. But it was also the year when things began falling apart. Jim Young had been approached by Graham Painter with the thought of turning his brilliantly performed design, Flap Martin's *Namu*, into a yacht for export. A company, New Zealand Yachts, was formed with Graham Painter, accountant Murray Tracy and J H Young Boats Ltd as its shareholders, and premises taken in Barry's Point Road, Takapuna. The aim was to export one to two wooden NZ37s a month to California, a market that was already fibreglass-oriented. That under pressure from US buyers the yacht's specifications were increased without an adequate price increase meant the NZ37s were being sold at a loss. That management failed to pick this up until it was too late was a tragedy. It took Jim Young's business with it.[13]

> Jim was a neat guy who made a lasting impression on me. Considered a bit of an outlaw, he was both an innovator and a really great designer. He did everything from catamarans to monohulls, from launches to swing keels. Witnessing his open-thinking approach to things was a very big lesson in my development. He also went out of his way to create an opportunity for me. He knew I wanted to be a yacht designer so he got me doing lofting when his Australian draughtsman went back home. I also learned how to build real boats properly. Sad as it was, the collapse of New Zealand Yachts turned out to be a blessing in disguise for me. With the future looking wobbly and with another eighteen-footer design-and-build having just come along — this time for Wayne and Noel Fleet — I decided to leave.[14]

At that stage Bruce had just enough money (£150) to buy an Austin A40 Countryman. More as encouragement for their socially retiring mate than from hard evidence, his sailing buddies named it 'The Passion Wagon'.

With six feet of headroom in the Tudor Street basement and just enough length to house an eighteen-footer hull, Bruce had the work space to begin building on his own account again. At the far end was his mother's sewing room. With ample light from a large window it looked like the perfect studio. And when Ileene returned to work after a three-month summer cruise and was sent packing by her employers, the 21-year-old landed the perfect secretary.

Bruce knew there wasn't much money in what was still regarded as a part-time activity; net income before tax from his first year as a full-time designer had been $500. But if he could build a little on that he'd be able to pay board, put gas in the tank and shout a few after-race jugs for the boys. What mattered most was that he was doing what he believed in — even if it meant, a decade on from *Resolution*, he was back doing it in the basement of the family home.

Chapter 6
Lord Sandwich

Bowler, who is only 21, has been hailed as one of the greatest open-boat skippers ever to come to Australia.
The Sun-Herald, Sydney[1]

A 12-footer is 12 feet long, has a 25 foot mast and a 500 square foot spinnaker. They are almost impossible to sail.
Bruce Farr[2]

Henry Cooper was a headmaster who selected his school's yachting representatives out of a hat. Sailing was a recreational pastime; it was not a real sport like rugby or cricket. His method of selection meant Bowler and Blakey missed out on defending the Auckland secondary schools title they had won the previous year for Auckland Grammar. This was a head who could also refuse to accredit most of a top-stream class their University Entrance because their extracurricular activities flew in the face of authority. Disgusted, Russ Bowler stormed out of school.

Having to sit the national examination was a thing of dread for most secondary school pupils. The attrition rate was high. But so was the level of determination in the youngest Bowler sibling. He worked hard and knew his stuff. Still, it was with enormous relief that he read his name among the *New Zealand Herald*'s small list of successful candidates in 1963. More than one in the eye for Henry, it was a defining moment for the sixteen-year-old. Never again would unthinking authority hold sway in his life.

Rather than studying law like his dad, Russ chose to follow the career path of his sailing buddy and brother-in-law. In 1965 he embarked on a four-year civil engineering course. It was conducted by the New Zealand Institution of Engineers, whose exams were set by its British counterpart. The course was arduous: one two-hour night and two four-hour nights, plus half a day per week of lectures at the Auckland Technical Institute. There was a high drop-out rate. But studying while he worked for the Ministry of Works meant

Russ could continue with his yachting as well as establishing his independence. And if he sold his BSA Bantam he might even be able to buy a red MG.

***

The 1951 success at Glendowie of Dave Mark's Pennant Class *Pathetic* convinced a growing number of enthusiasts that a twelve-footer class of unrestricted design had enormous potential. By 1955 there was enough support for the Q Class Owner's Association to be formed at Tamaki Yacht Club. No one was surprised at the exciting racing that ensued. Nor, given the experimental nature of the class, that the next significant step in indigenous dinghy design would take place within it.

After winning the Sander's Cup at Bluff in 1966 — with younger brother Jim and Tony Bouzaid as crew — Don Lidgard sold his fourteen-footer *Ecstasy*. Bereft of a boat for the following summer, he bought a well-performed Q Class, *Topsy*, from teenage track star Ross Duder. It was about this time that Joint Industries' manager Maurice Cosgrove talked to Don about the benefits of a new product his company was producing. Don and Tony Bouzaid mixed the contents of the two drums they'd been given and poured it into *Topsy*'s stripped-out hull. They sat back, watched the substance expand, 'then dug it out like a Maori canoe'.[3] *Miss Hycel* might have been competitive had it not been for the false wooden floor, the large dollops of fibreglass used to strengthen its hull and the serious delamination that followed launching.

'You'll never win a race in that, Don. Bloody thing's too heavy!'

Warwick Goldsworthy's words were like shards of PVC to the plumber-sailor's heart. Nevertheless, they stung Don into action. The following winter he and Tony turned *Miss Hycel* over. They covered it with a parting agent, resin-soaked fibreglass cloth and the new Joint Industries' product: inch-thick sheets of high density Hycel polyurethane foam. Polyester braces were set across the hull and two skins of eight-ounce cloth added to the sides. After glassing, the structure was entirely rigid. *Query* out of *Miss Hycel* out of *Topsy* had just been born. So had the country's first foam-sandwich dinghy. The year was 1967. Don Lidgard:

> To explain just how bad it looked, it would go along with its bow out of the water. But we had achieved an incredibly light boat. Tony and I could lift it on three fingers, something unheard of for a boat of that size.[4]

And successful too. In a field comprising three-times winner John Chapple and the promising young team of Bruce Farr and Rob Blackburn, Don Lidgard and Tony Bouzaid decisively won the 1968 Interdominion sailed on the Waitemata Harbour. It was only the fourth time in its twelve-year history that the Silasec Trophy had gone to New Zealand.

Although the experienced pairing of Lidgard and Bouzaid had worked a treat, at only 100 lb with all fittings attached *Query* had had a massive 30 percent weight advantage over its rivals. At least that's how Bowler and McGlashan read it. They also saw that *Query*'s rig could do with modification. As could its lines. There was much to ponder before Russ put pen to paper. It was time to talk things over with civil engineer John Chapple. Russ Bowler:

> John's boats were always interesting. From *Flamingo* to *Sopranino* to *Montrose of Penrose* you never knew what he'd come up with next. Towards the end of a season there was a period when you could jump from boat to boat and have a bit of a sail. You could always tell a Chapple design — very light on the helm, the rig a little further aft. And they always went nicely through the waves. He designed and built exquisite boats. And a neat guy with it. You could examine a boat with John and watch his mind get totally wrapped around the whole thing. He'd even draw the way to put non-skid on the gunwale or grip on a tiller. He showed me all his plans, explaining them in detail. He was both generous as a person and a terrific designer. So yes, there was definitely a Chapple influence in *Jennifer Julian*'s hull.[5]

In early 1967 Russ Bowler drew the lines of what would become Australiasia's most successful dinghy of the decade. It was round-bilged, drawn with a much deeper vee than was current at the time. In fact the vee ran all the way aft, producing much straighter lines than the standard hard-chined hulls. This straighter shape, Russ hoped, would enable waterline length to be maximised for upright upwind work while still providing good planing areas when heeled for downwind high-speed sailing. That, at least, was the theory. Russ and Don had hoped to have it ready for the 1968 Interdominion. However, the unexpected complexity of the construction — no one in New Zealand had built a foam-sandwich round-bilged hull before — meant it wasn't completed until after that contest had finished.

They had begun by building a standard mould of close-centred stringers. Over this they built another mould using eighth-inch Ivory Board. That was when the first problem reared its head: how to get strips of one-inch-thick foam to adhere to a round-bilge shape? With a brand new system of construction and no one to ask, they needed inspiration. They found it in the Sunday roast. With Don above the hull and Russ underneath, they spent a month in their spare time passing a needle back and forth, stitching the foam to the cardboard mould with string. Over this they laid two layers of fibreglass cloth and polyester resin. Sanding and fairing was followed with a resin sealer coating. After curing, Russ disappeared beneath the hull again, this time to pull out the string and cut away the cardboard. Turned over, the shell looked as it should but wobbled like a jelly. It needed two more inside layers of glass-fibre cloth before they had what they wanted: a monolithic flexing unit lighter than the lightest wooden hull. With sufficient buoyancy in the foam there was no need for tanks. Incoming water would sweep through the clean interior and out through the open transom. The aluminium space-frame, designed to pick up the structural loads of the mast and centre-case, gave them additional clutter-free strength. NASA-inspired but unique in its application, it was an idea of brilliant simplicity. It was also a concept Russ would store away for later application. The mast, another from the Bowler Pohutukawa Factory, was a three-inch diameter aluminium tube stiffer but six pounds lighter than a standard spar. And at one quarter of the cost it was also light on the pocket. The pocket-luff main, designed by Russ and made by Boyd & McMasters, was loose-footed and zipped over the mast to reduce windage. All these features were Bowler innovations. And all were firsts for the class.

Unsure of what they had on their hands, but eager to find out, the brothers-in-law launched their unnamed twelve-footer with neither primer nor paint.

> It was blowing a 20-knot northerly for our first race. Heats were being sailed for the national Q Class title following the Interdominion twelve-footer series just won by *Query*. We got to the start of one of these just as the gun went off. We went on the wind, climbed out on the two trapezes and drove straight through the fleet like a catamaran! I remember seeing John Chapple looking down, probably for weed on his rudder. Then we went past Don in *Query* as if he were standing still. We were first to Northern Leading. But no sooner had our big main driven us around the mark than we lost control and swam.[6]

Bowler and McGlashan spent the next few months finishing off their weapon. Ready mid-winter, it still didn't have a name. Not, that is, until prominent Auckland retailer and yachtsman Evan Julian suggested he could fund their sail wardrobe in exchange for naming rights. Julian, who had purchased the top apartment of the Bowler waterfront complex in Kohimarama, had not only been following his young neighbour's career in the papers. He had been glued to his binoculars over the winter months, watching the boys practising. The ex-World War II Hurricane fighter pilot knew a speed machine when he saw one. And a deal. Here was a boat worthy of carrying both his daughter's name and the name of the family store.

With its clean lines and simple rig *Jennifer Julian* may have looked an easy boat to sail. It wasn't. True, its deep vee gave it great speed and power upwind. But that same vee, run all the way aft to prevent 'mining', made it unstable downwind. A new technique would have to be found if they were to spend the rest of next season on, not in, the water. It took the rest of that winter to work out how to roll the hull on its side to get more shape in the water and to 'scallop' down waves at speed. Although it meant straight-line running off the wind was a thing of the past, they now had stability on what had been their weakest point of sailing. K kept score. It was eighteen-nil before her brother and husband could take their wonder boat out without taking a bath. In fact, so successful were the survival techniques they devised that winter that they would become standard in the class and influence future generations of small-boat sailors.

Having learnt how to drive their boat, Russ and Don were still unsure of its real potential. Only once had they been able to measure its performance. Their next opportunity would be the weekend before the national selection trials. Starting fifteen minutes late in light conditions, they sailed past the entire fleet as it sat seemingly anchored behind a southern leading reef. Their staggering 22-minute win had the pundits shaking their heads. Everyone was talking. *Jennifer Julian* was not only quick. Clearly, the foam-sandwich wonder was the outstanding prospect for the forthcoming Interdominion. Don McGlashan:

> It's not just his way-out thinking. The big thing about Russ is that he's not afraid to fail. Take that boat. Not only were its shape and construction hugely experimental. It had no foredeck and virtually no transom. Things like that were almost unheard of then. Of course there were the obvious — his famous space-frame and the pocket-luff main. But there were also the unseen things — the balance he achieved between the rig, rudder and a wider-than-normal centreboard. That meant we could perfect the art of twin-trapezing. Plus he kept experimenting right down to the minor details. He came up with this idea for securing fittings to a foam sandwich hull. He'd mix up a paste of micro balloons and resin, soak it in a strip of glass, roll it like a bandage and insert it into holes formed through the foam. That's why, when other experimental boats were falling apart, ours never did.[7]

When they lined up for the Q Class trials to select the national team, *Jennifer Julian* showed devastating speed. It finished top boat of the trials with three wins and a second. Finishing 1079 points behind, with three seconds and a first, were Bruce Farr and Rob Blackburn in their second *Miss Beazley Homes.*

***

October 1968 through January 1969 was a watershed in the life of Geoffrey Russell Bowler. In that first month he contracted hepatitis, sat the last of his engineering exams and won the national Q Class trials. Turning bright yellow transformed him into a social pariah. But that was the least of his problems. From running thirty miles a week he could hardly walk around the house. He was sleeping fourteen-hour days. For ten days he couldn't eat.

Barely recovered by January, he had less charge than a battery on the blink when the team arrived in Sydney to contest the Interdominion. It was a summer fierce with heat. Russ collapsed following a capsize during practice. After rigging the boat before each heat of the series, he would pick a shady spot and, exhausted, fall asleep. Then there were the sharks. Not the ones in the harbour; the ones on the ferries. Out of Russ and Don's peripheral vision would come these floating monoliths. If they weren't threatening to swamp *Jennifer Julian* with their wake they would demonstrate an uncanny knack of shadowing it upwind to disturb or take its air. Only then did the pair realise just how much their four-minute fifteen-second win in the first heat of the series had upset the Sydney bookies. And on top of that there was the wind — at times so dry you could feel it skinning your eyeballs.

Still, they managed to hang on. Russ, as if from memory, barely clinging to the wire; his powerfully built brother-in-law ever watchful and encouraging. Their first heat win proved that they could do it. But could they survive the 30 knots of the third and the 102-degree heat? Just, as it turned out, although they thought they were finished when Russ fell into the harbour. One-handed, Don hauled him out. A capsize avoided, the dip in the water and adrenalin rush worked a big recovery. In their smallest winning margin of the series they had edged out the Australian skiff *Pol* by a mere 30 seconds. A win by a collar. But in a heat when only 12 of the 24 fleet survived, to finish with maximum points was as good as you could get.

> The boat wasn't the reason we won. Nor the skipper. Nor the brilliant crew work. We won because we stayed sober for the whole tournament! Russ wasn't allowed to drink because of the hepatitis. So after each race we kept each other company down at the end of the bar, sipping lemonades.[8]

Accolades from the Australian press flowed so thick and fast it seemed the champion helmsman had put himself up for adoption. 'Russell Bowler is another dedicated sailor who joins that exclusive club whose members include the ... Jim Hardys ... Bob Millers ... Peter Manders, Paul Elvstroms of the yachting world,' announced *Australian Seacraft*.[9] 'Russell Bowler ... is one of the finest small-boat skippers to visit Australia in recent years,' trumpeted the *Sydney Morning Herald*.[10] And two days later: 'Richard Lee, publicity officer for the NSW 12ft Skiff Association, today described his performance in winning four out of the five heats as "phenomenal".'[11]

Indeed it had been remarkable. *Jennifer Julian* had finished with a 1/1/1/1 series, losing the discarded second heat by only eleven seconds.

But Bowler and McGlashan weren't the only well-performed New Zealanders on Sydney Harbour that January. Bruce Farr and Rob Blackburn had finished second New Zealand boat and fourth overall with a fine 7/4/5/2 result. And on Middle Harbour, contesting the Javelin Interdominion, were Peter Walker and David Hutchinson. *Rangi*, in its first season in the water, sported a swivelling aluminium mast from the Bowler 'factory'. Built from three sections of tubing with double diamond struts and adjustable jumper stay, Russ had also helped them tune it. For them to win four of their six heats and take that title as well was icing on the cake.

Praise from Australians can taste better than a beer in the outback. Nothing, though, tasted quite as sweet as the Bolly of a letter which had arrived that January at 227 Tamaki Drive, Auckland. 'Dear Sir,' it read, 'I am pleased to advise you that you were successful ...' It not only meant admittance to the New Zealand Society of Civil Engineers. It was Russell's ticket to the future.

Don went home to K. But Russ and Peter Walker decided that they still had things to do before they left Australia. Like crossing the Nullarbor Plains, for instance, and contesting the inaugural Cherub Worlds to be held the following year in Perth. It didn't take them long to agree on a strategy. Their first purchase was a map of Australia; the second, an old Ford Falcon. With their heads full of dreams, their bags in the boot and their water bottles beside them, they gassed up 'Joanna' and hit the Hume Highway south.

Back in Auckland there was no holding back Evan Julian: '... the store takes fashion honours and the yacht makes a clean sweep in the INTER-DOMINION 12FT YACHTING CHAMPIONSHIP held recently on Sydney Harbour,' shouted his advertisement. 'A combination of bold design and superb sailsmanship. Come to Jennifer Julian ... and see the magnificent Silasec Trophy on display ... tribute to the fastest twelve-footer in the world. Then hop into Jennifer Julian for fashions that take line honours every time. Like the stunning see-worthy bikinis you won't see on every deck, because they're exclusively made for Jennifer Julian.'[12]

Don even managed to sell the boat.

The team was unanimous. Januaries don't come much better than that.

Chapter 7
Wood is Good

I greatly deplore the ill effect that measurement has had on yachting ... These high-sided pot-bellied tubs with straight sheers and long ends may be all right for the mentally deficient ... but they never will appeal to the eyes of ... the true sailor.

L. Francis Herreshoff, American yacht designer[1]

Specializing under rules produces horrible boats. Rule-makers have done more harm to yachting than religion has done to mankind.

Jim Young, New Zealand yacht designer[2]

How to make sense of shape, its history and its voice? How to separate the dreamscape from the seascape? Originality from resourcefulness? Creativity from functionality? The artist from the artisan?

If history were the crucible of speech you could say it in a word: 'Aerodux'.[3] It is, of course, more complex than that. Even a micro-history is a fabric of intricate weaving: the warp of time, the weft of character, the multicoloured dyes of event and invention. True, it was resorcinol and urea formaldehyde resins that first blew life into this reveille of change. But before that dawn there had been a hundred years of history in the colony. Of scows beached in tidal inlets winching logs on board, or carting sheep against sharp currents. Of trading schooners running before squalls, their canvas-covered cargoes dark below deck. For without the Mathesons of Matheson Bay and the Scotts of Scott's Landing — and the many builders like them — the economy would not have had mobility. Without the work-horses of the sea there would not have been the thoroughbreds.

There would not have been the Baileys and the Logans, who between them set the standard for keel-boat construction and design in the southern hemisphere. For over half a century they dominated racing on either side of the Tasman, their yachts winning races as far afield as South Africa and England. Even today the names of their progeny are more than echoes: *Viking, Prize, Akarana, Moana, Rainbow, Ariki, Rawhiti*. All were synonymous with elegance and speed. Yet whether they had clipper or spoon bows, slack or firm bilges, fantail or tapered counters, whether driven by yawl or cutter rigs, they were all called forth by voices from the North: Herreshoff, Watson, Nicholson, Fife.

The northern hemisphere designers remained beacons to the Collings and the Wilds, the Woollacotts and the Robbs. And those who produced yachts of distinction in that second great wave of building — the Lidgards, the Wilsons and the Woolleys, the Brookes, the Salthouses and the Robertsons — in turn paid homage to the van de Stadts, the Knud Reimers, the Rhodes and Giles, the Carters, the Sparkman & Stephens. Some produced straight-out copies. Some the real thing. It was not that these gifted sculptors of wood lacked imagination. It was that those letting commissions in a tightly reined economy wanted to be sure of what they had before it hit the harbour.

The cost of a keel-boat precluded experimentation. The centreboarder was well suited to a culture not given to elitism. There were no naval architects in the Dominion. No repositories of locked-up information. Everyone was a specialist who specialised in everything. The colony had survived on a free-flow of information. In the society of the sea it was no different; ideas spilled like water from a tap.

Experimentation was the backyard builder's theme. Even when they couldn't see beyond copper-fastened planking they could still make their craft stand on their feet. From the Napier Patikis flying across Ahuriri Lagoon to the skimmers planing down the Waitemata, the shape of speed was at the centre of the centreboarder dream. In time, urea formaldehydes, cold-moulded laminates and plywood sheets made these fast yachts even faster. And if their newer, lighter construction fell apart under loads that were too heavy, they were rebuilt over winter and made to travel even faster. Centreboarders were not only value for money. For their length they were feet faster than a heavy-displacement keeler. There was the rebel in them. They broke all the rules, including Froude's.[4] And like Richard Pearse's revolutionary monoplane, they made it possible to fly in the face of mathematical odds.

But how to spin the two dreams together?

Nathanael Herreshoff gave a hint of how in the early 1890s when he first separated rudder from keel. His fin keel yachts were much faster than their rivals. As was the Arch Logan *Sunbeam*. It not only proved that a low centre of gravity produced more power than weight in a long-run keel. It also showed there was power in reaction. Denounced as unwholesome and a vicious rule-cheater by the establishment it threatened, the wickedly fast

30-foot linear rater was sold to Australia in 1900. Self-interest had drawn first blood. It anaesthetised the dream until 1949, when van de Stadt drew a plywood chine yacht, *Zeevalk*, with a split underbody.

In Auckland, Jim Young pondered the problems of this suspect configuration. But unlike their northern hemisphere counterparts, Young and his compatriots were free from the constraints of designing to a rule. In 1953 he produced the ultra-light *Fiery Cross* with its pendulum keel swinging to weather. Jim Lidgard replied with *Aerial*. Then Young again in '55 with his fin keel-spade rudder *Tango*. Yet while they brought the dream back to life, their hulls were inspired more by Francis Herreshoff and Knud Reimers than indigenous shape or thought.

It would be another seven years before an amateur designer produced the lines that wove the two dreams together. When he drew *Patiki*[5] — a 34-foot keel-boat with a ten foot two inch beam — for Peter Colmore-Williams, Bob Stewart turned away from all that his narrow K Class *Helen* represented. He peeled back memory in search of inspiration. With its dish-like hull, *Patiki* would be a synthesis of the centreboarders which pre-dated it: the mullet boats, the Logan Patikis and skimmers and the Emmies he had sailed in his youth. Built in 1959 by John Lidgard, *Patiki* was cold-moulded with three layers of quarter-inch kauri skins, resin-glued and fastened with copper nails over stringers. Sheathed in fibreglass, it was light, strong and fast.

Two new equations made a splash alongside *Patiki* when it was launched in 1960: beam equalled power; weight equalled money. The gentleman-designer and urea-formaldehydes had produced a fast, cheap alternative to the traditional keeler. But with more than a hint of Illingworth in its masthead rig and Rhodes in its curved after lines, the boat that would become known as the Stewart 34 'was perhaps not the New Zealand original or true native that accepted history would have it'.[6] In terms of voice it was more a Kiri te Kanawa than a Howard Morrison. A diva of a dream, but in the end one that neither Townson, Spencer nor Young could quite tame. Townson produced two stunning versions before his eye was diverted by the lines of Olin Stephens' racers. Spencer had success with his sheet ply race-boats *Saracen* and *Infidel*, but seemed to lose his footing on the light-displacement playing field. Young's *Namu* was arguably the most refined of this new breed of yacht. Sadly, its adaptations foundered with that visionary export business. Jim Young:

> There are so many things that are obvious in hindsight you wonder how on earth they weren't thought of in the first place. When you think about it, a Stewart 34 is a beamy boat and when it lays over that's when the rudder becomes smallest. Therefore, if you're going to have a beamy boat you need a deep rudder. Originally it didn't. Those sorts of things weren't obvious then

although they should have been. There were text books available before the War on yacht design procedure. But generally speaking it was shrouded in mystery. There were lots of ideas that nobody was absolutely clear on. For example, there's no question today that if you can make a boat lighter it will go faster. But in those days it was thought that if you made a boat too light it might not sail at all. Or if too beamy it might not go upwind. And back in the 1950s, using beam to get stability then combining it with light-displacement was not at all clear. So the Stewart 34 showed us that the old adage — that length gives speed — needed to be modified to the extent that when length is restricted, it is stability through beam that gives speed.[7]

Just as religious myths have been used to create the notion of gender superiority, so a yachting myth was born that said long ends were better than short. And just as patriarchy had enshrined its concepts in laws and tradition, so those at the heart of yachting's establishment locked theirs up in a series of rules that proved the myth (and enhanced the value of their waterborne investments). The reality was that monohulls were caught in an evolutionary oxbow. A circular time-warp of physics and mathematics, it went something like this. Long ends were required to give additional buoyancy and waterline length in order to produce the stability needed for the vast amounts of canvas required to drive their excessive weight and large underwater bodies through the ocean. And when this shape began to change, the rulemakers changed the rules again.

The Stewart 34 might have fired a warning shot across the bow of its heavy-displacement long-ended rivals. But it would take a design genius and an Armageddon of a battle before an entire history of muddled thinking would be undone.

***

The two-year period spent in the Tudor Street basement was to be one of the most fruitful of Bruce Farr's career, and perhaps the most important. It was here that he would design and build his first world champion eighteen-footer, his Qs and Javelins, the 50-lb Moon Man dinghy you could row onto a plane, and his revolutionary single-handed trapeze yacht: the 3.7. And it was here that he would draw *Titus Canby* and *Moonshine*, his first keelboats. Here too he would meet a blue-eyed beauty from Brisbane. Come from Trade Consultants to collect his business accounts, she would leave with his smile branded on her heart.

During this time his habits of work, already well established, would be cemented for life. He would approach each new assignment as if it were for Alan or himself. He would produce construction drawings of minute detail, build half models of his keelers. He poured hours into projects. He worked long days and often long into the night. And after the planning, drawing, building and finishing, inevitably there would be a functional work of art ready to be rolled out from the basement for his client. Building boats produced cash. It continued to give him invaluable hands-on experience. And like the easel to the artist, it provided access to the one activity that kept him going creatively: designing yachts. But more than anything else the time at Tudor Street was a time of experimentation, its learning curve the steepest of his career.

The intensification of this process had begun with the move from single- to two-handed dinghy racing. The first *Miss Beazley Homes*, a development of his Moth designs, had low freeboard forward, a false floor and a centre of gravity that was too high. In spite of being fast through the water it 'mined' a lot, especially when it blew. Which meant its crew did a lot of swimming in their first full season together. A different design for their second season in twelve-footers was required if they were to have any chance in this highly competitive class.

But that didn't mean new materials. Like a proselytising convert the young designer-builder purveyed his belief that the properties of wood could do the business better than exotic materials. 'Wood is good' became his mantra of the moment. Which meant that while Bowler and McGlashan were stitching strips of foam to a cardboard mould, Farr and Blackburn were moulding two skins of laminated pine over a wooden one. While the 21-year-old trainee engineer was stiffening his hull with a space-frame, the 19-year-old designer-builder was fastening kahikatea ribs and stringers to his with copper rivets. While Russ and Don were shunting sleeves of aluminium against a pohutukawa tree to make a mast, Bruce and Rob were gluing strips of pine between a curvature of nails hammered into the wooden shutters in the Blackburn family's basement. Yet for all its adherence to traditional boat-building methods, the second *Miss Beazley Homes*, built in a corner of Jim Young's shed in the winter of '68, was fast. It also sported the first gybing bowsprit, a Farr invention, and carried a version of the Chapple gybing centreboard. But above all, it was the stable platform needed to carry their 220 square feet of working sail and 500 square foot spinnaker. Rob Blackburn:

> It's interesting to think that *Beazley 2* and *Jennifer Julian* were conceived at almost the same time. Yet they were totally different concepts — ours more traditional, theirs vee-bowed with a deeper hull form and, to my way of thinking, more along the lines of subsequent designs. I remember Bruce had a lot of respect for that boat and what Russell had done with it.[8]

Bruce's time at Tudor Street was also when he turned his attention to sails. He read voraciously. He conducted his own research. He wrestled with aerodynamics. He spent hours discussing spinnaker shapes with his sail-making friend Ross Guiniven. Nor was it long before he'd stitched the nuances to the basics and was telling sailmakers what they needed to do to make fast sails. And when friends like Hamilton's Peter Hutchinson asked him to run an eye over theirs during a race, Auckland sailmakers would receive a rash of orders the following week. He could drive his crew nuts. If Rob thought their sails perfect, the famous Farr focus would decree that they had to go back to have an eighth of an inch taken out of the fourth seam up ... and this much out of the leach of the number two while they're at it! Sailmakers soon recognised who was behind the growing flood of pin-pricking alterations to Q Class sails. Despite the angst, it was this attention to detail that eventually made Farr and Blackburn winners on the water. Fortunately they'd received a once-in-a-lifetime main from Nalder of Nelson, so well cut and sewn that it lasted three whole seasons with only minimal alteration. If there had been a prize for the promotion of peaceful coexistence — a must when travelling at speed in a barely controllable hull — this sail would have won hands down.

Yet in spite of his wooden masterpiece, wood wasn't always the won-der-working material Bruce claimed it to be. *Miss Beazley Homes'* pine rudders snapped with regularity at the waterline. Nor did his wooden masts always last the distance. But when they eventually decided to try an aluminium one, Bruce spent so much time shaving it with a sure-form plane that he removed metal from around parts of the weld and left it with some holes. Rob Blackburn again:

> Amazingly, that one never broke. But he was a fanatic about having too much weight aloft and getting the right heeling moment. The three years I spent with him in the twelves we argued quite a lot. It used to brass me off. I was the one who had been through Auckland Technical Institute and studied mechanics, the strength of materials and columns. Yet invariably he'd be right. And if he made a statement about something he'd defend it through and through. He upset a few people and was perceived as dogmatic. Yet his thinking was always crystal clear. And if he made a mistake he'd admit it, then head off in his new direction.[9]

It was during this period that Bruce and Rob travelled regularly to Hamilton Lake to race with the 30-strong inland fleet. On a couple of occasions they pitched their red and white striped 'circus tent' on the edge of the lake and slept in it overnight. It was a natural locus of post-race 'debriefing'. Pete Hutchinson gleefully recalled:

> This particular evening it was quite cold. Everyone had had one or two too many, and Bruce was sitting around with the rest of us, his sleeping bag up to

his waist, sounding forth about power-to-weight ratios. Today, with his glasses, you'd call him a nerd. But then, because he was young, no one really knew just how bright he was. One of the older blokes, who'd had it up to here by then, gave him a second warning. But there was just no stopping him. I'm telling you, Bruce insisted, if you want to sail a boat properly you've got to have the right power-to-weight ratio. With that, three of them lifted him up bodily and threw him into the lake.[10]

1970 was the year nearly half the New Zealand twelve-footer team came from the Hamilton Yacht Club. It was also the year Farr and Blackburn, with a 1/1/1/3 series sailed on the Waitemata, won the Interdominion title left vacant by Bowler and McGlashan. For the second time in as many years Mike Chapman came second in Pol. Miss Beazley Homes' three-year-old mainsail amazed the Australians, as did its home-made Oregon mast. Ever the forthright pragmatist, the winning skipper commented that you don't change sails just for the sake of changing them. And as for the only wooden mast in the 24-boat fleet, well, it's like your hull and fittings — if you've made and tested it yourself, you know it can be trusted. In defeat, Sydneysider Chapman was as observant as he was gracious. It's your do-it-yourself attitude, he said of the Kiwis, that makes you the more innovative, and that counts for more than anything else in this class. He was also full of praise for the new Silasec Trophy holders, for their crew work, and especially for Farr's ability both on and off the wind.

But Australian respect for the young designer didn't stop there. When he went public with his plans to design a single-handed twelve-footer, the 3.7, the contest's rulemakers quickly moved to ban it. The catamaran Kitty Jim Young had designed for Glendowie's John and David Peet had outsmarted them once before, in 1958. It wouldn't happen again.

Having finally won the Big One, Bruce and Rob thought it time to part company and pursue separate interests. Bruce threw himself into design-and-build commissions. Now a qualified engineer with the Auckland Electric Power Board, Rob decided it was time to take a well-earned break from the rigours of dinghy racing. Time to ease up on the body and think of something more sedate — like leaning into the windward rail of a nice little cruising keel-boat, an arm slung over the tiller and a fist full of the brown stuff. It would be an understatement to say that Bruce was not a happy man when Rob announced he'd be building a Spencer Serendipity.

'What about one of mine?'

'Apart from that JOG competition you didn't win, Bruce, you haven't designed a keeler yet!'

In the end, though, it was the camaraderie — which had begun at Browns Bay, and taken them countless miles around New Zealand and across the Tasman together — and their respect for each other that won the day. Not that it was a freebie. Rob Blackburn parted with $500 for the design, a small fortune in 1970. Initially he received neither lines nor plans for his little offshore cruiser-racer. Not even a receipt. But he did get a table of offsets and sections drawings. And that was enough to enable him to approach a local builder, borrow his shed at the top of Kowhai Road, Mairangi Bay, and, with Bruce's help, set to work lofting.

***

How often they arrive out of the blue, those events that signpost time. A battle (Hastings). A discovery (the Pacific). A thought ($E=mc^2$). A recording (Sgt Pepper's). A fraud (the Winebox)[11]. A crash (in Paris). Some leave the granite of fact and vaporise into myth. Others are rewritten. Most, especially those in yachting, are debated and debated. But the one called *Titus Canby* can be bedded down in fact.

When the 20-year-old designer conceived Rob Blackburn's yacht in 1970 at 11A Tudor Street his primary aim was to create a keel-boat that would be fast and fun to sail. And cheap, for his cash-strapped mate. Given their experience in centreboarders and with material costs roughly proportionate to displacement, there was only one way to go: as light as possible. Designing it to a rating rule was not part of this equation. It was designed as a correct boat first. Only later were the numbers massaged to make it fit less badly into the still-unpublished International Offshore Rule.

Although from a friend, although a commission additional to his main line of work, it was the opportunity of a lifetime. Which was why, from the outset, Bruce worried it to bits. This in spite of all he'd inherited. The art and mathematics. His family's practical approach to things. A mind not cluttered with blocked-off avenues of thought. An enviable clutch of local light-displacement designs he could turn to for inspiration. And all around him was the evidence of that Kiwi approach to solving problems.

But no one could have guessed what was to follow, especially those who were laughing in their beards as they watched this high-wooded, knuckle-bowed, sawn-off ugly duckling float off its cradle at Westhaven. They had no idea its diminutive splash would make waves all around the world. That its first gun would be the first shot fired in a design revolution unprecedented in offshore yachting's history. Nor that this revolution would escalate into a full-blown war between hemispheres north and south and last two decades.

Not even Bruce Farr knew, as he handed Rob Blackburn the details, that he had just calculated the biggest change to offshore yacht design since the Wizard of Bristol in 1891.[12] It was all so obvious. You build it strong but light. You sharpen up the bow so it cuts through the chop but keep it broad and flat aft so it planes. Drive it with a fractional rig and you have dinghy-like flexibility. It sounded so simple. Yet so completely was *Titus Canby* the sum of Bruce Farr's thinking, his sailing experience and one hundred years of heritage of a nation's yacht design, that for yachties all over the world it would change the shape of speed.

Chapter 8
King of Hearts

Telegraph Office
12 Jan 1970
Claremont 6010

Lt ... Russell Bowler
65 Bayview Tce
Claremont Perth

Congratulations To You And Peter Thrilled With Result
Please Confirm Using Our Sails Stop Urgently Require
Best Photo For Advertising World Wide
                        Boyd and McMaster[1]

Dust and distance define the Nullarbor. Travelling its breadth is like crossing an ocean of dust. The reddish-brown powder behaves like water and gets into everything — your eyes, nose and hair, the dash, the boot, your suitcase.

'Joanna' bumped along the Eyre Highway that heat-soaked summer of '69. From coastal Ceduna to Eucla, interstate meeting point of the first Overland Telegraph, to Balladonia in the west, there was little to break the monotony save the saltbush shrubland and the occasional belts of myall and mulga. No emus. No herds of roaming 'roos. Just dust and more dust swirling in cannonades as they left the rutted road to travel on the smoother surface beside it. Massive water tanks, positioned at thirsty intervals, reminded them they were not alone. At night they would park beside the road, light a fire, throw butter in a pan and devise ways of getting food into their mouths before the flies did. They would hold casual conversations three hundred yards apart and, as the heat abated, curl up in 'Joanna', in pools of sweat, trying to get some shuteye. The first time they heard the distant rumbling it took an hour and a half to arrive, the massive road train exploding past them in a blaze of lights.[2]

When they arrived on the other side of Australia, seven days after leaving Sydney, Russ and Peter Walker found a city jumping out of its skin. Gold was still worth plenty. But now there was nickel. Lang Hancock had not long flown over his iron ore mountains and cut that famous royalty deal through his company, Hamersley Iron. And a young arriviste named Alan Bond had not long wooed and won an ebullient redhead of monied parentage.[3] The English migrant had been painting signs and houses; now, with the next-best thing to a banker in the family, it made more sense to buy the houses and get others to paint them for you. But Bond wasn't the only one changing Perth's landscape. High-rise buildings were shooting up faster than mushrooms in a Waikato paddock, and restaurants and wine bars growing like date palms around an oasis. And that, on both accounts, suited the young New Zealanders admirably.

Their first stop was Basil Wright's. A key player in the Cherub Movement, he welcomed Russ and Peter with open arms. So did his sons, both Cherub sailors. No questions asked, the Wrights' was to be their home while they found jobs and an apartment of their own. Russ joined an engineering consultancy firm and Peter architects Forbes and Fitzharding, the latter a prominent member of Perth's sailing fraternity and soon to be involved with Bond's early America's Cup bids. Only months out of Auckland and the two young friends walked into undreamt-of responsibility, designing and overseeing multi-storeyed building projects. They also walked cashless into banks and walked out owning brand new cars.

In the early days there were times when they missed home and the activity that had gone on around them. Russ missed Lynda. They wrote often, unsure of when they'd meet again. But Perth was a good place to be. Its sailors were competitive and fun-loving. And after joining the Mounts Bay Sailing Club, Russ and Peter were soon absorbed into its activities. There was a Wednesday night series through the summer, weekend races, and the Wrights' frequent pool and barbie parties. As well, there were the pubs and wine bars to fall in and out of as they each began to spread their social wings.

The dinghy community, as well as the two young Kiwis, was also looking ahead to the inaugural Cherub Worlds. As the first world championship of any kind to be held in Western Australia, it was an event the state capital was looking forward to as well.

Russ and Peter may not have had the tools with which to build a wooden Cherub, but they did have ten years of Cherub experience between them. They knew what a Spencer Mark VII could do. And after half a season sailing as guests on the Swan River they knew that too. They knew where it was flat and where, above the massive dredging holes, it coughed up short sharp waves. They also knew that the course for the Cherub Worlds would be an Olympic one, not the usual circular type sailed on weekends by the various clubs competing for space on the river. That, they figured, would mean more chop than usual and the need for a hull with more spring than the straight runs currently being built by the Aussies. They also assessed the effect of their combined weight on a hull that would weigh less than the bodyweight of one of them. Only then was the displacement curve set and, with Peter at his shoulder, the diagonals plotted and the drawings completed by Russell. Like the first *Jennifer Julian* there would be sufficient buoyancy in the foam-sandwich hull to obviate the need for tanks yet still give them a calculated buoyancy figure in excess of that required by the Cherub class rules.

Employing the same methods of construction as before, they literally stitched the hull together. Then they faired and glassed it. They stepped the mast eight inches further aft than usual. And strengthened the shell with alloy tubes that tied the centrecase, chainplates and mast together. From the beginning they involved the local measurer; with the only foam-sandwich boat in the series, they wanted no controversy. Although not required by the rules, they decided to place a canvas cover over the area where a foredeck would normally have been. The second *Jennifer Julian* was not a piece of furniture. But with its clean interior it was light and easily worked. Simplicity is what they'd aimed for and simplicity is what they'd got — and at less than half the cost of their rivals' boats they had also got it cheap.

The Australians knew they were up against it when *Jennifer Julian II* won the first of three selection races for the Western Australian team to compete in the national champs. Sailed out of the Nedland Yacht Club, only sixteen of the 52 boats competing were eligible for selection. The win was all the more surprising to locals because it relegated the favourite and three-times national Cherub champion, Gordon Lucas, into second place and the fancied John Cassidy Jnr into third.

The Australian national championship began on Boxing Day 1969. With five-foot seas and a sou'wester gusting 35 knots the conditions were tailor-made for the sailors from the Shaky Isles. They won comfortably but were under duress thereafter. The Aussies weren't happy with non-bloody-nationals winning the first heat of their national title in a $400 boat after the small fortunes they'd spent on their own. With the wind dropping, the New Zealanders were fortunate to snatch second place in Heat Two. Sailed in a fluky easterly — the conditions that would prevail for the rest of the week — they only beat *Jazzer* when forward hand Tony Barnes fell into the river ten yards from the finish. In the end, after the fifth heat and discarding worst performances, the brand-new *Ace III* and *Jennifer Julian II* were tied on 6437 points. *Ace's* two firsts to *Jennifer Julian's* one was all the officials needed to award the national title to Gordon Lucas and Philip Arnold. Off went Lucas and Arnold to celebrate their unprecedented fourth national title win. But they hadn't counted on the wily Kiwis and their experience of protests. Softly spoken, but with the hard logic of their academic training, Bowler and Walker approached the committee.

'Given that we've sailed this contest under International Yacht Racing Union rules, we believe the correct method for breaking a tie is to add up the four placings used to compile the original points and award the contest to the boat with the lowest total score.'

'Oh?'

It was the last protest to be heard, just half an hour before the prize-giving was due to commence. The protest committee agreed with the Kiwis' logic. The race officials, adding and re-adding the scores, agreed with their arithmetic: *Jennifer Julian*'s 1/2/2/3 did add up to two points less than *Ace*'s 2/6/1/1. And so it was that the race committee chairman, at 8.25 pm, mounted the podium and announced to the glittering gathering that they had made, well, a rather unfortunate error. It was tough on the unsuspecting Lucas and Arnold, just returned to collect the winner's trophy from the Governor General. But they would have a chance to turn the tables on the New Zealanders when the world championship began on January second, in two days' time.

For those two days Bowler and Walker left *Jennifer Julian II* outside the committee room, a precautionary and penitential offering to the class rules and the whims of officialdom. They wanted no eruption of debate around their revolutionary craft once the world series began. They even replaced the canvas foredeck with a plywood one. Knowledge of the rules was one thing, adherence to their spirit quite another. Especially now.

Held on Melville Water the day before the Worlds, the invitation race was robbed of extra interest when *Jennifer Julian II* dropped its triple sleeve mast soon after the start. It did, however, produce a taste of things to come — a close finish between Lucas, Wilmot and Brian Wright.

The main event, a seven-race series, contained a high-class field of 48 boats: two from the United Kingdom, seven from New Zealand and the first 39 from the Australian national championship. Yet in spite of the numbers it was obvious from the beginning that it was going to be a three-way battle between Lucas and Bowler, representing WA, and the gifted sixteen-year-old from New South Wales, Jamie Wilmot.

Only a minute separated *Ace*, *Jazzer* and *Jennifer Julian II* throughout the first heat, sailed in a twelve- to eighteen-knot sou'wester. While Lucas and Wilmot showed greater downwind speed, it was Bowler's ability to pick the shifts that gave him and Walker the edge. With each boat holding the lead at different times, *Jennifer Julian*'s five-second win from the fast-finishing *Jazzer* was to be the closest of the series. *Ace III*, in spite of having capsized once, was not far behind in third. In Heat Two, Bowler's genius on the wind saw them break from the pack, pick up a puff near the Perth Flying Squadron and finish twenty seconds ahead of Lucas's *Ace III*. Two wins from two heats and the expats' prospects were looking promising. More outstanding windward work saw *Jennifer Julian II* gain six places on the final triangle of Heat Three to finish second. And with *Jazzer* finishing fifth and *Ace III* twenty-fourth, it now looked as if the title would be going to the West Australian Kiwis. But three things were to work against them: continuing light airs, the young Sydneysider, and a sudden loss of coordination. In the fourth heat, a five-knot sou'wester puffed *Jazzer* and its lightweight crew into first but left *Jennifer Julian II* gasping in fourth. Second at the wing mark in Heat Five, Bowler and Walker capsized sensationally in the middle of their jibe. Like wounded ducks they lay in the path of 46 Cherubs, their crews all screaming 'BUOY ROOM!' as they bore down on them in fifteen knots of air. It not only scared them witless. By the time they had righted their boat they were last. They improved to nineteenth, but this would have to be the result they dropped when their best six finishes were totalled. It was the wake-up call they'd needed; there could be no more mistakes, not with *Jazzer*'s two wins in the last two heats and an overall lead in the series. Even then the drama wasn't over. Just before taking to the water for Heat Six they discovered that their top rudder gudgeon was broken. Patching it with lashing and tape, they did their best to forget the problem as the sea breeze grew to a vigorous 20 knots. Which was when things began going their way again. *Jazzer* retired early when Jamie Wilmot suffered a severe attack of asthma. And unlike *Jennifer Julian II*, which had been drawn specifically for these conditions, the extreme straight sections of the Aussie boats were not handling the choppy flat runs well. Several boats, including *Ace III*, capsized. And with *Jennifer Julian II* again taking the lead on the windward leg, Bowler and Walker breezed in with their biggest winning margin of the series: 1.30.

It was now down to the final heat.

Gang warfare was in the air when Bowler and Walker found themselves surrounded by a host of Sydney Cherubs during the pre-start manoeuvres of Heat Seven. The unwelcome attention turned the normally passive helmsman aggressive. Executing a couple of turns, he extricated himself, came up on the wind and hit the start line ahead and just upwind of *Jazzer*. When Wilmot and Barnes attempted to break away by heading for the wrong side of the river, Bowler and Walker covered. And they kept shadowing the Sydney pair until the latter had effectively sailed themselves out of contention. The Kiwis then broke away and worked their way through the fleet to snatch fourth place. With *Jazzer* ninth and *Ace III* eighth, the series was *Jennifer Julian*'s. Ashore and surrounded by admirers, the Kiwis popped the champagne corks. And as the cameras clicked away and reporters fired questions, Bowler addressed the gathering. After thanking all those who had made the inaugural Worlds such a memorable event, he paid tribute to his crew.

'We came here together, built the boat together, lived together and sailed together. It was a real team effort.'

But no one was that surprised. Between them Bowler and Walker had, in the last year, won two interdominion titles, the Australian Cherub title and now the Cherub Worlds. In a brand new country among new-found friends they had proven themselves among the very best small boat sailors in the southern hemisphere.

The New Zealand team was quick to recognise that the inaugural Cherub Worlds spelt the end of the Spencer Mark VII. *Jennifer Julian II*, with its low chine, slight rocker, moderate flat run and wide transom, was the shape of things to come. That was how the New Zealand team manager saw it when he announced they would be taking its plans home for manufacture. The Australian praise was rather more direct. 'She's a foaming beaut!' wrote Jim Sharples in *Australian Seacraft*. 'Australian and World Cherub champion, *Jennifer Julian*, is a foam and glass unsinkable that has knocked the small-boat sailing world for a loop.'[4]

It was at the end of the Worlds that Lynda Smith arrived in Perth. The reunion was, to put it mildly, impassioned. But with Ces and Jean in Perth supporting their adored but still unmarried son, it wasn't exactly a cake-ride for the couple. In fact, their ingenuity reminded them of how it used to be making out in the red MG. And of the time a policeman shone his light through the window and told them to move on. Which, as they untangled themselves in the back of the Fiat 850, was the thought that once more crossed their minds. Maybe it was time to be doing just that.

***

The decision to go to London was an obvious one. After a decade of competitive dinghy racing Russ needed a break. A mecca for young antipodeans, London was particularly so for a newly qualified civil engineer. The large consulting firms that attached their names to the giant projects in Australasia and around the world were all there. Adding one of their names to his CV would be more than helpful when he returned home.

Leaving Lynda was the tough part. But Lynda was philosophical. If Russell wanted to wander off, that was fine by her. She had settled into a flat with friends, had a good job and a social life of her own. Perth was fun and she wasn't moving. At least not yet. There was also a sense in which this leave-taking had become a part of their relationship — unspoken recognition, perhaps, of their need for independence if they were to contemplate a future life together.

In a visible demonstration that no hard feelings existed from their fierce on-the-water rivalry, Russ leased a North London flat with Gordon Lucas and Philip Arnold.

Things weren't so congenial on the work front, however. Coming from the fast promotions and big money of Perth it was hard to adjust to the prevailing attitude of the large English firms. 'You need us' translated into lowly jobs and thin pay packets for those on the bottom rungs of the engineering ladder. It was an entrenched hierarchical attitude that had no appeal for Russell. But at Peter Heath and Partners, a small consulting group in New Bond Street, he eventually found what he wanted: interesting work with reasonable reimbursement.

For a lad from Down Under there was much to immerse himself in culturally and an amazing variety of warm bitters to try. Together with Gordon Lucas, Russ also made contact with some of the 250 Cherub owners in England and accepted an invitation to go sailing on breached gravel pits. They admired the use of space and found themselves in awe of the English sailors' equipment: sophisticated wind gear, wet suits, watches and compasses. But coming from a windy part of the world with miles of sea to stretch out on, it was hard to get excited about drifting around on tiny patches of water.

They also came up with a number of schemes to offset the effects of the low-wage high-cost-of-living regime they now found themselves in. Settling on the design and manufacture of sails, they bought bolts of fabric and an old Elna, and ripped up the carpet in their flat. Their reputations and the superior shape of their sails ensured they sold the first few sets. But neither attribute helped their margins much and, unable to compete with the British sailmakers' prices, they reluctantly gave their fledgling venture away.

Six months later, Lynda and Barbara Kendon arrived, squeezing into the Collingdale flat. At a turn life became hilarious and one of delicate manoeuvring, especially when Jean arrived for a visit. There were the inevitable trips to Europe and a half-hearted attempt at the 1971 International Moth Worlds in Cannes in his *Wooshmobile*. To supplement his income, Russ began doing quantity surveying work at night. That enabled him to trade up from his £10 Anglia to an Alpha Romeo, and helped him and Lynda to move into a tiny Kensington apartment. So did Lynda's job as model-receptionist for West End leather couturier Henry Lehr. Soon she was introducing Russell to a side of London life he didn't know existed.

It was on little more than a whim that Russ walked through the doors of the United States embassy and, much to his surprise, was approved for immigrant status. With six months to accept, by the time five had rolled by he had itchy feet again. His last-minute decision left little time for planning. It also meant the economic necessities of life (and Colonel Sanders) would determine his travel arrangements. A £30 charter ticket to the East Coast was all he could afford for himself, a 'freight forward' passage to the West Coast the only way to ship his Alpha. Following a farewell party the night before his departure, Russell found himself eating Kentucky Fried Chicken in a Hampstead Heath gutter with someone he'd never met before. Look this lady up when you reach New York City, the stranger told him.

Which he did, staying with her while he arranged with a car delivery firm to drive an Oldsmobile across America. The cost of his five-day stay was a promise to carry a large bunch of artificial flowers across the continent to his hostess's sister. The journey gave him the opportunity to reconnect with the friends his mother had made in Indianapolis all those years before. But with thoughts of his Alpha and high-rise projects on his mind, Russ politely declined their offers of work and hospitality and kept moving west. He also wanted the experience of arriving in a strange city knowing no one. Which, after dropping off the car in Los Angeles and running the gauntlet of shifty-eyed cowboys eyeing his bouquet and slim body, is exactly what he got. Arriving in San Francisco by Greyhound, he found a bed above a Chinese restaurant, delivered the flowers, and went looking for a job. It took 56 interviews before he was invited to join Deingenkolb & Associates and was assigned to high-rise foundation work. Not only was the work in the earthquake-prone area invaluable, but the pay was excellent. It enabled him to pay the freight on his Alpha and take an apartment. And when he came across New Zealanders working in the same field as himself, he was able to undergo the interview, required after three years of practical experience, to obtain his New Zealand Society of Civil Engineers papers.

As she had in Auckland and Perth, Lynda stayed on in London. There was still a lot to see and do. There were the Muira showings at the Prêt-à-Porter and elsewhere in Europe to help arrange, the rich and famous clientele to attend to. London was swinging. Oil-rich Arabs were buying up property like they were playing Pick-up-sticks. There was clubbing most nights after work, *Hair* to see, and birthday celebrations with joints instead of candlesticks and sundry parties to attend. But after six months even this began to pall — at least enough to prompt Lynda into buying a charter ticket to Los Angeles. Russ, on the other hand, had been thinking seriously about extending his earthquake experience by working in Japan. As it turned out their reunion in San Francisco did more than change his mind. It transformed the engineer into a romantic and sent him, during a candlelit dinner, down on one knee. They thought about getting married at a drive-through chapel in Las Vegas but knew it would break Ces and Jean's hearts. And so it was, in January 1973, that Lynda Smith and Russell Bowler, for the first time in five years, boarded a plane together. And, much to the delight of their families, married three months later in Auckland.

Chapter 9
The Mind of Design

Titus Canby

Designed by: Bruce Farr
Constructed by: Rob Blackburn
Year of construction: 1970–71 (15 months)
Year launched: 1971
LOA: 26 ft $7\frac{1}{2}$ in
DWL: 23 ft 8 in
Beam: 9 ft (max at 70 percent aft)
Stern: 6 ft
Displacement: 4820 lb
Ballast: 1740 lb
Draft: 4 ft 10 in
Sail area: 298 sq ft (main and foretriangle)
Hull thickness: $\frac{1}{2}$ in
Hull construction: Glass over ply. 3 skins of $\frac{6}{32}$ in Luan ply laid vertically with joints staggered. 7 stringers astride at 10 in centres. Bulkheads double-sided $\frac{1}{4}$ in ply on $1\frac{1}{8}$ in framing
Deck construction: Single layer $\frac{3}{8}$th ply on foredeck beams of 2 in x $\frac{7}{8}$ in at 10 in centres
Rig design: $\frac{7}{8}$ fractional
Mast: $5\frac{1}{4}$ x 4 Foster sections. Newly fabricated
Rigging: Single forestay and backstay. Cap shroud, single lower shroud per side. Single spreaders raked aft
Engine: 8 hp Kubota[1]

Cost:

Initial (excluding labour):	$6000.00
Extras (including life raft):	$1000.00
Total (in full racing trim):	$7000.00[2]

'When I think about *Titus Canby*, what blows me away is that you went from dinghies to a keel-boat and got it right first time. Age 20–21. John

Spencer tried to explain. Any good keel-yacht designer in New Zealand, he wrote me, began in the very competitive dinghy classes of those days, both designing and sailing them. That was helpful but it didn't explain process. I knew something significant had happened when I sailed *Redeemer* for the first time; going from a Herreshoff to a Farr 10^{20} was like going from a steamroller to a Porsche. In fact the only similarities were that they both floated and were propelled by the same natural force. But was it the quantum leap I imagined?'

'Well ... in a sense John is right. You do have to go back to centreboarders to understand its beginnings. The first yacht Rob and I sailed with a jib and spinnaker was a twelve-footer. With their massive sail area and short length they're probably the most challenging boat in the world to sail. We worked hard and went through a huge learning curve during the three years we spent in the class. But in terms of designing *Titus*, I was just doing what I had always done — trying to design a fast yacht that handled well.

'At the time it was the natural thing to do in New Zealand, to build your own keeler. And it was the only way Rob could afford one. He'd settled on a Spencer design — fairly narrow, good downwind but not sparkling upwind — when we began kicking around the idea of finding a better way of doing it. There were two things that stuck out in our minds. Could you get better windward performance from a light-displacement keel-boat without destroying its downwind and reaching potential? And could we build a round-bilged boat for not a lot more effort and cost than a chine plywood one? If we could it would look better, perhaps sail better, and have a better resale value. So that's how the original idea came about. And we decided we could probably do those things if we put more beam into the boat without adding much more weight.

'I went through a lot of soul-searching. I thought constantly about the concept, trying to figure out the displacement-length ratio up to which you could get enough weight into the boat to give good stability and power upwind but above which you would fall out of the downwind surfing-planing regime and end up with a slow boat.

'But there were a lot of other factors too in terms of the shape. I was convinced from the work I'd done that you needed a fine bow to get through waves. Earlier light-displacement keel-boats with good beam hadn't done that well. That is, although they had power, they didn't get good speed in waves in all conditions. And that's what we'd learnt in small boats — if you made the boat fine in the bow it got through waves a lot better. The downside to that was it made for worse handling downwind, particularly in big waves, and caused cranky behaviour when reaching and running. But I'd also learnt that if you shifted the centre of buoyancy further aft it counteracted that tendency and the boat became easier to handle. To some degree the answer was inspired by the Twelve-Metre yachts of the day. It seemed logical to me to go from a raked bow to one with a knuckled stem and U-shaped sections. That way I could get reserve buoyancy in the bow for reaching downwind in waves without making the bow too full for upwind work. It also helped to dampen pitch. That, I now realise, was a landmark decision — going for a fine-bowed boat that not only got through waves easily but extended its length in waves because of the small bow overhang.'

'What about your JOG design? Did *Titus* come out of that?'

'No. That design was more like a standard keel-boat of the day although it was lighter, had a bulbish keel and a separate rudder and skeg. One of the things that was different about it came from my reading books on aeronautics. In order to compensate for the added volume of the keel I reduced the volume in the bottom of the hull where the keel joined it. Although commonplace now it hadn't been done before then. I carried it to an extreme, to the point where there was almost a hollow in that area of the hull. This helped to lift the centre of effort of the keel close to the bottom of the hull. I did the same for *Titus Canby*, though not to the same extent. So although the two were far apart conceptually, the JOG design was a good stepping stone.'

'If you'd been thinking that intensely about appendages before your first keel-boat commission, it must have been a critical part of your thinking by the time you put pen to paper.'

'I worried a lot about the keel. If a keel is too small the boat gets blown sideways. If it's too big the boat trips over itself and ends up going sideways anyway. But somewhere between the two there had to be a shape that worked. The approach I took was that the keel needed to be a sensible plan-form shape. That was essentially an intuitive decision although I had been observing keels for a while. At the time, in order to get the lead low down and reduce wetted surface area, keels in New Zealand were being cut away in the middle and made big at the bottom. But that seemed to me to be an inefficient shape. In retrospect that's another thing we got right: a plan-form shape that is correct for producing lift. So I went for a relatively parallel-sided keel with a small bulb at the bottom. I'd also planned to have a trim tab on the boat. That seemed a logical thing for the keel although it made fastening more difficult. Which is why we settled on the central keel bolt scheme, which Rob engineered, rather than just using conventional bolts. This two-inch steel rod ended up going from below the lead's centre of gravity, up through the deadwood into a bulkhead three feet above the cabin sole. So even with something as simple as a keel we were doing things quite different from the norm.'

'And the rudder?'

'Hang it over the stern like a dinghy! By doing that it meant it would be economical both to build and maintain.'

'The rig? Did that also come from dinghies?'

'Definitely, although only after I'd read a book on yacht design and done the mast calculations — a first attempt to engineer something as opposed to guessing it. So it was very conservative. But we knew we wanted it to be fractional, without runners and not too tall a top mast. And that meant swept spreaders. That decision, though, was affected by other things as well. There was the efficiency of a big mainsail in terms of being able to depower the rig and the ease of handling the smaller headsails you get with a fractional rig. It was also a question of cost. Small-headsail–big-mainsail rigs are a lot cheaper than the configuration of masthead rigs with their huge headsails, the norm back then. That's because you have only one main — the sail you don't duplicate — and multiple genoas and spinnakers that you do. There's also a flow-on effect from that. With the loads being less on a three-quarter rig, the winches and tackle don't need to be as big and that becomes another area of cost-saving. But although a fractional rig was the right solution, we ended up being conservative. The rig on *Titus* was relatively low-aspect ratio — wide for its height — and the mainsail quite small for a fractional rig. It was a first tentative step. Later boats, including *Titus Canby* under Ian Gibbs' ownership, got more extreme fractional rigs once we were convinced it really did work. So I actually designed the rig and chose the spar sections, which was an unusual practice for designers then. However, I'd been putting as much effort into detailing the engineering of my eighteen-footer rigs as I had their hulls so it was a natural thing for me to do. And it's been a hallmark of our work ever since.

'The next step was to build a half model. After drawing the lines I thought: Oh ... this is different from anything anyone else has done! I wonder what it looks like? But once I saw its three-dimensional shape I felt comfortable with it.'

'And the method of construction?'

'The internal structure was built not too differently from the way I was building dinghies at the time, a lot of stringers over wide-spaced frames. These were big stringers — roughly an inch and a half square — and framed about every three feet. *Titus* was criticised by the traditionalists as being too light. But with what we know now — which is a lot more than we knew then — it was, in fact, a very robust boat. So even though the skin may have been light, set as it was over a strong internal structure, it was perfectly adequate for keeping the water out and dealing with the water loads.

'In terms of its general structure there's no question it was inspired by John Spencer. It was John who started the three-foot frame-spacing-with-stringers method of construction. But with my approach the stringers were more closely spaced and the skin a lot lighter. So instead of Rob putting big plywood sheets onto the boat — as he would have done with a Spencer design — he used eight- to twelve-inch strips of ply laid at right angles to the stringers. And when that was done we covered it with a fibreglass skin. As a structure it had about twenty-five percent more hours and cost in it than the sheet-ply method. But the extra cost and labour were worth it. And that's an approach we still take today: if it makes it more efficient in terms of weight we'll make a structure more complex.'

'One of the keys to *TC*'s success was that it fitted the IOR. Was it difficult to adapt it to the new rating rule?'

'When we conceived the boat in 1970 the RORC rule was being phased out and the IOR phased in. But that didn't happen until 1971 so there was little we could study. However, we knew certain things. We knew that they placed a high emphasis on beam, and both were nasty rules to design boats to. And I also knew what we'd done at Jim Young's. So while the rules had no influence on the basic design, for purposes of competition and resale we did decide to massage the hull shape a little.[3] I put a bit more beam into the topsides amidships. I placed the sheer carefully so the girth stations at the bow fell around the stem knuckle. I pushed it a little in the midship depth area where the measurements were taken. Lastly, I cut off the stern almost vertically to give it a lower rated length. All of which became important after the Half Ton level was struck internationally and we did, in fact, manage to squeeze *Titus* into a Half Ton rating.

'It was much the same approach we took twenty years later with our first IMS boat. We didn't run a single IMS VPP test while designing *Gaucho*.[4] We understood the essence of the rule but didn't study it to see how we could beat it. We just designed a decent boat relying on our years of experience then got it rated on completion to see what it was. But the similarities of effect were the same. *Gaucho*, for the time and market, was very fast for its size and brought about some changes to a rule that should never have been made. Likewise, people couldn't believe we could get *Titus* successfully rate to a rule designed to exclude light-displacement boats. And that's what startled the New Zealand yachting world. *Titus* was three to five feet shorter than all the other boats, half their weight and about one-third their cost.'

'So even your modern boats have their roots in *Titus*?'

'Not all the same features. But the same general style, yes.'

'And the pivotal design elements — those things you had to get right in *Titus* and did — were?'

'The displacement–length ratio: a boat light enough to surf in sea conditions and plane in flat water but with enough weight in the keel to be powerful upwind. And moving the centre of buoyancy aft by putting more volume in the back of the boat. That was critical. It allowed us to put the keel further aft and, along with the keel, the rig. Match a further-aft centre-of-gravity position with the upright centre of buoyancy and you have a boat less inclined to drop its bow when it heels. Which is what allows you to have a much finer bow for piercing waves. For its time *Titus* would have been considered to have had an extreme-aft centre-of-buoyancy position. It's still a hallmark of our boats. Even our IMS designs have centres of buoyancy that are as far if not further aft than we had on *Titus Canby*. It's one we're comfortable with although classical naval architecture would say our centre-of-buoyancy positions are too far back. What's interesting, looking back from this distance, is to see that so many of the things I did in that inspirational moment — or whatever it was — are still valid today.'

'And you can't explain it any more than that you just happened to get the fundamentals right at the right time?'

'No. It was largely intuitive, almost more the product of an artist than a technician. And the product of hard work. I had no other serious design job then — apart from designing and building dinghies — that consumed me. So I put an enormous amount of time into just worrying about what was right. And by getting a feel from observing other boats and where the barriers were for them. It was a case of something good coming out of a pretty big effort. That was my mode of operation then. And it's still what happens today. As a design team we put a huge amount of effort and detail into every project. I think that's a key. If you think about things longer you'll find a better solution.'[5]

Chapter 10
A Tearaway Called Titus

Tradition is the consensus of the dead.
Felipe Fernandez-Armesto,
Oxford-based historian and writer[1]

I'll let you into a secret not too many people know. If you'd put a bone in front of that boat it would have barked. The only reason it did any good was the sheer brilliance of its incredibly talented crew.
Rob Blackburn, builder/
original owner *Titus Canby*[2]

The shed, old but defiant, sits on its elevated site among developed blocks of an Auckland subdivision. The engineer-builder begins at night, after he and his yacht-designer mate have translated the table of offsets into the lofted lines of a life-size yacht. His shadow moves across the old brick walls, stalks the cobwebbed beams. Moths career in the half light above him. For all its weathered dilapidation the structure seems secure. Until he looks up and runs an acquisitive eye over the roof's wooden trusses.

With its designer married to his Javelin and eighteen-footer commissions, he sets his mind to doing everything himself. He turns the bolts. Wrestles with the deadwood. Makes the moulds for George Abbot to cast the fittings. He even removes sections of the shed's structural partitioning. They'll do for the jig and temporary frames on which he'll mould the hull. Later on he'll cut and position the laminated stringers and roll the narrow plywood strips around the round-bilged shape. Three skins with staggered joints, the ply bellying out of control, forming into nightmares once his head hits the pillow. Weeks hunched over sanding boards loom large before him, the muscles at the back of his neck knotting at the thought.

He works sixteen-hour weekends in the Mairangi Bay shed, every night after work. Visitors are few, his dedication total. Already it's a labour of love. It has to be. His social life has just been laid to rest under a mountain of Luan ply.

Stories, apocryphal and true, grow along with the hull. Materials, it is said, are vanishing from building sites on Auckland's North Shore, lead from the roofs of local parish churches. But there's no question what's happening to the leftover lead from certain Auckland Electric Power Board jobs. High in antimony, it is hard, strong and perfect for a keel. Nor is there any doubt about the shed's kauri lining. Progressively it is becoming the elegantly patterned joinery in the body of the hull. That is not so bad. But when temporary framing for the cabin roof is needed he finally attacks the trusses. Paying his cousin a call, John Coady draws attention to the shed's dangerously sagging roof.

'Rob, reminds me of that bloke in the "Bringing Up Father" cartoon strip. You know, always borrowing and never returning things!'

'It's called double-entry cost control. The builder gets debited with reduced demolition costs and I get credited with a cheaper interior!'

But the 'borrowing' isn't over yet. The engineer-builder snaffles an old transformer case from work and braises a one-inch steam pipe to its inside floor. Then he's throwing in the lead and lighting a fire beneath it. Which is when he makes his biggest mistake of the project — not checking the temperature of his metallurgical brew. As he turns the pipe on its thread and the molten metal flows, he hears a sickening pop. Despite the concrete reinforcing the glass-fibre plug splits open. Suddenly he's leaping beyond the smoking delta, grabbing a spade to chop at the rapidly cooling metal. If it's not in pieces small enough he'll never get it off the floor and back into his makeshift vat.

And when it's over he collapses against the topsides. That's it, he thinks. No more. I'll just patch the bloody keel up. It'll mean no trim tab, a trailing edge wider than the specs, and an irate designer. But after fourteen months he's wasted and that's the best that he can do.

As the project nears completion, friends begin to gather. Mostly from the Moth Class days, there's more than altruism in their actions. Future brother-in-law Garry Roberts calls to make sure 'Blackarse' is still alive. John Mann, a sparky with the Power Board, oversees the wiring. Commonly known as 'Elton', if he's going cruising at night he wants to be sure the navigational lights will work. 'Hydraulic' Jeanes turns up to check on the keel bolts he's turned and, while he's at it, the builder's sister. The designer makes it his mission to cut out the hatches. He wants to know they'll open if he's down below and they hit something solid at sea.

***

About to make yachting history, *Titus Canby* nearly didn't make it. The first night of the inaugural cruise they anchored in South Cove, Kawau Island, cooked a handsome dinner and sucked a bottle of Scotch to death. Next day they dropped the pick in Mansion House Bay and watched the ferry take Bruce back to the mainland. Pity he was busy because this was the life, and the proud owner's pay-off for fifteen months of hard labour: settling in with mates to knock the scabs off a few while the sun sank over the pines. No matter that they were on a lee shore (a term not yet part of their vocab) and the bottom of the bay littered with decades of yachting detritus. Their 26 lb anchor was big enough to hold their little boat in anything, including a rising nor'easter.

'Aren't we closer to shore than we were five minutes ago?'

'Does that mean we're getting closer to the bar?'

'No. I think that's called a reef.'

'Best we start the motor.'

But these were sailors used to fast-planing dinghies. They had never been backwards in a yacht before. Nor had to check on something as mundane as a tender. Clunk went the Kubota when the skipper threw it into gear, the propeller shaft and painter jammed together tighter than a granny knot.

'Best we hoist the sails.'

'Best we had.'

But even under main and jib they were still going backwards.

'The dinghy, Rob, the bloody thing's sunk!'

'ELTON — QUICK, GRAB THE BREAD KNIFE!'

The painter cut, Rob pointed his brand new boat at the rocks. He needed some way on badly; without it he couldn't steer. But now free of constraints, *Titus* leapt to his command. Steady on the helm, he waited until he could see the oysters on the rocks before he called 'ABOUT!' Like men possessed the crew cleated home the sheets. Their sails cracked and filled. And like the thoroughbred it was, *Titus* raced away from danger. They may have missed the reef by feet. But what they couldn't escape was the applause from the crews still anchored in the bay. The *TC* crew each took a bow then, blushing like boiled crays, slid around the point for the night.

For the next five days, while they waited out the Christmas storm, they anchored *Titus* in the relative safety of Ladies Bay and themselves in the more traditional shelter of the Mansion House bar.

***

There are red faces among top ocean-racing crews this week. A 26-footer with the crazy name of *Tituscanby* ... has won the South Pacific Half Ton Cup (contested in New Zealand) against 19 of the hottest Half Tonners south of the Equator ...

The winner carries no instruments and when Farr was asked how they navigated among the rocky islands of the Hauraki Gulf in a 180-mile event raced throughout the night he said: 'We just looked for the next light and pointed the sharp end at it.'

But it was, in truth, brilliant racing and it has rocketed Farr into the top ranks of New Zealand designers, alongside other light-displacement experts such as John Spencer.[3]

It was a mad panic to get ready for the Schweppes inaugural South Pacific Half Ton championship. The sails gone over. The running gear checked. Safety equipment bought, begged and borrowed. There was also the unexpected complication of *TC*'s measured rating being higher than the designer's calculations. In a way it wasn't surprising. Rob Blackburn had over- rather than under-built his boat. But its stability was definitely higher than the righting moment for the rule and that would have to be fixed. The weekend before the race *Titus* went on the hard. Rob was far from happy. He'd sweated blood over the keel; now he was having to cut lead out of its trailing edge and plug the ensuing 'window' with wood. It also meant finding eighteen pounds in lead ingots to fasten under the foredeck. Yet even with this bow-down trim the bargain-basement cruiser still didn't rate. That sent sailmaker Roscoe Guiniven snipping with his sheers, removing precious square footage from the spinnaker and jib. And when that still wasn't enough they had to take a hacksaw to the spinnaker pole. Finally, late on Thursday night, *Titus* received a provisional rating. Hand-calculated by measurer Jack Allen, it would have to be confirmed by computer before the fourth race.

Friday, 31 March 1972. At 1100 hours Blackburn, Guiniven, Jeanes, Coady and Farr crossed the start line in the first race of the five-race series. Their nearly five-minute win in that forty miler stunned their critics into silence. Huddled below decks over post-race rums, the other crews were talking.

'Can a yacht which looks so different be a breakthrough?'

'From a kid who grew up in Leigh?'

'Built by an amateur in his spare time?'

'Should we now regard it as favourite for the series?'

'Hell, is history being made?'

They were not unreasonable questions. The nineteen other boats were not only designed by well-known names: Davidson, Salthouse, Tarabocchia, van de Stadt and a young Paul Whiting. They were being sailed by some of the country's most experienced offshore sailors, including those who had just won the Southern Cross Cup from Australia. There was Roy Dickson helming Tony Bouzaid's *Blitzkrieg*. Alan Warwick on *Swooper*. Bob Salthouse on *Cavalier II*. Jim Lidgard on *Tramp*. Peter Colmore-Williams on *Conquero*. Noel Angus on *Janna*. Peter Willcox on *Pretender*.

The dinghy boys, however, were still feeling their way with a yacht they'd spent precious little time getting to know. They were still trying sail combinations. Still learning its idiosyncrasies. Thirty-knot winds in the second race with only one large spinnaker didn't help. Nor did their spectacular wipeouts. But once they got their sail combinations right, and with the designer on the helm, they managed to climb back through the fleet on the beat back home to finish a close second to *Conquero.*

Although holding an overall lead, their crowning moment was yet to come. Race Three was an overnight race of 180 miles. No one on board *Titus Canby* had been on an ocean race before. It was a race in which navigation was expected to be a key. A race in which the experienced Dickson and Warwick were expected to take charge. Having helped Chris Bouzaid and *Rainbow II* to that historic One Ton Cup victory in 1969, they could surely do it now. But by the time the fleet had reached the first mark — Flat Rock off Kawau Island's eastern shore — the ex-Moth boys had *Titus Canby* in front of Alan Warwick's *Swooper* by a tidy five minutes. Sailing by the mountains rather than their compass wasn't the most efficient way of heading up the coast, and by the time they reached Sail Rock they had slipped back a place. Ocean-racing practice of the day said you sailed down the rhumb line and set the sails most suitable for that course. For a crew whose spinnaker was too large to carry on the tight-reaching leg back down to Channel Island, that was now impractical. They'd have to find a new way of sailing if they were to regain the lead and stay ahead of the fleet. Their reaction was instinctive — they'd chase the wind rather than setting sails to catch it. Just as they had when sailing twelve-footers, they reached up under genoa until the course opened up enough for them to set their spinnaker and angle back down. And when they got too low to keep carrying it against the decreasing angle of wind, down it came, until they'd reached up high enough under their genoa to be able to reset it again. Watching the leader's crazy antics the fleet thought *Titus Canby* uncontrollable or its crew lost or drunk. They had other thoughts when positions were reported by radio and the dinghy sailors had made Channel Island twenty minutes ahead of their nearest rival. It was a lead to which they clung tenaciously on the beat north to Flat Rock and the reach across to Gannet Rock. And as they tacked up Rangitoto Channel early the following morning they knew they had it won. True, they had sailed a far from perfect race. But the ocean race was worth double points and, provided they kept out of trouble for the remaining two, the series should be theirs. Something special was happening. The designer knew it. The builder-skipper and the crew all knew it. And so did the media.

Funny how you remember all this detail when it happened so long ago. The beat up Rangitoto Channel in a freshening breeze in the morning at the end of the long ocean race. The next boat a binoculars-job. It was a really exciting time — having done the boat, gone into ocean racing without any experience and won with relative speed to burn. It was a huge moment, for the series and for us. I think Peter Montgomery might have even been bubbling over the airwaves at the time.[4]

He was. Having heard him on the radio announcing that *Titus Canby* was rounding North Head with no other boats in sight, Jim and Ileene Farr had run from Tudor Street down to Torpedo Bay. There they watched the little 26-footer slicing through the Waitemata, hurrying towards the finish line. So unexpected, and against all predictions, it would remain for them the highlight of their younger son's career.

After winning the ocean race by fourteen minutes the dinghy sailors were quizzed by a retinue of offshore veterans genuinely bewildered by their success. A few beers later they were none the wiser.

'Who is your navigator?'

'Don't have one.'

'Why did you sail a zigzag course from Sail Rock to Channel Island?'

'We have a nice little Japanese compass. But right underneath it we have a nice little Japanese motor.'

'Why were you sailing so close to the Waiheke shore just before dawn?'

'Had to go in to ask the way.'[5]

The following Tuesday they received computerised advice that *Titus Canby* was under the 21.7-feet rating maximum. Their confidence and faith in the system restored, they won the eighteen mile Olympic-course Race Four by a massive thirteen minutes.

But the series wasn't over yet. A tough and gifted competitor, Roy Dickson was about to hand the dinghy sailors the offshore sailing lesson of their lives. In the fifth and final race he shaved marks closer than a builder with a chisel, changed sails more often than a model changes clothes, and stuck closer to his opponents than a Bedouin to his blanket. It was a hard-fought battle over 78 miles with the brilliantly sailed but slower Davidson edging out the new sensation by a minute seven seconds. *Titus*, however, had already done enough to secure the series from the much heavier *Blitzkrieg*,[6] with the Salthouse Cavaliers battling it out behind them.[7]

***

Within two years Rob Blackburn would sell *Titus Canby* for $13,000 to engineer-developer Ian Gibbs. Renaming it *Tohe Candu*, Gibbs, Blackburn, Guiniven, Kevin Lidgard and Phil Edgar would sail it to another South Pacific Half Ton Championship victory in 1974. Later that year the pioneering Gibbs would again demonstrate the yacht's potential by beating 40

British Half Tonners at Cowes before sailing in the world Half Ton championship at La Rochelle, France. Although *TC* only finished eighth, its entry and performance — the first time a New Zealand-designed, built and crewed yacht had been sailed in a major world championship — was portentous.

While Laurie Davidson, Paul Whiting and expatriate designer Ron Holland would all produce sensationally fast yachts in the years ahead, it was Bruce Farr who had broken the mould. In thinking outside the square he had stepped ahead of his talented compatriots and the rest of the world. Everything after *Titus Canby* would be a variation on a theme: Farr's. He still had refinements to make to the hull and appendages, and work to do on the fractional rig. But the inaugural South Pacific Half Ton Championship had proved that his concept worked. In fact it had exceeded all expectations. And it did so because its designer, without regard for convention, had, like Bob Stewart before him, gone back to his roots. Bruce Farr had produced what he loved most — a yacht that was fast and fun to sail. In so doing he created the first light-displacement keeler which sailed to the point of excellence on all points of sailing and successfully competed within an international rating rule. By refusing to sacrifice speed for rating or concept for conformity, he had, with his first keel-boat design, completely revolutionised the sport.

Chapter 11
Flying Eighteens

Keeper of the Secrets this season is the man who stands head and shoulders above the rest in the 18 ft skiff world, designer Bruce Farr. The off-season for Bruce exists no longer since his recognition as king of the 18 designers ...
Jenny Farrell in *Sea Spray*[1]

The 'weird one' of the fleet is Benson and Hedges. It's hard to say if she is a lot of aluminium tube held together by plywood, or a lot of plywood held together with aluminium tube. Her crew includes some of the smartest men in the business and, by the time the trials start, they could be a real threat.
Tip-Over Ted in *NZ Boating World*[2]

By 1973 the orders had reached such a level that the Farrs' Tudor Street basement could no longer cope. This called for a visit to Gil Galley. Known as the Godfather of Devonport for the way he pulled strings for local youngsters, the automotive electrician quickly arranged factory premises in nearby Fleet Street. Bruce signed a two-year lease, installed a row of benches in the sunlit office, and employed boat-builders Charlie Webley and Robin Williams. They were soon hard at work on several OK Dinghies, two post-*Joshua* Javelins and the first Farr export order: an eighteen-footer for 'Radiogram', Andrew Gram III of California.

With Bruce now able to spend the bulk of his time drawing, ideas grew at a rate of knots. An earlier wave had produced *Joshua*. An ultra-light construction of laminated ply, Andy Ball's Javelin was designed to improve performance by concentrating weight in the centre of the boat. What had emerged from Tudor Street was a hull drawn around the minimum class restrictions. It had a minimum beam across the chines, a maximum vee, and a flat centre to the hull with a lot of round on the outside. With its mast placed further aft than normal and weight-saving devices throughout, it was designed to plane early in marginal planing conditions. *Joshua* and its twin *Afakazze* were not only quick upwind. They showed phenomenal speed reaching. And sailed by two talented crews — Jock Bilger and Murray Ross, and Andy Ball and Peter Newlands — both boats went on to dominate national and interdominion Javelin contests and the Sander's Cup. This would become a signature hull for the young designer and provide key elements in the design equation of his next generation of eighteen-footers.

***

Not since the first World Eighteen-Footer Sailing Championship in 1938 had the contest been won by a New Zealander in Australian waters. Gordon Chamberlain had sailed his Emmy *Manu* to victory in 1939. Jack Logan had skimmed *Komutu* to a title win in 1950.[3] Peter Mander had won twice in *Intrigue* in 1952 and 1954. And Bernie Skinner, the last New Zealander to win the Jas J. Giltinan Trophy, had done so in his double-chine *Surprise* in 1960. But all were victories on the Waitemata.

Bruce had designed and built his first *Cool Leopard* for Gary Banks in 1967, and designed him another in 1969. Both had shown potential. And in the minds of the cognoscenti, Wayne and Noel Fleet's *Miss UEB*, the third Farr eighteen-footer, was a potential champion. It had finished runner-up in the 1971 Worlds and won its race in the Long Beach Sea Festival in August 1971. Accompanying the skiff as coach and reserve mainsheet hand, Bruce had also used the trip to take a close look at the latest Twelve-Metre yachts, with their U-shaped forward sections and knuckle bows.[4]

In the 1971 world championships Don Lidgard sailed his first *Smirnoff*, a foam-sandwich boat he'd built to Jim Young's design. But he was unhappy with the stability of its narrow-bodied hull. Impressed with the speed and all-round ability of the Fleets' runner-up, he arranged for his sponsors to pay the Devonport designer $1600 the following winter. He received a ground-breaking boat in return.

Where Bruce had it over everyone else was that he came up with the complete solution. Up until then designers hadn't put whole boats together. It was generally accepted that a designer would draw the boat, a builder would build it, and a third person would rig it. But Bruce not only drew the hull. He positioned the centrecase and designed the rudder, centreboard and rigs. And, what was unheard of then, he designed three centreboards for us — one for heavy airs, one for light airs, and this amazing gybing centreboard that would make the boat jump to windward. In fact it was so effective that I put my pelvis out once when I tried to pull it up before it had straightened in the centrecase! He also designed the sails, although I made them, and Don and I both had input into the rigs. But it was a whole new approach. And you can still see that pattern today: Bruce's attention to detail as a yacht designer is second to none.[5]

What Bruce Farr had produced for the lightweight *Smirnoff* crew was a stable platform with a simple internal layout and cantilevered sides to get their weight out wide. Another in the emerging breed of Farr three-handed designs, the lightly glassed plywood hull produced good speed upwind and blinding bursts of sustained planing downwind. Perfecting the Bowler-McGlashan technique of tacking downwind, *Smirnoff* avoided the straighter but slower flat-running technique of the Aussies. In fact, it was so well balanced and well sailed by Don Lidgard (helm), Jim Lidgard (forward) and Ian Harrison (mainsheet), *Smirnoff* was untouchable on the waters of Brisbane's Waterloo Bay in 1972. In conditions ranging from ten to forty knots, they won the invitation race and the first four heats. Only a capsize and second placing in the fifth cost them the winning of all five heats. In the process they buried four-times world champion Bobby Holmes, who could only manage fifth in his Bob Miller 'wedge'.[6]

'It's the worst thing you could have done.'

That's what Helmer Pedersen told Don Lidgard the following year.[7] That was after Don had told Bruce Farr what he could do with the $1500 invoice he'd sent directly to his sponsors, Gilbey's, for repairing a hole in *Smirnoff's* side. It didn't matter that three Farr-designed skiffs had filled the first three places in the 1973 Eighteen-footer Worlds sailed on Sydney Harbour. Don Lidgard was not going to be 'the filling in the sandwich'.[8] And nor, as far as he could tell, had Rob Muldoon[9] let the rate of inflation get so out of hand that a repair job cost as much as building a brand new boat the year before! That winter he got on the blower to John Chapple. It was a call that would open the Farr Fleet Street doors to Kim and Terry McDell.

Bruce Farr had always given freely of his time. But with a small business to run, two employees to oversee, and often waylaid by casual calls for advice, he finally started to charge. If lawyers, accountants and plumbers could put a dollar value on their time, why not a yacht designer? The answer was simple and in the negative. Yachting was a pastime. You paid for what you got and as little as possible for the thought that went into creating it. Hell, sometimes you could even get lines plans from recognised designers without having to pay a brass razoo.

It was, in fact, the first public airing of a leitmotif that would be played again in Bruce's career. Without warning he'd found himself in no-man's-land, a lonely frontier of ideas where an emerging professionalism was doing battle with a virulent strain of egalitarianism. And the still-loud echoes of a proud colonial past were not blending well with the high-pitched

A sailor and a painter. Bruce Farr's great-great grandfather, Capt Matthew Thomas Clayton (1831-1922) with wife Ellen and family, Manurewa, circa 1914. At age 88 he said: 'My hand is getting shaky ... but my nerve is as good as ever. I could take a ship round the world today if my legs would only hold out.' *(Jim and Ileene Farr.)*

Capt Clayton's 1905 depiction of 'The *Endeavour:* Capt Cook beating into Poverty Bay 1769' was used on the reverse side of the New Zealand One Pound Note. *(Jim and Ileene Farr. Used with permission of the Reserve Bank of New Zealand.)*

'Landing of Lieutenant Governor Hobson at Waitangi', 1840, oil-on-canvas by M.T. Clayton. *(Auckland City Art Gallery Toi o Tamaki.)*

The Farr-Jacquart connection

Eighteen year-old Odile Jacquart (Bruce Farr's French grandmother), Roubaix, France, circa 1914. *(Jim and Ileene Farr.)*

Bruce Farr's Scottish grandfather, Harry Farr (standing), Paisley, Scotland, 1917, just before leaving for the Western Front as an artillery observer. *(Jim and Ileene Farr.)*

The Clayton-Edmond connection, Auckland, 1949. Standing: Colin Clayton (Ileene's brother), Elsie Clayton (née Edmond, Ileene's mother), Jim Farr with Alan, Ileene Farr (née Clayton) with Bruce; Front row: Duke Clayton (inset, Ileene's father), Isaac and Cecelia Edmond (Ileene's grandparents). *(Jim and Ileene Farr.)*

Alan (4) and Bruce Farr (2), Orakei, Auckland, 1952. *(Jim and Ileene Farr.)*

Bruce Farr (6), 1955. *(Jim and Ileene Farr.)*

Smoko at Claude Greenwood's boat yard. Jim Farr (left) with Bruce (11) and Claude Greenwood, Whangateau, Leigh, 1960. *(Jim and Ileene Farr.)*

Bruce Farr was eleven years old when he designed and built the 10 ft 6 in *Resolution* (Design 1), seen here under construction at the Farr family home, Mathesons Bay, Leigh, 1960. *(Jim and Ileene Farr.)*

(right) The Pennant Class *Resolution* under sail with its designer at the helm. *(Jim and Ileene Farr.)*

The launching of *Resolution* at Mathesons Bay, Leigh, New Year's Day, 1961. *(Jim and Ileene Farr.)*

Bruce Farr planing across the Whangateau in *Mammoth* (Design 3), his first Restricted Moth. It was in *Mammoth* that he won his first New Zealand national championship (March 1966). It was a big series for the young designer-builder as he collected all three trophies: the national title, the junior national title and the designer's trophy for the boat with most points skippered by its designer.
(Bruce Farr.)

Alan Farr with the first Farr Mk II Restricted Moth, *Behemoth* (Design 4), circa 1966. Alan was Bruce's first client. *(Jim and Ileene Farr.)*

Bruce sailing *Dynamoth* (Design 6), his Mark III Restricted Moth, circa 1967. Note the twin-spreader high-aspect ratio rig and lightweight, stringered hull.*(Jim and Ileene Farr.)*

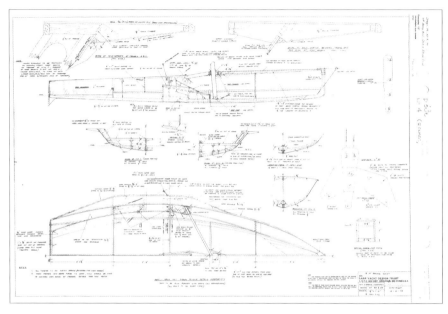

Design 40 and 50. Deck and internal construction drawings for an eighteen-footer racing yacht. *(Bruce Farr Yacht Design.)*

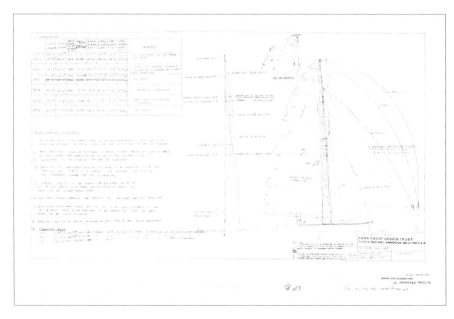

Design 50.
Sail plan of an eighteen-footer racing yacht. *(Bruce Farr Yacht Design.)*

The four Farr-designed eighteen-footer world champions

(All photos courtesy of Bruce Farr Yacht Design.)

Don Lidgard's *Smirnoff*, Brisbane, 1972.

Terry McDell's *TraveLodge*, Auckland, 1974.

Bobby Holmes' *TraveLodge*, Sydney, 1973.

Dave Porter's *KB*, Sydney, 1975.

Titus Canby (Design 26), the yacht that launched Bruce Farr's keel-boat design career and fired the first shots in the light displacement revolution. Built and owned by Bruce's sailing friend Rob Blackburn, it would later be purchased and campaigned by engineer-businessman Ian Gibbs as *Tohe Candu*.

Titus Canby with its designer on the tiller. *(Jim and Ileene Farr.)*

Titus Canby's crew of dinghy sailors introducing a whole new way of sailing keel-boats. *(Jim and Ileene Farr.)*

Relaxing after a hard day's sail on Hamilton Lake. L-R: Graham Nye, Bruce Farr, Peter Hutchinson and Rob Blackburn. *(Peter Hutchinson.)*

Cecil, Jeanette, Katharine, Jean and Russell.*(Jeanette Holley.)*

Russ Bowler, circa 1947. Soon he'd be exchanging horses for boats. *(Russell Bowler.)*

Russ (5), rowing on Lake Rotoma in the 9-foot clinker dinghy he converted into a sailboat to sail solo across the lake.*(Russell Bowler.)*

Harry Highet's P Class has trained many of New Zealand's champion yachtsmen.

Members of the Kohi P Class crowd, Kohimarama Beach, Auckland, 1960. L-R: Malcolm McGill, David Hutchinson, Vernon Davies, John Petrie, Wayne Dillon, Russ Bowler, Ena Hutchinson (front). *(Norm Beetson; Ena Hutchinson.)*

A champion in the making: Russ Bowler's first outing in his first P Class, *Marika Jnr*, Auckland, 1960. *(K.L. Fish, Auckland; Russell Bowler.)*

The P Class regatta in full flight on Kawau Island's Bon Accord Harbour, 1961. *(Phil Hutchinson; Ena Hutchinson.)*

A big haul for Russ Bowler and *Tranquil* at the end of the 1963-64 season, including his first national title: the Tauranga Cup. *(K.L. Fish, Auckland; Russell Bowler.)*

N.Z. **YACHTING AND BOATING**

INTERDOMINION CHERUBS • OLYMPIC REPORT • WORLD KEELERS

DEC.-JAN. ISSUE. 1968-69   35 cents

Twice national
Cherub champions:
Russ Bowler,
Don McGlashan
and *Mecca*.
(Russell Bowler.)

Australasian dinghy of the decade: *Jennifer Julian*. As skipper of his twelve-foot
round-bilged foam-sandwich design, Russ Bowler was 'hailed as one of the
greatest open-boat skippers ever to come to Australia'. *(Russell Bowler.)*

Beach-bound, the elegant Lynda Smith. *(Russell Bowler.)*

Peter Walker about to boil the billy. The Bowler-Walker trial run up the NSW coast in their FJ Holden the year before they crossed Australia. *(Russell Bowler.)*

Packing up *Joanna* for the journey across the Nullarbor to Perth. *(Russell Bowler.)*

The shell of the revolutionary *Jennifer Julian II*, the first foam-sandwich Cherub. *(Russell Bowler.)*

About to become Australian National Cherub Champions. Note the canvas foredeck. *(Russell Bowler.)*

Russ Bowler, Peter Walker and *Jennifer Julian II*, Australian and inaugural World Cherub Champions, January 1970. *(Independent, Russell Bowler.)*

Russ Bowler and Peter Walker sailing to their World Cherub title on the Swan River, Perth, January 1970. *(Russell Bowler.)*

The narrow round-bilged body of a national champion: the Bowler designed-and-built *Benson & Hedges*. *(Russell Bowler.)*

Trapeze-artist trio: Russ Bowler with Graham Catley (mainsheet) and Simon Ellis (forward hand) sailing *Benson & Hedges*. *(Russell Bowler.)*

Russ Bowler's *Benson & Hedges* slipping through to weather to claim first place in Heat Two of the 1977 eighteen-footer World Championship sailed on Auckland's Waitemata Harbour. *(Russell Bowler.)*

notions of intellectual property. It was hard to know it then, but the gossip about the Farr charge-out rates doing the rounds of the Auckland Sailing Club was really a groan of transition. It was a painful journey to be making, this going from a society where you cooperated to survive to one where you paid for that cooperation in order to compete against those outside it. And it was never more so than for those fiercely independent members of the community Bruce Farr served. Sponsorship was available. But in New Zealand that only paid for a small portion of the costs attached to an eighteen-footer campaign. Besides, these were funds raised by the organiser of a campaign, not the designer. And they were not to be frittered away on calls for advice or repairs. He may not have been a very tall poppy, but the secateurs were already out and snipping for Bruce Kenneth Farr.

There was, however, no such ego-pruning on the other side of the Tasman. There, during the 1973 Worlds, the young designer had joined the spectators on the ferries, offered guarded advice to 'Joe the Bookie', and watched as Bobby Holmes (*TraveLodge NSW*) and Dave Porter (*KB*) sailed his designs into first and second places. In recognition of his efforts (and as the designer of Hugh Treharne's four-hander *Booth Holden* a.k.a. *Boobs for Holding*) the inimitable Aussies dubbed him 'Loose Bra'. They also bought six of his seven new eighteen-footers built in 1974.

***

Crouching under the foredeck of his father's X Class as it sailed over the Waipou bar is Kim McDell's first yachting memory. Tearing down to Auckland each Sunday to race as bail-boy with six others on his father's eighteen-foot skimmers *Patricia* and *Sluefoot* is another.[10] As was the long drive home, sitting among the wet cotton spinnakers he'd have to dry and tie with wool in preparation for next weekend's racing. Before Roy McDell sold his businesses and the family moved back to Auckland, Kim's younger brother Terry had become Northland P Class champion. As an eighteen-year-old he would also win the National Zephyr title. Kim was the athlete. A brilliant schoolboy half- and one-miler and one of the country's elite in the golden era of New Zealand athletics, he walked from the sport in 1967.[11] Athletics' loss was yachting's gain.

Looking ahead to their second year in eighteen-footers, the McDell brothers knew they would be up against it if they were to win the 1974 world title. In the fleet would be five-times world champion Bobby Holmes and former world champions Hugh Treharne and Don Lidgard. Dave Porter and Mike Chapman, both former interdominion twelve-footer champions, and Nev Buckley, twice Australian sixteen-footer champion, were also lining up. Kim and Terry McDell had no doubt who should design their new *TraveLodge*. They had seen professionalism, not a paucity of spirit, in Bruce

Farr's methods. The Devonport designer would respond in kind, with the gift of his most valuable asset: time.

The McDells had already decided to approach this campaign with an Olympic-style build-up. So after driving to Devonport to order their new eighteen-footer they went out and bought two second-hand Farr 3.7s.[12] That winter, while Charlie Webley was building *TraveLodge*, they plunged into a daily regime of road-running and the honing of their sailing skills in the single-handed trapeze yachts. No matter the conditions, they were out on the Waitemata practising after work and racing every weekend. Kim McDell:

> From the word go we established a very good relationship with Bruce. He was totally professional and easy to work with; you could always talk things through with him. We had complete confidence in him and that made the whole exercise of building our first *TraveLodge* a pleasurable and exciting experience. I hadn't worked with designers before. Nor had anyone built me a boat. So the amount of detail involved was a real eye-opener. You could tell from his drawings that he'd thought about absolutely everything. He'd designed the lot.[13]

Terry McDell was on his honeymoon when the Auckland eighteen-footer fleet knew that *TraveLodge NZ* was throwing out a serious challenge. With Kim as forward hand and Peter Brook on mainsheet, Bruce helmed his new design to victory in the Auckland Sailing Club's Mark Foy event. There was still work to be done when the McDells and Brook were beaten in the Auckland Champs by the Chapple-designed *Smirnoff*. That was December 1973, the month they hit upon the idea of a summer training camp in the Bay of Islands.

> We took the boat, all the gear, our three wives and Bruce to the Bay of Islands that Christmas. Our base camp was Terry's parents-in-law, the Lewises, who had a place in Jack's Bay. We sailed every day, damned near, with Bruce in the runabout telling Terry, Peter and me what we were doing wrong and how we could improve. Up there he again showed just how clever he was. And when we broke the mast on one of the fresh days, he set to and designed a new medium rig for us. Built by Don Baverstock, it was that rig which really brought the boat alive.[14]

With the married couples and Bruce all sleeping in the one big room, and the men on the water every day, it wasn't long before the minds of the women began to wander. Not that far, as it turned out, though quite far enough for the still-shy designer who would spend most nights looking for the pants to his pyjamas after the three couples, giggling, had gone to bed.

He survived, with good grace, as did the crew the three weeks of intensive practice he put them through. And so successful was their training camp that when they arrived back in Auckland they found themselves on a completely different level. They were pipped by *Smirnoff* in the nationals that February, but following further after-work practice sessions with Bruce they won four of the seven heats in the New Zealand selection trials and headed into the 1974 world championships full of confidence.

Although nine of the thirteen entrants were from the young designer's board, what soon became apparent to the McDells was that the other new Farrs, with the exception of *KB*, had been altered by their owners or not built in strict accordance with the designer's specifications. That would show in their performances — just as the McDells' faith in Farr and Webley would prove decisive. Four minutes ahead at the second mark of the first heat, there was no catching *TraveLodge NZ* after that. As *Smirnoff* had two years before, so the McDells and Peter Brook sailed a notch or three above the rest of the fleet. They won the invitation race and the first four heats to secure the world title. A hundred small craft followed them across the finish line in that title-clinching heat. But like the Lidgards and Ian Harrison, they failed to win the fifth and final race, finishing a disappointing sixth in the fluky conditions. Still, it was a brilliant victory. And sweet. Their perfect 400 score had given them the Giltinan trophy and New Zealand's first win in home waters for fourteen years. And for Roy and Pat McDell, who had lived and breathed the eighteen-footer class from its earliest days, their boys had given them one of their proudest moments of their lives.

> Bruce was obviously a major part of that win for us. He has such an incredible feel for boats. I mean, he could jump on *TraveLodge* and know straight away if there was something holding it back. It might have been that the leach was too tight or the jib a bit full or the mast not bending enough. Whilst we prided ourselves on being good boat handlers, he was in a different league from us in terms of what made the boat go fast theoretically.[15]

But Bruce had not only been a major part of the McDells' win. By the time his eighteen-footer designing days were over, with Dave Porter's win in *KB* in 1975, Farr designs had won an unprecedented four world championships in a row, two each with Kiwi and Australian crews.

***

For Bowler, long recognised as one of New Zealand's most talented small-boat sailors, the way to the top of the flying 18s has been a hard one ... From the outset [he] has sailed boats of his own design ... refus[ing] to follow the fashionable Farr designs which then dominated the class.

It was a struggle at first. Wiseacres were saying that Bowler's sponsors acquired the best sailor and the worst boat! But they persevered. Results improved year by year and in the 1976–77 season with a radical new hull they began to really shake things up.[16]

Newly married, and working for Kingston Reynolds Thom and Allardice, an 80-strong Auckland group of consulting engineers, Russell wasn't sure he had the time to return to centreboard sailing. But there's nothing quite like the sight of an eighteen-footer fleet doing battle to stir a dinghy sailor's blood. For the 26-year-old they were still the 'biggest, fastest, angriest, nastiest pieces of equipment'[17] on the Waitemata. And with former rival Don Lidgard often leading the way and old hand John Lasher still competitive, the thought of getting back into the game began growing in his mind. That turned into a dozen letters of request and, finally, a $2000 sponsorship deal with cigarette manufacturer WD & HO Wills. With Don McGlashan and Peter Walker as crew, the pundits were soon picking the Bowlermobile as a serious threat to the more conventional Farrs. There was only one problem. By the time Russell had the funds he no longer had the time to build a foam-sandwich hull. He settled instead on an aluminium-framed plywood shell for his first *Benson and Hedges*. The boat showed flashes of brilliance, but more often than not it finished with a busted bow or broken gear or right at the back of the Auckland fleet.

So that first year was a stunning plunge back into reality. Bruce's boats, based on a shape he'd developed over a number of years, were being very successful. And I was still guessing a lot, shooting the corners and not being very disciplined about it. We tried to get a sandwich system working the next year but were unable to match the weights of the finely developed wood construction going on then. It wasn't until my third boat — a very quick piece of equipment — that I finally got it together. And not until my fourth that I was able to get it down to a hundred and eighteen pounds and better the wooden boats for weight.[18]

Before leaving for London Bowler had been Australasia's leading light in the field of exotic dinghy design and construction. He had learnt a lot about earthquake engineering in his three years' absence from the sport. But apart from the speed at which disaster can strike it had little relevance when it came to designing and sailing an eighteen-footer. As well, Bruce Farr's skiffs had attained a place of pre-eminence achieved by no other designer in the 35-year history of the class. Losing McGlashan and Walker as crew the following season didn't help either. And with the birth of Gareth in 1975, there was no longer just Lynda and himself to think about. Russ Bowler not only had a lot of ground to make up if he was going to be competitive again. He had just entered a crucial cross-over phase in his life.

Bruce Farr, on the other hand, had crossed over at sixteen. It meant he was still living at home, had no tertiary education or the security of a recognised profession. But although he continued to struggle financially, he was also free to dedicate every waking moment to the development and refining of design ideas.

The different approaches of Farr and Bowler were the product of life choices. Having designing as a pastime meant you could experiment to your heart's content; too many failures might mean the loss of a sponsor. But one failure if designing was your livelihood could mean the loss of your entire client base. That meant Bruce took a more conservative approach, immersing himself in the detail of construction and the massaging of shape to attain maximum speed and stability for his customers. It also meant, more often than not, he'd arrived at the heart of a problem before he'd put pen to paper. Conversely, Russ might still be working on an aspect of design after he'd launched his experimental craft or lost its rig in a wave. An order book full of various commissions helped Bruce discover that the fourteen-foot Javelin was a much better boat than a Q Class on which to base an eighteen-footer design. With 30 percent less sail and two more feet in length, the Javelin was a boat you sailed, not just survived in; it had more inertia and things happened more slowly. It would take two spectacular failures — a deep-veed version of his original *Jennifer Julian* and a flat-hulled shape that sailed on its bilges (and which he finally carved up with a chainsaw) — before Russ would get to the root of the same theoretical problem. He knew the total resistance in water was made up of the same factors. But it would take two seasons before he discovered they were in different ratios; that a twelve-footer's resistance was composed more of wave-making resistance and an eighteen-footer's more of wetted surface. That sorted, by the time he had designed and built his third hull, tuned his big rig to perfection and established a smooth working partnership with Graham Catley (mainsheet) and Simon Ellis (forward hand), he had produced the fastest skiff in the country. Keith Chapman:

He was not just a major talent on a tiller and a talented designer who created quick boats, whose rigs and hulls were always very different. He was also generous with his time, always happy to share what he'd figured out for himself or learnt from others. He was someone you could bounce ideas off, someone who would stimulate you into thinking about aspects of a hull and rig that hadn't crossed your mind. All the young dinghy sailors looked up to him. The first time I met Russ was when I was building one of his *Jennifer Julian* Cherubs. Then, after I had graduated to eighteen-footers, I asked him if he could give us a hand to figure out why ours wouldn't go. And that race completely changed my thinking. I'd read the books and seen the movies and thought you had to be a gorilla to sail an eighteen-footer. But with Russ on the

helm that day we were in a completely different world. It was, 'Ready to tack,
old chaps? Tacking now.' There was no yelling, no shit-fight when we tacked,
no loading things up. It seemed unbelievable to me that someone could be so
relaxed and calm yet so precise in one of these flying machines. Not surprisingly, we went on to win that race.[19]

Russ Bowler's ultra-light, round-bilged *Benson and Hedges* was seen as a
radical departure from the Farr chine skiffs which had, since 1972, dominated eighteen-footer racing on either side of the Tasman. Although the boat
immediately showed up as a light-wind flyer, a jammed spinnaker halyard
while leading the fifth heat cost the Ciggy Boys the 1977 national title and
the winner's prize of a Sitmar Pacific cruise. It almost cost them the boat.
Screaming in on Auckland's Compass Dolphin, the container terminal rocks
came precariously near before they finally freed their giant sail. A close second to the McDells that February made them all the more determined to
reverse the tables in the New Zealand team selection trials the following
month. Not only did they have four wins and a discarded second placing.
Such was their superiority they could even capsize at the last mark of the
last heat, right their boat and, without being headed, win by over a minute.

That Easter Russ Bowler led the five-Farr six-boat New Zealand team
into the 1977 Worlds sailed on the Waitemata Harbour. It was a championship *Benson and Hedges* should have won. A hot tournament favourite,
they were leading Heat One when they sailed into a hole in the lee of North
Head and, in great embarrassment, capsized to weather. With what was to
follow, they needed a win, not an eighth, in that heat to win the title for
their country. They won a thrilling second heat from *Colour 7*. A sixth and a
third in the next two races saw them enter the final heat in close company
with tournament leader Dave Porter and the 1976 interdominion twelve-
footer champion, Iain Murray. But their mast gave way in the big seas and
30-knot winds and that was that. Eighteen-year-old Murray, on the other
hand, seven minutes behind Porter's *KB* at the last mark, ordered his spinnaker hoisted. With the young trio hanging off the transom, *Colour 7* flew
from Rough Rock to Orakei wharf to snatch a famous victory from the hap-
less *KB*, still tacking downwind under genoa. Although 1977 would be
a bumper year for Bruce Farr in the ton classes, it was the first time in
six years that not one of his designs made the top three finishers in an
eighteen-footer Worlds.

The following year, after winning the national title, Bowler, Catley and
Ellis went to Brisbane determined to bring back the Giltinan Trophy.[20] They
were full of hope and had good reason to be. Not only was their Smoke
Boat performing to its brilliant best in light airs, they now had it going in all
ranges of wind. And so it proved in the first heat when they played the
freshening conditions to complete advantage. But with four nautical miles to

go the stitching of their jib peak parted and they were left with only their main. That it took Dave Porter's *KB* three and a half miles to catch them was indicative of their pre-disaster lead. Their second in Heat One was followed by a DNS in Heat Two when a shattered mast fitting forced their withdrawal. Then came the protest. Things had been confusing enough as a result of the dated constitution and an amalgam of rules being used to run the contest. But all hell broke loose when Russ Bowler protested the race committee for the way it postponed Heat Three. His request for a hearing turned down, the Kiwi team threatened a boycott if the protest wasn't heard. The local officials threatened to pick up their buoys from Waterloo Bay if it were. Outside it was a scorcher. Inside, that April, it was even hotter as members of the Queensland Yachting Association (QYA) and the Brisbane 18 ft Sailing Club almost came to blows. Finally, and to the relief of all, the QYA officials acknowledged Bowler's right to a protest and temperatures cooled. Ironically, the resailed third heat disadvantaged the consistent Dave Porter who, following just the second tie sail-off since 1938, had to accept the runners-up trophy for the second year behind Iain Murray. And although dominant in the light, Bowler, Catley and Ellis had to settle once again for third place behind the two New South Welshmen.

A second to Iain Murray in an open world championship sailed in Plymouth, England, later that year didn't alter the obvious — the sport Russ Bowler adored was demanding more and more of the commodity he had less and less of: time. With a young family and a blossoming career — including, as its structural engineer, designing and building Auckland University's School of Architecture — something would have to go. But it wasn't until he received a proposition over lunch from Peter Walker, now working for Bruce Farr Yacht Design, that he began to see the possibility of turning his passion into an income.

Chapter 12
The Seminal Vessel

... styles, whims and fads are often carried too far, for the human animal is a myth-loving creature.

L. Francis Herreshoff, American yacht designer[1]

I would finally like to suggest that the reason ... small boats do not carry masthead rigs has nothing whatsoever to do with problems of staying such a set-up ... but is because small boat skippers have discovered the advantages (through evolution) that a lower-stayed controlled-bend rig can offer ... Surely these same principles, applied intelligently and with some modification and refinement, can work safely and efficiently in an ocean racer. If ocean racing is not prepared to follow the path of evolution, it may well come to the same end as the dinosaur.

Bruce Farr[2]

Brave words from a designer who was still only 24. Not one to rush into print, Bruce Farr's comments were typical of what he said when he spoke his mind. They may have sounded heretical. But he was no iconoclast. He simply said what he believed, and proved it on the water. With his first keel-boat he had done what no one had done before him: 'marr[ied] the two seemingly incompatible characteristics of light-displacement with agreeable upwind form.'[3] *Titus Canby* had proved itself better balanced than Bob Stewart's *Patiki* (1960), more refined than van de Stadt's *Zeevalk* (1949), more radical than Jack Giles' *Myth of Malham* (1947), and a more complete breakthrough than Nathanael Herreshoff's *Dilemma* (1891). Yet its significance had been missed by New Zealand's yachting establishment. Sure, the youngster had designed a rash of dinghy title winners and won a Half Ton National. But it was a big step up from there to designing real ocean-racing yachts. Bendy rigs on big investments? Not likely. That was anathema to a fraternity convinced that their country had neither the talent nor the technology to foot it with Sparkman & Stephens in the IOR design game. It was a conviction born of myth. The conviction was explicable: money. But why and whence the myth?

From its earliest beginnings yacht racing had two main problems to overcome if it were to flourish as a sport: management and measurement.[4] Firstly, there needed to be racing rules if races were to be conducted in an orderly manner. The second problem was the more vexing. How could yachts of different dimensions compete on an equal footing in the same race? To answer that question meant finding a way (or ways) of estimating the difference between the speeds of different yachts, differences that could become handicap corrections applied to finishing times. But because many things influence a sail-boat's speed — most notably waterline length, displacement, beam and sail area — it would not be an easy task.

The problem was first addressed when the Royal Yacht Club used weight to group yachts competing in the Cowes Town Cup of 1829.[5] While calculations to do with sail area, girth, displacement and time allowances would all be added to the tonnage formula, weight would remain the basis of English measurement rules thereafter. The Europeans raced under various rules, including the Scharenkreuzer Rule which classified yachts according to their sail area.[6] In the United States the 1882 Seawanhaka Rule remained the dominant measurement rule for twenty years, although most large yacht clubs had their own methods of measurement which they changed from time to time.[7]

The first attempt to standardise this plethora of measurement rules was made by the New York Club in 1901.[8] From designers all around the world it chose Nathanael Herreshoff's Universal Rule which, for the first time, took displacement into account.[9] Then in 1906 the Yacht Racing Union called together representatives from sixteen nations for a meeting in London. This produced the International Rule, the first to include scantling requirements, and created the International Yacht Racing Union (IYRU), the sport's governing body.[10, 11]

But neither the Universal Rule nor the International Rule were suitable for passage-racing (offshore) yachts. Realising the limitations of its length-based rules,[12] the Cruising Club of America (CCA), sponsor of the Bermuda Race, adopted a rule in 1923 developed by engineer-sailor Wells Lippincott.[13] In an attempt to accommodate the differences between cruising and racing yachts Lippincott had structured his rule around a 'base boat', giving penalties to factors which increased speed and credits to those which made a yacht go slower than his theoretical model.[14] Two years later, following the 1925 Fastnet Race, the Ocean Racing Club[15] formulated its own rating rule 'to bring together by time allowance when racing together in open water, yachts of the widest possible range of type and size.'[16] Although they existed on either side of the Atlantic, two offshore rules were one too many in 1957, the year American yachts first competed in the RORC-sponsored Admiral's Cup. Designed to the CCA Rule, they attracted a higher rating (or greater penalty) under the RORC Rule than under their own.[17] A phone call from designer Dick Carter to Olin Stephens in 1965 was to change all that. Two years later the CCA and RORC met in London and the International Technical Committee (ITC) was formed.[18] Instructed to amalgamate the two rules by placing emphasis on the British method of hull measurement and the American method of measuring rig, the ITC, with Olin Stephens as its chairman, announced the new rule in November 1968. First used for racing in 1970, the International Offshore Rule (IOR) was to have a profound effect on the yachts it rated — and Bruce Farr's career.

In one sense it was to be of great service to the sport. 'By the mid-1970s, there were more than 10,000 certificates around the world; for the first time an owner could have his boat built and rated in New Zealand and be able to race in Florida or the Solent with a valid rating. An IOR certificate was international currency.'[19] But in other areas it was a disaster. It contained no scantling requirements. And like measurement rules before it, the IOR was soon producing hull and rig distortions as designers sought to gain advantage for their clients (investments) and themselves (reputation).[20] And therein lay the problem. To win one of the IOR's ton regattas, now the world's premier offshore yachting events, it was too easily forgotten that you had to have a boat which chased the rating advantages the new rule offered. That is, if you had a heavy boat with a substantial mid-point beam, a pinched-in stern and a masthead rig — in essence, a Sparkman & Stephens design — your rating went down and your chance of winning went up. Soon these characteristics were being interpreted as those needed to make a yacht go fast. The problem was that no one was asking the critical questions: were performances following trends, or were trends just producing yachts with lower ratings? Soon the distorted features of IOR yachts came to be seen as normative and preferable to those designed on sound sea-going principles. The myth had taken root. So much so that by the mid-1970s there was a strong, almost fanatical belief that unless your boat was fat, heavy and had a tiny stern, it was not just incapable of winning an IOR regatta — it was a dog.

It was a myth that needed rewriting; had already been rewritten on the windy, shallow reaches of the Whangateau. Prophets, though, have never done well at home. And this one might have been missed had it not been for those who understood his principles and rescued them from a life confined to dinghies. Yet it was like osmosis, this slow filtering of ideas through the membrane of reaction. Hostile at worst, disinterested at best, the establishment chose denial. It would be left to a group outside their yacht club doors — the risk-it-all dinghy boys — to jump aboard these apologies-for-boats before the centreboard explosion of the fifties and sixties could become the offshore revolution of the seventies. If the powers that be had only glanced in Farr's direction from 1971 to 1974, they, like the rest of the offshore-racing world, would have had no option but to take a long hard look in 1975.

***

There has to be a happy medium somewhere, a compromise between having to sit out on the weather deck [of a light-displacement yacht] 24 hours at a time, wet and miserable and just human ballast, and sitting down to leeward [in a heavy-displacement yacht] sucking on a can of beer thinking of dancing girls and swaying palms.
Chris Bouzaid, New Zealand yachtsman[21]

... having had the doubtful pleasure of pushing one of these lightweight go-go machines to windward in 50 knots-plus off Cape Barrier, with wind against tide, I can only admit to being most impressed with her handling of the conditions and with her performance against so-called orthodox yachts.

We must remember that, being so much lighter, they don't knock themselves around as much as a heavier-displacement yacht. And when the downhill part comes, not only are they a joy to handle ... the speed is exhilarating.

If we don't go sailing for pleasure, like this, it is time we gave it up!
Roy Dickson, New Zealand yachtsman[22]

Titus Canby had been the breakthrough. But there were to be four more meticulously cared-for gestations before the boat that would give Bruce Farr worldwide recognition was ready for delivery.

George Knightly's was the first. Like Rob Blackburn, George had been bumped and bruised in the wild twelve-footers and wanted something more cruisy when he retired from the class. In fact Bruce was still working on the *Titus* lines when George asked him to draw a 33-footer. George also suggested he help him update his old plywood clinker *Mislead* and race with him in the 1971 Leander Trophy.[23] Although their planed-back mast exploded below the hounds while he and Bruce were screaming down a reach — and George saw the trophy slip through his hands yet again — it didn't divert him from the bigger plan: the moderate-beamed cruising yacht growing in his backyard. With a displacement of eight thousand pounds and a finer bow than *Titus*, it would prove extremely fast upwind. *Moonshine*, launched in 1971, was classic Farr: a yacht that could surf downwind in waves and plane in strong winds and flat water. Without an IOR calculation in them, these would be the lines plans to which Bruce would return almost a decade later when he sat down to draw the 10^{20} for Sea Nymph's Kim McDell and Peter Gribble.

Following the 1972 Half Ton Nationals, accountant Terry Harris commissioned a Quarter Tonner of similar concept to *Titus*, but with one overriding instruction: it had to be designed to the IOR Mark III.[24] Uppermost in Bruce's mind was improving light-weather performance, which meant a much larger rig in proportion to length. He now had the confidence and permission to install the first true three-quarter rig on an ocean-racing yacht. It would prove the critical ingredient in his alchemy of design. Others had tried, but failing to push the envelope they'd reverted to masthead arrangements.[25] The difference, though, was not just commitment to concept; Bruce Farr had refined these rigs on his many dinghy designs over a wide range of conditions in both New Zealand and Australia. He knew they increased efficiency by giving greater control, through mast bend, over the biggest portion of the rig, the mainsail. He had also calculated that they would produce more power for a keel-boat than the small-main big-genoa set-up of a masthead. Launched in January 1973, *Fantzipantz* was more than just a *Titus* clone. It was the prototype refined, the seminal vessel. True, it had the knuckled stem and U-shaped slab-sided bow sections. But it was finer forward, flatter beneath the waterline, and had more beam aft. As well, the reverse curve of its hull had been incorporated into a deep fin keel. With its fast lines and rakish rig the little boat looked a winner. And so it proved, taking out the South Pacific Quarter Ton title that year with Farr on the helm.

While the ocean-racing establishment still looked askance at anything that wasn't a Sparkman & Stephens, one of their number was more than a little impressed with the yachts being drawn in Devonport. John Senior, a North Shore builder and Commodore of the Royal Akarana Yacht Club, asked if the *Titus* concept could be enlarged — considerably. He'd noticed the extra room-for-length achieved in the little Half Tonner and thought the light-displacement concept could work well in a larger cruiser-racer. Immaculately built in three skins of kauri by Brin Wilson, his *Kailua* was 42 feet long, had a beam of almost thirteen feet and displaced nearly 16,000 pounds. It was by far the biggest commission Bruce Farr had had. But even before Brin had begun, Auckland furniture manufacturer Graham Eder had let a commission for the same design, Design 39, to boat-builder Graham Wheeler. Begun before but launched after *Kailua*, *Gerontius* epitomised the ultra-modern fast cruising yacht. It had a different cabin top (to Eder's design), a plush interior, a quadraphonic sound system as well as a lighter twin-spreader masthead rig designed to improved performance. If the pair received only grumpy glances from those who disapproved of their wide-bodied shape — and an indifferent 'interesting' from Olin Stephens, in town to visit *Corinthian* — they thrilled their owners with their turns of speed, their space below deck and relatively low cost for boats of their length. *Gerontius* hadn't been designed for the Admiral's Cup. But the famous regatta wasn't far from the minds of its owner and designer as it floated to its lines just before Christmas 1974.

***

> *Angry French fishermen might wreck the hopes of New Zealand yachtsmen who are at present joint leaders of the world Quarter Ton yachting series off Deauville. They are blockading the harbour. For the second day they have kept the yachting fleet bottled up in Deauville basin by sending a task force of eight trawlers to blockade the exit of the locks into Seine Bay. The fishermen are protesting about the increasing pollution in the bay that is destroying their fish harvest ... Crockett said he couldn't argue with their protest, only the timing.*
>
> Charles Cooper in *The Press*[26]

Things seemed to happen when Farr and Crockett got together. If they weren't talking wider decks for an old eighteen-footer,[27] they were removing 12 millimetres from the rocker of the old *Miss Beazley Homes*, now *Harvey Tiles*. Good enough to make the national team but never quite able to keep his fast but modified boat under control, Murray Crockett spent so much time swimming he became known in twelve-footer circles as Harvey Fish. That experiment might not have worked all that well. But he was still a passionate believer in what his buddy could do.

With design commissions just sufficient to support his still spartan lifestyle — although it had, since the middle of 1971, progressed beyond beers with the boys to include pharmacist Anne Clark — when the Fleet Street factory lease came up for renewal, Bruce decided to stop building boats. Drawing them was what he loved most and that, he now determined, would be the irreversible direction of his career. With his mind made up, he moved into a Kohimarama flat with Murray Crockett and set up an office downstairs.

The main reason for starting Alpha Marine was simple: generate more interest for Bruce. He just wasn't getting the exposure he deserved. I'd seen *Titus*

Canby and was convinced he could really go places. Yet in spite of his success in the Half and Quarter Tonners few were taking him seriously. Ocean racing was still very conservative, still locked into Sparkman & Stephens and those old clunker types, and Bruce hadn't made the progress he should have. It was partly his own fault; he was pretty introverted and had little idea of self-promotion. In spite of all that he'd achieved, you had this sense of him struggling for direction. Perhaps it was because I was a couple of years older — I'm not sure — but I looked at *Fantzipantz* and thought that that was the way to go. The only problems I saw were in terms of the market. I thought it too upright in the bow and stern to be acceptable as a production boat. So I pushed him to pretty it up.[28]

Which is what happened. Eleven inches were added to the overall length, the keel repositioned, the sail plan increased, the coach roof lowered, the bow and stern tapered a little more, and the rudder hung under the transom. It was the work of a gifted artist, the 'subtle matching of lines and camber'[29] producing a timeless classic. They called the Mark II version of *Fantzipantz* the Farr 727.

In need of space for lofting, Bruce drew the lines full-size in his mother's sewing room, then he and Murray lofted them in Jim and Ileene's basement. A heavy-engineering draughtsman by day, he would leave Fletcher Bernard Smith at five to work on the plug till midnight. Having built nothing bigger than a bookcase, the outcome was a miracle. Not only was it completed in nine months, nothing was off-square. Admittedly, Bruce, who lent a hand most days, drove Murray to it; if something didn't fit it was ripped out and replaced. And with Air New Zealand engineer Bob Farrell and both their fathers helping with the finishing, Murray and Bruce duly delivered the plug from the Crocketts' Mt Roskill garage to the site of its first public showing: the 1973 Auckland Boat Show. There it attracted attention from the local cognoscenti, including Roy Dickson, before being sold to Noel Angus.[30]

When the third party — an engineer-investor — dropped out it was Murray who picked it up. Murray has a huge amount of enthusiasm and commitment to things. It was his enthusiasm and sheer will to do it that got Alpha Marine formed and the 727 thing going. So there I was, suddenly a partner in a boat-building business. And by 1974 Murray was pumping these things out in fibre-glass from an old bus barn in Devonport Gil Galley had organised for us.[31]

***

1974 was Decision Time for offshore yachting in New Zealand. Should they send a team for the first time to the Admiral's Cup? Chris Bouzaid had shown what could be done by winning the One Ton Cup twice, sensationally so in

1969 when his victory captured national front page news along with Neil Armstrong's moon walk.[32] His *Rainbow II* had been designed by Sparkman & Stephens and his 1972 winner *Wai-Aniwa* by Dick Carter. But the established designers were now under siege. *Titus* had given more than a hint of that in 1972. Expatriate Kiwi designer Ron Holland had designed and built *Eyghtene* and with it won the US and World Quarter Ton Cup in 1973. And that same year American Doug Peterson, a Holland friend, had produced the radical One Tonner *Ganbare*. The choice was clear. The New Zealand Yachting Federation (NZYF) could take a fix on a local star and follow it over the horizon or coast home between port and starboard markers to trays of gin and tonic.

Tension was running high during the New Zealand Admiral's Cup trials of 1975. There were big bucks, big boats and big egos at stake. And when the three Sparkman & Stephens designs — Evan Julian's *Inca*, All Black great Ron Jarden's *Barnacle Bill* and Russ Hooper's *Corinthian* — jointly protested *Gerontius*, challenging its measurements, emotions reached boiling point. Matters weren't helped when the selection committee reserved the right to choose the team's third boat irrespective of trial placing. Finishing one-two meant *Inca* and *Barnacle Bill* were automatically selected. But should the third-placed *Gerontius*, with its absurdly wide transom, be taken instead of the latest from Sparkman & Stephens? *Corinthian* even looked a classic, all glistening navy-blue and bristling with sophisticated gear. Could a canary-coloured cruising boat, built at a quarter of the cost from two-year-old lines plans be taken seriously? Having helmed *Corinthian* into fourth, Chris Bouzaid resigned from the selection committee, unhappy with its methodology and convinced the yellow peril's rating would make it uncompetitive in international competition. Dickson stepped off *Inca* for personal reasons but was openly critical of team selection policy. John Lidgard posited that if a team were to go at all it should include a Kiwi design. As the politicking peaked, *Barnacle Bill*'s skipper let fly in typical Kiwi speak. 'We have a chance in Cowes because we've won against big odds before. Let's have faith … in the guys slogging their guts out to do well.'[33]

Ray Haslar's plea struck just the right chord. But a problem solved produced another: how to finance three large yachts and crews for an event on the other side of the world? It was a big ask in 1975 and the team was to sweat every sponsorship cent. To the point of humiliation for some, as the burliest rode trikes along Auckland's Quay Street and those with the best pecs posed nude for goss-mag *Truth*. Bruce had to sell everything, including his car, and move back home in order to make the trip. For a self-employed designer it wasn't just the cost. There would be three months without income and the prospect of starting all over again when he returned home. And no Anne Clark; she had taken herself off to Medical School in Dunedin. For the first but not the last time in Bruce's life career would come before love.

There was drama aplenty in the 1975 Pacific Quarter Ton contest run by the Panmure Yacht and Boating Club. The fleet was substantial: 31 entrants from fourteen different designers including seven 727s. Stiffest competition for the Crockett-Farr-owned *727* again came from Roy Dickson. He won the first two races on *Genie*, the first production 727, but was disqualified from the intermediate race for not having his second anchor in the stowed position. An unpopular decision, it gave *727* the race and threw the event wide open. But that was just the beginning. In a night of wild winds and high seas during the long ocean race, the Fleet brothers fired off flares when their lightweight *Cruntch* blew out its rig and began taking on water near the Ahaaha Rocks. *727* may have led the fleet into the back end of Waiheke, but with Murray Crockett heaving up all over the cockpit floor and Peter Walker having flashbacks of his brother-in-law's very recent drowning, the decision was made to lower the main. Too bad that Roy Dickson had foamed up beneath them as they were rounding Gannet Rock and Murray Ross, on gybing out of Navy Buoy, had blasted past them with spinnaker flying in *Stan's Family Jewels*. By then, though, they had the contest won and could afford to play it safe. Farr's precision helming and the quality work of Mike Menzies, Walker and Crockett had already given them a 2/1/1 result. And with a fourth in the last they comfortably took the series. It also gave Bruce Farr his seventh national title, his sixth sailing a boat of his own design.[34]

That win convinced Murray Crockett that a tilt at the world title was on. But with Farr, Menzies and Walker already committed to *Gerontius* for the Admiral's Cup, replacement crew would have to be found. Graeme Woodroffe, an OK Dinghy international, was seconded as skipper and Roy Dickson offered the navigator and co-helmsman jobs. Rob Martin, a Walker cousin and New Zealand yachting rep, filled the remaining berth. Renamed *45° South*, the boat was shipped to London along with Bruce Eadie's *Genie* and from there trucked to Lymington. A win by *South* in the Wight by Night gave a hint of things to come. Two days later, in windless conditions, the little boat motored under outboard across the English Channel. Arriving at Deauville, the crew settled into a four-level four-bedroom apartment with their wives and girlfriends and prepared for things to come. Roy Dickson:

I was in love with the 727. It was a beautiful boat, easy to handle and a joy to sail. I'd been used to sailing heavy-displacement keel-boats before that. But the 727 was a real gem. We had some fantastic rides in *45° South*. We did the Coupe de France out of Deauville prior to the Quarter Ton Worlds. Racing against Half Tonners in the one-hundred miler, we were fifth around the first mark. But on the spinnaker run towards Cherbourg it really blew and we just took off, beating all the Half Tonners around the leeward mark. It was fantastic. We not only brought the Cup home. Between *Genie*, skippered by Bob Farrell, and ourselves we won all three races, taking both line and handicap honours.[35]

Contested by a fleet of 43 yachts from fourteen nations, the 1975 Quarter Ton Worlds began on Bastille Day. The night before there had been fireworks displays and a lot of red wine drunk. But the pyrotechnics were not just outside the Kiwis' apartment. In bed with his wife in the early hours, the skipper had come under siege from a plaintive female voice from the footpath below.

'Woodee! Woodee! Es-tu à la maison? Woodee! Woodee, es-tu là ...?'

Embarrassing enough for the fun-loving Woodroffe, it would be considerably less unsettling than the French fishermen's dramatic plea for attention.

Patchy winds and a fierce tide turned the first race — a 30-miler round-the-buoys — into a lottery. It was only after the wind freshened on the last leg that the two Farr boats were able to break from the pack and finish one-two. In the second race, an Olympic course, the three-quarter rig came into its own as the winds touched 25 knots. Two wins from two starts and a tournament win was in sight. Lack of wind, however, caused a delay of several hours to the start of the 100-miler, and by nightfall *45° South* and *Genie* had slipped to twenty-fourth and twenty-fifth respectively. A night breeze of eight knots enabled Dickson to coax *South* through to sixth by daybreak. But when Woodroffe found himself on the wrong side of a 50-degree wind shift early the next morning, they were suddenly back in fourteenth, one place behind *Genie*. Which was how the two Farr 727s finished, ghosting across the line in the almost non-existent airs.

The Quarter Ton crews woke on the morning of the fourth race to find the entrance to Deauville basin blocked. Now there was apprehension — not just the colourful language of French protest — in the air. If the contest were abandoned, Italy's *Charlie Papa*, also light-displacement, would win on a countback. The Kiwis sweated the two nerve-wracking days it took the fishermen to end their blockade. Then the officials got to work; the Olympic race, which the New Zealanders expected to dominate, was abandoned, and the ocean race shortened to 185 miles. But as it turned out the protesters had done the Kiwis a favour. During the two days of the blockade there had been no wind; now, as the yachts milled around the start line, the wind was coming in at a steady ten knots. Woodroffe made amends for the intermediate race with a brilliant start. Leading after the first beat, he was never headed after that. The little Farr flew across the English Channel in gusts of 25 knots and literally sailed out of sight on the way back from the Nab Tower, the Isle of Wight mark. And although the wind dropped on the spinnaker run from Le Havre to the finish line, *45° South* still managed to finish 24 minutes ahead of its nearest rival.

Even the locals were enamoured of the fun-loving Kiwis as they walked and rode, jubilant, around Deauville, the Quarter Ton Cup jauntily displayed in the wicker basket of their bicycle. They had a lot to celebrate. Never before had their country won an international keel-boat contest in a boat designed, built and crewed by New Zealanders. And with *Genie* finishing sixth overall, the champagne tasted even sweeter.

The English had dismissed *45° South* and *Genie* as little more than trailer-sailers. That wasn't surprising given they displaced only 2700 pounds and relied on the deployment of crew weight for stability. This was a radical departure from the traditional approach in which a keel-boat's ballast provided the entire righting moment. And with only three winches on board, the traditionalists wondered how the two boats could be sailed with any degree of control. Even the experimental French had been sceptical of the Farrs' high-aspect ratio keels. But that was then. Now everyone knew the little boats were unstoppable in a breeze. And in flat running conditions, two crew sat forward of the mast to lift the wide stern sections clear of the water. Revolutionary concepts, by the end of the tournament they had opposition owners and designers all shaking their heads.

Meanwhile, back in England, *Inca*'s disqualification at the start of the Fastnet had seen the New Zealand team slip from fourth to fifth. But that was fifth out of eighteen nations, a commendable result for a first attempt at the Admiral's Cup. And the designer who could barely afford to be there had confounded the critics yet again — *Gerontius* had finished top New Zealand boat.

Long before the campaign camps had been packed away they were humming with debate. Was the Farr approach the direction of the future? Between Admiral's Cup events, and despite an injury to his back, Bruce had managed to make it to Deauville. Dockside, Canadian Bruce Kirby, designer of the phenomenally successful Laser, confessed that he'd thought a light-displacement approach to Quarter Tonning wasn't on. It was a statement that begged the question: was a radical departure from the heavy-displacement approach a dangerous path to be taking? The chief IOR measurer, Major Robin Glover, didn't think so. He was not just impressed with *45° South*'s clean lines; without any hull distortions in concession to the IOR, he considered it would have done well under any measurement rule. But his optimism hid an ominous problem. While the wide variety of designs successfully competing in the Quarter Ton contest proved the IOR was working, what would happen if the Farr approach rendered obsolete the expensive heavy-displacement boats of the larger ton classes? It was an issue that needed addressing. If not, the value of these investments might shrink to the size of sail bags.

But despite this sudden surge of success and attention from the yachting world, Bruce remained the level-headed person he'd always been — and just as caring. At the end of the Fastnet, while Peter Walker and crew sailed *Gerontius* back from Plymouth, Bruce drove Peter's wife Annie and daughter Frith the two hundred miles to Lymington, where he had two days' work to do before flying home. And as he'd been unable to book a hotel room in advance because of the holiday season, they decided to share accommodation. Annie Walker:

> I'd turned up in Lymington with my baby on my hip to support Peter and the team. But unable to stay with him at The Angel, we ended up at Mrs Bridle's Bed and Breakfast. Which is where Frith and I returned after leaving Peter and the boys in Plymouth, and where, over those two days, I got to know the real Bruce. He was just like one of my brothers, sensitive and thoughtful. He would even leave the room when I got up and was preparing to wash and dress my baby girl. He would talk to Frithy about those all-important animals on the Frostie cornflake packet. And when he sensed I needed a break he would take her walking in her pushchair. He would even come shopping with us. He was such a good friend. So caring. I really valued his quiet empathy.[36]

***

Still only 26, Bruce Farr had just booked his ticket to fame. Unwittingly, he had also set himself on a collision course with the northern hemisphere establishment. If they couldn't have his prophetic head on a platter, they would do the next best thing — turn him into an *enfant terrible* from whom all respectable owners of decent ocean-racing yachts needed protecting. And if that didn't work, they would validate their own myths by leaning on the rulemakers, doubling as their designers, to change the rules.

Oblivious to all of this, Bruce took himself off to Madison Avenue on his way home. There he was cordially greeted at Sparkman & Stephens and given a fifteen-minute tour of the famous premises. A world champion designer but still the boy from Leigh, Bruce Farr had no idea that beneath the cordiality lurked something he'd only briefly seen at home: the shadow side of international yachting politics.

Chapter 13
A Fine Prospect

... speed is the result of driving power overcoming resistance.

L. Francis Herreshoff, American yacht designer[1]

[Light-displacement yachts] are easier to sail, more fun to sail, they are cheaper to build and equip and they have more room potential. All-round, they are probably a better style of boat than what the rule has given us as the norm. People in New Zealand haven't spent 50 years developing this style of boat because they are worse than anything else.

Bruce Farr[2]

With a signal from the Lizard lighthouse keeper — 'YOU ARE FIRST' — Olin and Rod Stephens burst upon the international offshore yachting scene in July 1931. They were about to record a massive line-honours win in the transatlantic race in the yawl Olin had designed and in which their father had invested.[3] Pressed by his parents to enrol at MIT, Olin had left, ill, within a year. The family anthracite business, Stephens Fuel Co, held no interest for him. Instead, he worked briefly as a draughtsman for Henry Gielow then Philip Rhodes before being invited to join a young yacht and marine insurance broker, Drake H. Sparkman.[4] This was the beginning of a glittering career for the self-taught designer. It would include co-designing and co-helming the J Boat *Ranger* in its 1937 defence of the America's Cup, designing small craft for the United States Navy during World War II and the Korean War,[5] seven other successful defences of the America's Cup[6], winning the inaugural Whitbread Round the World Race,[7] and, at the peak of his Admiral's Cup career, having 42 percent of the 1971 and 1975 fleets built to his designs.

At a dinner given by the Cruising Club of America the night before that famous transatlantic race, George Roosevelt remarked to Uffa Fox that a boat as fragile as the Stephens *Dorade* should not be allowed to cross the Atlantic.[8] Nearly five decades later the Cruising Yacht Club of Australia would echo those same sentiments when three Farr retractable-keel keel-boats were about to cross the Tasman to contest the Southern Cross Cup.[9] This would turn out to be just one of many similarities in the careers and personalities of the two bespectacled designers: their passion for drawing boats; their mathematical minds; their eye for line; their love of sailing; their gifted helming; their abnegation of formal training; their ability to make boats go fast upwind; their periods of total dominance in offshore yacht design; their many unpaid hours of input into the sport; the potent structures of their businesses; their love of art; their attachment to fast European cars; their respect for each other. There was, as I was about to learn, only one major difference between them: how they saw the shape of speed.

I had travelled to Annapolis, New York and Vermont in search of the shape of speed. In his cabin on Putney Mountain the octogenarian designer, gracious and sharp as a tack, told me that, for him, making yachts go fast to windward was the key. Being the first designer to take tank-testing seriously had also had great benefits.[10] I had Olin's words running round my head, diskettes of information from the Farr office, and conversations with yachting's illuminati to mull over on my return to Auckland. But it wasn't until I sat in an Orakei lounge that I came significantly closer to understanding the shape of speed and where it could be found. The afternoon sun was streaming in through his lounge window when Peter Beaumont, architect, yachtsman and flight instructor, opened a couple of cold ones, threw his feet on the coffee table and let rip.

'It's like aeroplane design. If you've got the shape of speed, a structure that stays together and a crew that makes it sing, everything else is out the back door. The seventies was a very competitive time. There was Jim Young, Laurie Davidson, Ron Holland, Paul Whiting, Murray Ross. All great designers. But I think they all had respect for Farr because, deep down, they knew he held the shape of speed. Not just in hull form. He was the holder of the shape of speed both above and below the water, in rigs and appendages. Soon everyone was on the bandwagon. But Farr had led the way. And although challenged and occasionally beaten, he would come again and forge ahead with another innovation or refinement.'

'Why do you think he held it?'

'The shape of speed?'

'Yes.'

'I've never quizzed him on the subject because quizzing a designer on design is like quizzing an America's Cup team member on their campaign. Ask no questions, be told no lies! But my guess is that for Farr it's part mathematical, part experimental, part observation of things that go fast, and an interfacing of all three. If you compare the profile of the elliptical wing form of R.J. Mitchell's Spitfire you'll see similarities with Farr's early rudders. The propeller blade tips of the later Spitfire Mark XIV look remarkably like Farr's more recent rudder plan forms. And look closely at the 727's keel and you could be looking at an upside-down version of a dolphin's dorsal fin. I find all that significant. I doubt that Farr ever noticed the Spitfire or set out to copy the dolphin.[11] So without reference to each other, similar design conclusions were reached by all three designers: Mitchell, Farr and God.'

'And without copyright infringement! But how do shapes like these help yachts go fast?'

'Well, first you've got to grasp the notion that air and water densities change very little at sea level in the middle latitudes, and lift and drag formulae don't change as best we know it. So all a designer can do is to mess around within given parameters, small as they may be. That's how I think Farr may have solved a lot of his speed problems. And he did it early, eliminating, as best he could, that nasty by-product of lift which slows you down: induced drag.'

'Induced drag?'

'Yes. It's the result of water escaping around the tips of things like keels and rudders in order to equalise its environment. It's the same with air around sails and wings. Gliders solve the problem with high-aspect ratio wings. The 747-400 does it with a winglet on the end of each wing to stop the cross-flow of air around the wing tip. And Ben Lexcen reduced his tip vortex problem with a couple of wings on an inverse tapered keel.'[12]

***

Prospect of Ponsonby ... *displaced a mere 8510 pounds — almost uncharted territory back in 1975 — and showed a narrow beam of just under 12'. She also showed an amazing turn of speed, which is unquestionably a Farr signature.*

Michael Levitt, American yachting writer[13]

Prospect of Ponsonby*'s absolute dominance in all five races has not been seen in New Zealand since the peak of* Rainbow II's *career and it is all the more remarkable when considered that this series was the most competitive so far.* Prospect *easily beat such opposition as three* Peterson One Tonners *... as well as Sparkman & Stephens designs of various sizes and ages, and a couple of Australian Miller designs and several local boats.*

Jenny Farrell, New Zealand yachting writer[14]

Scratch the surface of art history and you become aware of the artist's restlessness. Rubens travelled widely for patronage and fame. Van Gogh exorcised the anguish of his soul in a country not his own. Gauguin travelled to the far corners of the world in search of light. McCahon travelled to the far corners of his faith in search of meaning. There was, it seemed, no difference between beaux-arts and the art of yacht design — you had to be committed to find both meaning and success.

Arriving home from his success in Europe, Bruce would soon be immersed in observing and, as crew, in working with his functional works of art: *Prospect of Ponsonby* in the Southern Cross Cup trials and the Cup itself sailed out of Sydney, and later on *Jiminy Cricket* in the Auckland Dunhill series. But first he bought a Datsun station wagon and drove across the harbour bridge to the old but trendy suburb of Parnell. He was not only looking for business advice. He was also thinking about premises more in keeping with his growing reputation than his mother's sewing room. At the ripe old age of 26 it was time to leave home for good.

***

Peter Beaumont and Bob Eastmond, former Canterbury Javelin Class champions, couldn't get close to the formidable pairing of Jock Bilger and Murray Ross in the 1973 Javelin Nationals. The 1972 Olympians were untouchable in their borrowed boat, the Farr-designed *Afakazze*, reputed to have cost two grand.[15] After the contest Beaumont wrote to the designer requesting something similar, preferably faster. But after having it built, he reconsidered; a move from Christchurch to the country's sailing capital seemed a smarter way to fast-track his sailing career. In Auckland he began crewing on Bruce Eady's *Genie* and struck up a friendship with Farr. And when Bruce would later need a pad near his new place of business they would end up sharing an old Remuera house with three young women, Stuart Finlayson, 'a piper of no mean account',[16] and Robert James Maynard Kitchener Lampwick Scott, an ex-airline pilot caring for oil riggers during their R&R tours.

***

Even before the 727 plug was finished, Alpha Marine, through the promotional efforts of Chris Bouzaid and Rob Fry, had received inquiries from Japan. Interest remained strong in New Zealand. But Deauville had opened the taps. After the Admiral's Cup and before returning from Europe, Bruce had arranged in Germany for a limited production of a semi-custom wooden Quarter Tonner, similar to but considerably more expensive than the fibreglass 727 and more oriented towards the lighter airs of the Mediterranean and the European lakes.[17] And Murray Crockett was now flat-chat finalising with companies in France, Japan and Canada to produce fibreglass 727s under licence. No such arrangement had been made with Australia until Paul Whiting got on the blower to Devonport. He'd just returned from Perth where he'd been promoting his popular Reactor.[18]

'How many 727s have you sold into Western Australia, Murray?'

'One.'

'Interesting. I've just been sailing over there and saw seven!'

It wasn't hard to nail the trail. The WA sailmaker and yachting personality, owned up the moment he heard the voice.

'Yeah, mate, flop-moulded it. Didn't think you'd mind. But look, wouldn't want to cause another quake in the Shaky Isles. How 'bout I send you boys some lolly in lieu of royalties?'

Crockett said yes, provided it was a bank draft drawn on a New Zealand-owned bank.

***

Noel Angus, owner of *Pinto*, the 727 plug, was so impressed with his little boat that he commissioned a One Tonner, *Prospect of Ponsonby*, in 1975. Bruce was still working on the last of its details in Lymington, between races on *Gerontius* and his trip to Deauville. He returned home to a finished boat and, in spite of his still-injured back — damaged in the Admiral's Cup build-up — to a vigorous sailing programme in preparation for the Southern Cross Cup trials.

Few now doubted that the 727 signalled something new in offshore yachting. But it was still a tiny boat. Would the design principles and sailing techniques that had given Farr his first world title work in yachts of longer waterlines and greater displacement? Was its designer really a major new talent or just another one-shot wonder?

As with his previous drawings, a huge amount of thought had gone into this commission before pencil met paper. Being able to adapt his balanced shapes to the IOR had been one of the keys to Bruce's keel-boat successes. Closer scrutiny might maximise these results. At least that was his thinking as he put a microscope to the rule and saw more clearly how to increase sail area for no increase in penalty by way of wetted surface or displacement. This would be a boon, especially in light airs where his boats had been vulnerable. And by drawing a stronger, lighter construction he was able to add a higher ballast ratio, making the design stiffer and more powerful in a breeze. While retaining his signature knuckle stem, entry was made even finer and more flare added to the topsides. With tumblehome a feature of many IOR boats, this made *Prospect* look even more odd. Flared topsides, however, made the beam wider than the waterline and meant greater use could be made of crew weight on the windward rail. This would add to the keel-held ballast and, ultimately, to the overall speed of the boat. But it was in the stern sections where the most dramatic changes were made. Not only was the maximum beam set well aft. The 36-footer (28 ft 6 in on the waterline) was given a long, broad overhang. Just clear of the water in plan form, its length when sailing would be significantly increased for only a minimum rating penalty. Like his keel-boats before it, Bruce drew it as a fast boat first. And like his first Quarter Tonner, his first One Tonner received a fractional rig, its swept-back spreaders obviating the need for running backstays and complementing the clean lines of its no-nonsense hull.

The name had been doing the rounds of Auckland yacht clubs for some time before Noel Angus picked it up. It was a typical Kiwi response to Englishman Arthur Slater's line of posh Sparkman & Stephens (S&S) yachts called *Prospect of Whitby*. It might have ended up *Prospect of Penrose* had Noel Angus known that within a decade the then downmarket suburb of Ponsonby would become the seat of soaring property prices, with film and TV directors and local stars all gliding along its café-lined streets behind Ray Bans and Gaultier shades. Despite its up-you name, the shoestring One Tonner 'with a lot of cruising compromises'[19] was virtually unknown when the first gun of the New Zealand Southern Cross Cup trials was fired in December 1975. Keith Chapman:

> When Bruce came out with *Prospect of Ponsonby* I was crewing on *Quicksilver*. Almost overnight the IOR boats of the world were made obsolete. *Prospect* was that dominant.
>
> In the first race of the Southern Cross Cup trials we were sailing out of the Rangitoto Channel and this little red boat was just sitting there; we were 41 feet long, it was 36, and it shouldn't have been keeping up with us! And as the race developed and sheets were cracked, away it went. It wasn't on for a boat of that size to be beating big S&S boats around a Balokovic Cup course. But that's what happened. I remember sailing up the Waitemata after the race, looking at this little thing, thinking it had to be some kind of Super Boat. Of course it was nice to be going overseas to compete with a yacht like that in our national team.[20]

Built in Tauranga by Brian Lonegan, *Prospect of Ponsonby*, with its designer as co-helmsman, blitzed the eighteen-strong trial fleet. Its 1/1/1/1/1 result buttoned the mouths of light-displacement's critics and left establishment tongues in neutral. Most noticeable was *Prospect*'s ability to break away from larger boats in a slop and light winds, conditions considered pre-tournament to be its nemesis. But this was a boat fast on all points of sailing — even with its full cruising interior and a crew given to taking its morning bacon and eggs on the windward rail. Richard Wilson's *Quicksilver*, built and campaigned in the 1973 Southern Cross Cup by his father Brin, and Cliff Johnson's Brin Wilson-designed *Tempo* made up the New Zealand team to contest the Cup.[21] The Kiwis thought they had the boats to win. They also had Roy Dickson. Team captain and *Tempo* helmsman, he had the crews exercising early each morning in Sydney's Rushcutters Park and navigators and skippers meeting on board *Quicksilver* to discuss tactics before each race. Keith Chapman again:

> On Wednesday night, after *Prospect* had annihilated the fleet, all these Ockers, who had heard by now of this magic little boat, came down to the dock to get a closer look. And while they were studying its slender mast — there among all these masthead rigs looking like bloody great tree stumps — Terry Gillespie was down below tweaking the backstay tensioner lines. They couldn't believe their eyes as they watched it bend back fifteen inches then suddenly jerk upright. That seemed to epitomise the Kiwi way. Here we were, grown up in rucks and mauls, doing things on the sniff of an oily rag, not afraid to drive our boats hard, racing against wealthy Americans, Aussies and Poms — guys who would drive down from London in their Rollers to pick up their boats. Combine that approach to the superb Farr designs we've had and I think that's why we became so successful as a yachting nation.[22]

Having won the first two races and taken second in the third, *Prospect* was hot favourite for the Sydney–Hobart — the famous 630-miler doubling as the fourth and last race of the series. It surprised everyone that it finished third, not first, on corrected time, one place behind Ted Turner's 1975 One Ton champion *Pied Piper*. But with an eight-inch crack in the mast the crew had had to nurse their boat down the Australian coast. And with their compass loose on its mounting, they'd travelled well east of the rhumb line and an extra 40 miles to Constitution Dock. All bluster and challenge, The Mouth of the South wasn't convinced that the Farr flyer was faster than his world champion Peterson. This in spite of *Prospect* finishing top yacht of the 33-boat contest with a 1/1/2/3 series and *Pied Piper* finishing second with placings of 3/1/9/2. Sailing for Great Britain and finishing third boat overall was Ron Amey's *Noryema*, the immaculate German Frers design which had finished top points scorer in the 1975 Admiral's Cup and was widely considered before the contest to be the world's top ocean-racing yacht. Fourth and fifth respectively were *Love & War* by S&S and *Rampage* by Bob Miller.[23] Suddenly, it seemed, the tide had turned and was on the run for the big names of yacht design. But this wasn't a mere repeat of yachting history, or the beginning of just another cycle. There was something more complete than that about this red Renaissance boat.

***

Olin Stephens' early designs, with their narrow deep hulls, had taken over from the Alden-style schooners as the preferred style of offshore yacht. In spite of being labelled overgrown Six Metres by their critics, they had produced stunning windward performances and, contrary to predictions, made safe transatlantic crossings. But they were heavy and lacked performance across the wind. In the mid-1960s Dick Carter's designs surged to prominence on the wave of his ideas.[24] Yet within a decade his and Olin's hold had begun to slip away. The new kids on the block were Ron Holland and Doug Peterson, who were putting into practice the dreams they'd dreamed together on the deck of George Kiskaddon's *Spirit.* While theirs was a substantial though brief domination of Ton Cupping, their 'style — of a deep forefoot, negligible rocker and pinched pintail stern ... — was a cul-de-sac. This style of boat exhibited great upwind form, but they were not great reaching boats and were positive bitches to get downwind in a blow. Their aptly named death roll was the result of big masthead rigs and beamy hulls with fine ends.'[25]

That was the difference. Gifted designers, yes, but each had painted pictures of a limited dimension. Farr had not only filled in the missing bits. His ideas had introduced a whole new way of seeing things. What had begun in *Titus*, had been refined in *727*, was now full-blown in *Prospect*, Design 51. For the first time in offshore yachting's history one designer had delivered a complete breakthrough design. And this in the premier class. With its arrow entry, wide stern sections and revolutionary $^{13}/_{16}$th rig, it went upwind, downwind and across the wind. The metamorphosis from the child mind dreaming to the tangible product of the adult mind was now complete. In just two regattas *Prospect of Ponsonby* had turned the best designs from the world's pre-eminent drawing boards into yachting postscripts. The Farr genius was in full flight.

Prospect of Ponsonby had added that edge of brilliance to the steady performances of *Quicksilver* and *Tempo*, eighth and fourteenth respectively in the 1975 Southern Cross Cup. But that was enough to produce a 28-point margin of victory and give the Kiwis the coveted Cup for the second time. To top it off, Noel Angus pulled a sale out of the bag the day before leaving for Hobart. Sadly, New Zealand would not see the little One Tonner again, and Angus would himself be tragically lost at sea two years later.[26] *Prospect*, however, would live on as the Farr 11.04. Between them, Glass Yachts of New Zealand and Compass Yachts of Australia would produce nearly one hundred production versions of this design.

***

Back home it was more of the same. The second but lighter version of

Prospect — Stu Brentnall's *Jiminy Cricket* — won the inaugural Dunhill Cup offshore series from the 50-foot *Anticipation*, the Brown/Moyes-owned Bob Miller design, with *Gerontius* third. A host of Peterson, Davidson, Frers, Mull, Spencer and S&S designs followed in their wake. In the 1976 One Ton Nationals, the 11.04s cleaned up, filling the first four places. Graeme Woodroffe's fibreglass *The Number* narrowly beat the wooden *Cricket* with Peter Kingston's *Rockie* third and the chartered *First Edition* fourth. Jim Young's *Checkmate* was fifth. 'All the heavier-displacement One Tonners, including Chris Bouzaid's [Peterson-designed] *Streaker*, Bob Stone's *Katena*, Peter Smith's *Warchild* and Bob Salthouse's *Crusader*, finished the contest as mere statistics'[27]; not one completed the stormy intermediate race or the 340-miler. In the South Pacific Half Ton trophy, had it not been for Peter Spencer's successful protesting of the sailing committee, Ian Gibbs would have won the South Pacific Half Ton title for the second time and *Tohe Candu* for the third.[28] As it was, the former *Titus Canby*, with its new fractional rig, suffered its first national title defeat by only a fraction of a point from Spencer's *Cotton Blossom*, an S&S design. A first, not a second, in this and the Quarter Ton Nationals[29] would have been a fitting way to round out the summer season for what was now the hottest name in southern hemisphere yachting.

'I can't see the point of designing or building heavy boats any more.'[30]

By the end of that Easter you could see what Farr meant. And to prove that not only Kiwis could sail his boats, *Why Why*, a 727, went on to win the North American Quarter Ton title that year and *Piccolo*, a stock glass-fibre version of Design 51, the Sydney–Hobart.[31]

Feted Bruce may have been, but there was a lot going on on the inside as well. The chains of shyness, once so crippling, were beginning to fall away. Perhaps the process was helped by the rush and social chaos of those with whom he lived. Although he remained devoutly focused on work during the week, come Friday and Saturday nights it was for Loose Bra, Pebo and their friends a time of simple choices: the yacht clubs or the pubs, and the occasional pretentious party. They lost count of those who seemed to have a key to their home of many rooms, those who would turn up in the middle of the night to raid their fridge or crash there.

On the business front success was coming in waves. At times it seemed Crockett and Farr were screaming down the harbour with everything up and nothing tied down. Alpha was in the process of turning out its sixty-three 727s and ten 727 Horizons, the cruiser version with its slightly higher coach roof and inches added to its freeboard. It was also gearing up to do a production Half Tonner (the Farr 920), thinking about a line of trailer-sailers and hearing unheard-of numbers being quoted by their Canadian licensee. It was not inconceivable that their 727 could become the Laser of the keel-boat world. Business was like Aladdin's cave. But while success spilled into their hands, cracks were appearing in the walls. Murray, hugely enthusiastic, was still learning about business organisation. And with the explosion of interest in both design and production, that vexed choice was once more looming large for Bruce: to draw boats or to build them? Then there was the tension, born not out of personality but out of the inexperience of two young Kiwis setting sail on the treacherous waters of overseas trade. Accepting a boat-builder's invitation to travel to Germany after the Admiral's Cup was a decision any young designer would have made. But Bruce was unaware at the time that Murray was making arrangements with Mallard to produce fibreglass 727s in France. And nor did Murray know that after selling fifty of the pretty little Farr, Mallard would go belly-up and not pay Alpha royalties.

Bruce had stayed in touch with Graham Painter ever since his ill-fated New Zealand Yachts 'adventure'. Painter, now running a marketing and consulting business in Parnell, said yes to a beer on Bruce's return from Europe. With a soft spot for yacht promotion, he offered the designer a corner of his premises in the Cadelfa building and assistance with administration. He'd also send a young woman to collect his accounts and see what they could do to promote his growing portfolio of designs.

Cathie Cochran, just seventeen, had not long begun an accounting degree when Steven Cross backpacked his way into her home town of Brisbane and her life. Two years later the dark-haired youngest of four sisters and the sandy-haired sailor were carrying their packs together around New Zealand. Early the following year, 1974, they travelled to England, marrying that December, and returning to Auckland in 1975 to settle. Yet in spite of the miles they'd travelled together, their inner maps weren't meeting — Steve was still searching, Cathie still growing. Nor was it any easier for this particular Englishman to understand a woman of fiercely independent spirit. On divergent paths through the rest of 1975, their tracks were about to disappear over different horizons. The following March Cathie dropped Steve and his 'trusty rucksack' at the end of Auckland's southern motorway.[32] It was during this unravelling of her personal life that Cathie made a visit to 11A Tudor Street.

I was Graham Painter's Personal Assistant and was also responsible for looking after Trade Consultant accounts. It was the summer of 1975–6 that I first met Bruce. He was still working out of his mother's sewing room in Devonport. Which is where I visited him, together with his accountant, Bruce Mellor, to go over his accounts that I was going to maintain. Instinctively I thought him a nice guy. He seemed unassuming, easy-going, had a sense of humour and, well, this wonderful spontaneous smile. However, I took no more notice of this new facet of my life other than to think, because I enjoyed sailing, that maintaining his accounts could be a lot more fun than tariffs, duties and import licences.[33]

Another visitor to the Farr sewing room that summer was a tall blond sailor, grown up in the nearby suburb of Bayswater. Like the designer, 28-year-old Peter Blake had spent most of his life in boats, racing in Ps, Frostbites, Zeddies, his parents' ketch and his home-built 23-foot keeler *Bandit.* And while he hadn't won much himself, he had done a lot of miles. He had been one of the watch leaders on Robin Knox-Johnston and Les Williams' 71-foot *Ocean Spirit* when it won the inaugural Cape Town to Rio race. He had been a watch leader on the British yacht *Burton Cutter* in the first Whitbread Round the World Race, and was thinking of an entry in the next that could fly the New Zealand flag.[34] Gathered into his challenge committee was an eclectic bunch of supporters: his father Brian, an advertising art director; his long-time mentor, Martin Foster, the New Zealand Yachting Federation's executive secretary; and Peter Mulgrew, the mountaineer-adventurer. Bruce promptly drew a 65-foot light-displacement flyer, ideal for the reaching and running conditions of the Southern Ocean. Lack of sponsorship interest saw the campaign founder. But another failed attempt in the following Whitbread[35] would see Blake back looking for more ideas he could call his own and peddle to those who'd listen.[36]

Chapter 14
A School of Salmon

I agree, however, that the tendency is probably going too far toward light-displacement, and these boats that go fast and hit seas hard ... I think the older boats were nicer cruising boats, and in the sense that they were evenly matched, were just as good for racing.

Olin Stephens, American yacht designer[1]

I hated [the IOR] because it encouraged bad features: like excess beam, excess displacement, small ends, low stability. Working within the limitations of the rule was both a challenge and a frustration at the same time. You could design this boat, and put in all this effort, and then you would think if I started with a clean sheet of paper, I could have a boat the same length that would be a lot faster.

Bruce Farr[2]

At heart a philosophical issue, these were the lines along which the hemispheric yachting war was about to be fought. Thrust reluctantly into the breach by the brilliance of his concepts, Bruce Farr would find himself fighting for the underdog (for his country to be heard in yachting's corridors of power) and for decency (for decent shapes, decent boats, decent speed, decent fun).

***

'If I'm going to win the World Quarter Ton title I need more grunt than Stan's *Family Jewels*[3] and more speed in the light than *45° South*.'

When, in reply, Bruce Farr offered Murray Ross a modified 727, the double Olympian and multi-national titleholder took a walk. He didn't have far to go. Next door to his sailmaking business was another young designer. He not only leapt at the opportunity. He'd go 50/50.

'You rig it. I'll build it.'

So began the Whiting-Ross partnership and an intense period of competition between a talented triumvirate. If Renaissance Italy had its Leonardo, Raphael and Michelangelo, Yachting New Zealand had its Farr, Davidson and Whiting. Over the next three years their gifts and rivalry would produce world titles with the fastest rated monohulls the world had then produced. Their lightweight flyers would beget their own school of yacht design.[4] It was revolution versus evolution, a re-exporting of ideas from the far-flung South Pacific back to the cradle of the sport. But as the curve of competition steepened, so the odds of risk lengthened. And as the battle lines hardened, few stopped to ask what would happen if the two should intersect. And those who did asked the wrong questions.

1947–48 was the summer 20-year-old Laurie Davidson launched a boat built to his own design. Its fine entry and lightweight construction were laughed at by the experts. The howls of laughter turned to protest when it cleaned out the Auckland M Class fleet, and officials moved to have it banned. But after being remeasured and receiving minor modifications, *Myth* was allowed to go on its winning way and so help save the Emmy from extinction. An accountant serving in the Air Force, Laurie leapt at the chance when he was offered a job designing boats at Ferro Cement Ltd. That was in 1969. Two years later, after the boat-building division was sold, he found himself designing boats on his own account.

Paul Whiting was fourteen and still at school when he designed his first keel-boat, the Reactor 25. He left school in 1966, the day he turned fifteen. Having presold three Reactors, he and his sister Penny set up business in the alleyway beside their father's factory. Later moving to Onehunga, Whiting Yachts would go on to produce one hundred Reactors and other bigger production designs all built to Paul's designs.

Before the advent of *Titus Canby*, the Farr 727s and 11.04s, both Davidson and Whiting had been locked into the northern hemisphere's concept of weight and shape, designing masthead yachts of heavy-to-medium displacement. But Deauville changed all that. 'Farr had started something with the 727 which set local designers thinking of ways to improve the concept.'[5]

Paul Whiting agreed. 'The closer we looked at the Farr 727s, the more we could see the advantages in the light-displacement and dinghy-style boat.'[6]

So it was not surprising that when he drew a boat to embrace Murray Ross's ideas, it was completely different from his previous work and had discernible Farr features: a knuckle stem,[7] slab-sided forward sections, a 'flat' behind the knuckle to the centre line, a small high-aspect ratio keel, flat runs aft and a lightweight hull.[8] It also sported a bendy three-quarter rig.

While these were mostly adaptations of Farr's ideas, the two deferred to quite different logic. Only after drawing a correct boat first would Farr massage the shape 'to make it fit the rule less badly'.[9] Whiting on the other hand, and on his own admission, 'chased everything possible in the rule',[10] drawing distortions in his hull to make it fit the IOR better. So instead of the usual pleasing lines emerging from his shed, his latest design 'brought forth the old comment of snakes swallowing apples'.[11] But brimming with gear and rig inventions from the man who took the helm, it proved virtually unbeatable.[12] *Magic Bus* not only won the national Quarter Ton title from a host of 727s.[13] With Murray Ross as skipper, Paul Whiting, Steve Allen and Steve Trevurza as crew, it went on to narrowly win the 1976 Quarter Ton championship in Corpus Christi, Texas.[14] For the second year in succession New Zealand had won the same world title in a boat designed, built and crewed by New Zealanders.

Also contesting that September regatta was a Davidson design. Although regarded as a designer of becoming lines, Laurie was of an older generation than his two compatriots. But after having his Half Tonners *Blitzkrieg* and *Tramp* trounced by *Titus Canby*, he too had taken a long hard look at Farr's light-displacement concept. He thought the 727 'a nice little boat'[15] but considered he could do one better. And not just in shape. If he shortened the hull the rule would allow more sail. That, he thought, would overcome the 727's weakness in light airs. He was also convinced that 'there was a loophole in draught measurement if you had a centreboard'.[16] With this in mind he conferred with American designer Britton Chance, himself garboard-deep in centreboard research.[17] Although not new, the board concept had been given a good offing by most recognised designers.[18] But encouraged by this encounter, Laurie, like William Webb-Ellis, picked it up and ran with it. Thus *Fun*, a design for Napier chemist John Bonica, received a lift rudder and lift keel, and the first daggerboard to be used in a rated light-displacement monohull.[19] A narrow-beamed trailerable Quarter Tonner, Laurie thought it superior to *Magic Bus*. But when he fell ill prior to the contest he was unable to take the helm and *Fun* finished fifth. There were compensations though. *Fun* sold for twice the *Bus* money, won the North American Quarter Ton title the following year, and led directly to Laurie's next race-boat commission and first World Ton title.

***

New Zealand's two leading One Tonners — Graeme Woodroffe's *45° South II*[20] and Stu Brentnall's *Jiminy Cricket* — went to Marseilles in 1976 convinced they could win the world's glamour level-rating contest. But they hadn't counted on a cedar centreboarder designed by Britton Chance. Nor had anyone else.

It had all come about because keel-boats with centreboards had for many years been commonplace in the shallow cruising grounds of the USA. With centreboards withdrawn they were able to ply the shallow waters of the eastern seaboard, and when extended (in a swinging motion through the bottom of their shallow-draught keels) to sail offshore with similar upwind efficiency as a fixed-keel yacht. This tradition meant that when the rules of the Cruising Club of America (CCA) and the Royal Ocean Racing Club (RORC) were folded into one,[21] centreboards remained allowable. But not just allowable. The International Technical Committee (ITC), subcommitee of the Offshore Rating Council (ORC) and formulators of the IOR, failed to differentiate between the bottom of the keel (in the keel-centreboard arrangement) and the bottom of the board. That is, the IOR saw the bottom of the boat as the bottom of the keel, not the bottom of the centreboard when extended from the keel. This gave a centreboard yacht added depth without further penalty being added to its rating for that extra depth. Was this ruling an oversight by the ITC? Or were there forces other than water trying to equalise their environment beneath these traditional cruising hulls? It seemed inconceivable that this august body, with Olin Stephens in its chair, could think that the ballast and beam required to make a lift-keel yacht as stable as its fixed-keel equivalent would offset its huge gains in draught. But that was the 'logic'. Olin Stephens:

> I think there has always been this problem with offshore racing that there have been, you might say, constituencies, and specifically an American constituency, which has always been highly negative to boats that seemed too light in displacement. The American wish, which has seldom been fulfilled, has been to find a mid-cruising, so-called wholesome cruising boat which could still win races. That's especially true of members of the CCA, which is an old organisation and a rather conservative one committed to cruising but always hoping that they could get a chance to go racing with their cruising boats. They had seen it happen, in a sense, time after time ... lighter and faster boats which [had] taken over and made it difficult, if not impossible, for them to win.[22]

Instead of making what should have been a simple technical adjustment or insisting on centreboards being locked in the down position while racing, the ITC deferred, deliberately or by default, to a powerful lobby group. Perhaps it seemed too small to matter, this failure of their formulaic scanning to see the wood for the water. But having pulled the bung on its professional ethics by choosing to ignore the centreboard loophole, the ITC had, at that moment, preordained a watery end for the IOR.

Politics aside, there were major design advantages inherent in the loophole should a designer wish to claim them. Given that a lift-keeler's ballast was carried internally, it would experience less pitch than a hull with ballast in a keel. It was easier to shape a wooden daggerboard than a keel made of lead, and a daggerboard's better hydrodynamic shape would improve a yacht's degree of lift, especially in the light. And with its board withdrawn when travelling before the wind, a centreboarder would reduce its wetted surface and resistance, so gaining a corresponding increase in optimum speed over its conventional-keel rival.

Then there were the rating considerations. Because a fixed-keel yacht got rated for the full draught of its keel and a centreboarder did not, the reduction in rating meant that the latter could carry more sail than its fixed-keel equivalent. That would be important in light airs or when running or reaching. Alternatively, if the option to increase the area of sail were not taken, the centreboard yacht would gain more draught for the same rating and thereby increase its upwind performance.

But while all this was legally permissible, the centreboard loophole was so obviously an error or an oversight by the rulemakers that few had bothered going there before Chance and Davidson. Substantial rating advantage for no design return had seemed so slick and improvident that designers had assumed that once the loophole was breached the ITC would plug it quick-smart. And once plugged, a centreboarder's value would be headed down the mine.

Things, however, hadn't been going well in the world of Chance. Based heavily on tank-test results and his own Velocity Prediction Programme (VPP), Britt's 1974 Twelve-Metre design, with its deep, truncated bustle and flat transom, had proved 'impossibly slow'.[23] 'The submerged step at her stern was supposed to let the flow of water separate as on a planing boat'[24], thus eliminating the quarter wave and making the waterline seem longer to the water flowing by. But neither Ted Turner — helmsman for the Hinman-headed syndicate — nor the sea were fooled by the hull's 'foolproof' lines. *Mariner* was a woofer and Britt's career headed for the rocks.[25] But here, in a different rule, was an escape clause big enough for a school of salmon to swim through.

***

When *45° South II* and *Jiminy Cricket* finished first and second in the opening race at Marseilles — in winds ranging from 10 to a blustery 22 knots — it seemed the New Zealanders would carry off the One Ton Cup and Farr the design accolades. They finished a respectable eighth and fourth respectively in the drifting conditions of Heat Two. But when fifteen skippers successfully protested the French Navy's setting of the wing and weather marks in relation to the start line, the gentlemen of Societe Nautique de Marseilles declared the race and all further Olympic courses abandoned. *Jiminy Cricket*'s forty-first in the resailed race was a disaster — the placing final despite Stu Brentnall's protest that the race wasn't a resail of the original Olympic course. *South* and *Cricket* were placed eighth and ninth at the end of the fickle-aired intermediate. They recovered brilliantly when the wind blew again for the fourth, finishing one-two. In the long ocean race, the fifth and final of the series, the Kiwi boats duelled for the lead all the way to the island of Levant in the eastern Mediterranean. Spot-on navigation by Dick Jones saw *Jiminy Cricket* find the Borha Mark in the middle of the Med while the leading boats — *45° South II, Resolute Salmon* and *Pied Piper* — sailed right on by in the hazy conditions. Brentnall, Bouzaid, Dickson and crew had to work hard from midnight to dawn as they headed back towards Marseilles in just two knots of air. But while they hung on for a stunning comeback win, eighteenth in the double-pointer was a body blow to Woodie and his team. In spite of being one of only two boats to win two races, *45° South II* finished fourth overall behind *Resolute Salmon*, Ted Turner's *Pied Piper* and the Scott Kaufmann-designed *American Jane.* 'Only in France,' muttered the Kiwis.[26] If the second heat had not been resailed, *Jiminy Cricket* would have finished first instead of fifth.

But in the end it had been a numbers game. The two Farr boats had won three of the five races. And if there had been more light-displacement designs, when the wind was up they would have filled the minor placings, pushing the heavy-displacement yachts further down the points table. As well, it could be argued, the variable conditions and shambolic organisation meant that little was proven in terms of design. But it did show that the Farr boats were superior in winds over fifteen knots, and had there been a moveable appendage factor *Resolute Salmon* might not have won. However, for all the emotion it caused, *Salmon* was a boat well suited to the conditions and gave Britt Chance the win he badly needed. Deep and fine forward but a little broader aft than traditional heavy-displacement types, it sailed exceptionally well in the slop and light airs off Marseilles. Downwind, though, its character was chaotic. Peterson, Holland and Kaufmann thought this was because of its high-aspect ratio rudder. Farr — observing, not participating in this regatta — was rather more considered. He thought that with the ballast scarcely below the centre of buoyancy, 'when the boat was upright there was very little initial stability.'[27] This produced oscillating problems when running downwind under spinnaker, 'which meant it started a rhythmic roll very easily ... [and] was out of control before the hull was stiff enough to support it.'[28] Farr concluded that in 'hard running conditions [it] would be a disaster and a type of boat the ORC would not want to encourage.'[29]

Doug Peterson thought 'the concept legitimate'[30] but in all other respects was damning. 'It is taking advantage of the rule and in the yacht design business, this is stepping on toes .... I bet you will see few people actually buying Chance boats because ... they'll be wary of a rule change to make the centreboard less of a proposition.'[31] Perhaps miffed that his defending champion had been beaten by a centreboarder built in Port L'Abbe, Doug ran up his ethnocentric flag. 'There was a big difference between the West Coast United States method of tuning rigs and sailing boats and that of the people from Europe. This may sound a bit egotistical coming from me as a West Coaster, but, damn it, we're just sharper than most other people.'[32]

His buddy, the savvy Ron Holland, was unerringly prophetic. 'If a board is penalised because it seems to make for a more successful yacht, then what should be done about wide-stern light-displacement boats?'[33]

Noticeable by their absence were designs from Dick Carter and Sparkman & Stephens — not one in the 43-boat 20-nation fleet.

***

When, at its annual meeting in November 1976, the ORC announced an amendment to the IOR, it was like a cold front preceding the southern hemisphere summer. Instead of addressing serious issues such as the non-existent scantling requirements, stability and moveable appendages, the ITC devised a change to the IOR that would target a characteristic found in only a small percentage of its fleet: wide sterns.[34] Effective April 1, 1977, Kiwi-style boats received a whopping half foot of rating penalty at a cost of twenty-plus square feet of sail. Centreboards, which already gave an estimated half foot of rating advantage over conventional keels, received not a mention. With a decisive stroke of its rule-making pen, the ORC had made uncompetitive nine to ten percent of the IOR fleet. Farr boats may have fired the opening shots. But the Big Guns of the northern hemisphere had won the opening battle in this escalating war.

This, though, didn't trouble the Kiwi School of Yacht Design or their growing number of supporters. On the contrary, the NZYF, in a public announcement of where it now stood, took the unusual step of naming a yacht designer as Yachtsman of the Year for 1976–77. 'Reading the citation to the award for Farr, Mr J.M. Foster, executive secretary of the New Zealand Yachting Federation, said ... it was remarkable that, although he had no formal draughting training, he had achieved worldwide recognition through his designing.'[35] After receiving the trophy from the Governor General, Sir Dennis Blundell, the quietly spoken 27-year-old thanked all those who had helped him achieve his success.

Nothing succeeds like success. And in New Zealand — at least for the moment — the yachting establishment was happy to extend the right hand of fellowship to the gentlemanly young designer. No longer the *enfant terrible* at home — more a leading light — there were lots of warm fuzzies for those standing in the growing arc of Bruce Farr's reflected glory.

Chapter 15
King of Speed

Farr breathed life back into the rule and improved every-body's boat speed. That's right, boat speed, not just rated speed.

Robert Perry, American yachting writer[1]

The fact that psychological and financial bankruptcy can be virtually guaranteed are only small prices to pay for experiencing one of the most exciting, exasperating and complex sports now being offered: [offshore yachting].

Mike Spanhake, New Zealand yachtsman[2]

If 1976 had been a busy year with its generous harvest of winners, '77 would be an *annus mirabilis* with a bumper crop of champions.

But first there were those to whom Bruce Farr Yacht Design would have to make a deep two-handed bow. If they were to regain the rating lost under the November IOR Mark III they had some tidying up to do to the after end of their alarmingly successful prototype: Design 51. But little did they know as they worked on Design 64 that they were about to unveil a yacht which would prove virtually unbeatable and see the hemispheric yachting war all but blow the sport right out of the water. Not that this was of the Farr office's making. With the failure of the ITC and ORC to close the centreboard loophole, they were merely walking through the open gate behind the early starters.[3] But that didn't stop the NZYF flexing its reactionary muscles as soon as the latest Farrs were launched. It would not only call into question the Farr office's ability to calculate stability. It would subject Farr designs — and those like them — to the indignity of haul-down tests. And no sooner would their ability to self-right be proven in these practical tests than the Cruising Yacht Club of Australia (CYCA) would weigh in with a ban on Farr lift-keelers competing in the Southern Cross Cup. And all that before the ITC rewrote Salem 1692 in nearby Newport. If Farr wouldn't confess to the witchery of his light-displacement god it would string him up on the gallows of their changing rules. It was time for the Parnell office to don its lifelines and grab the grab rails. 1977 was about to turn into one crazy wild ride.

*****

Bruce had first met Peter Walker in Sydney in 1969 when he and Rob Blackburn were part of the triumphant Kiwi contingent that saw Bowler and McGlashan make a clean sweep of the twelve-footer Interdominion, and Walker and David Hutchinson take the Javelin title. Walker had eventually returned from Perth and married Ann McGlashan.[4] With her and her family's support he had completed his BArch degree and later obtained a job at the Auckland Technical Institute (ATI) teaching Design Appreciation. He also raced on a number of yachts, including 727. It was at the beginning of 1975 that Bruce visited Peter and Annie at their Browns Bay flat and asked if Peter would crew on *Gerontius*. And when Bruce began casting around for an assistant following his return from Deauville and the Admiral's Cup, Walker was the obvious choice. An all-round talent on a boat, passionate about sailing and a romantic to boot, Pete jumped ship.

Bermuda-born Roger Hill was halfway through his NZCD (ARCH) when he came under Walker's tutelage at ATI.[5] He too was yachting mad, with an eye for line and a gift for draughting. A meeting was set up so Bruce could assess the young man's work. By then Hill was qualified and working for an Auckland architect and the Farr Parnell office had been operating for a year.

> My first impression on meeting Bruce was that he was very young. I can still remember the denim suit he was wearing. He looked at my drawings and thanked me for coming in. I didn't hear anything for several months. Then I got a call out of the blue. It was Peter. There's a position here if you'd like it. My father was incredulous. What do you want to do that for? You'll be out of a job in six months![6]

From Easter 1976, when Bruce had first moved his drawing board to Graham Painter's premises, a friendship over lunches and after-work drinks had developed with his blue-eyed assistant from Brisbane. By that December it was something neither could deny. But like two uneasy thoroughbreds Bruce and Cathie weren't yet ready to surrender. At least not fully. Two strong wills would see their maintaining separate interests and their own private lives for some time yet.

The office of Bruce Farr Yacht Designs was at the front of the suite leased by Trade Consultants at 470 Parnell Rd. They shared general office facilities but had two rooms to themselves: a small one for Bruce and a larger one for Cathie, Peter and Roger. Cathie was responsible for secretarial and accounting. Peter worked on deck layout, liaised with builders and customers, and helped set up and sail the finished product. Roger was full-time creating the working drawings of the lines plans Bruce produced. Roger Hill:

> He was very particular about having drawings done the way he wanted them. You could draw a geometry line of a cabin top or the end of a cockpit seat and he'd come along and move it an infinitesimal amount. But you soon learned to respect that; those tiny changes were often the difference. He also had a superb grasp of wooden boat-building technology. His ability to design a cold-moulded stringer hull is probably unsurpassed even to this day. He had an amazing ability to get the scantlings right — the frame size and the spacing. Everything was perfectly balanced. Everything worked and held hands. If you look at a set of his plans for the 3.7 for instance, it's made up of lots of little bits of wood all dovetailed in and frames with holes cut out. The woodwork and harmony of his structures was quite superb, just like a Stradivarius.
>
> Despite all that attention to detail, Bruce really was good to work with. He was never officious; he never played the boss. He was always kind. We developed a good working relationship and a strong personal friendship. It included all aspects of our lives. We worked together. We socialised together. He'd come sailing with us and we'd go sailing with him. We'd eat out together, go to yacht club functions together, party together. It was a close relationship. The office was like a family.[7]

When they began designing their first yacht in Parnell it never occurred to them that the forthcoming November rule change would not close off the centreboard loophole. Built in France by Tecimar, the fixed-keel Three-quarter Tonner was the first design they'd detailed for building in Kevlar-sheathed sandwich core. 'That gave better weight concentration to minimise performance loss in waves.'[8] Further weight was saved in the bulkheads by replacing ply with Nomex sheathed in glass-reinforced plastic (GRP). Better speed to windward in light airs was gained by way of increased sail area at the expense of length. In compliance with the recent rule change and in anticipation of further changes to rating gains through longitudinal distortion, the step in the stern was removed. It was Bruce's best all-rounder to date. Performing equally well in light and strong airs, *Joe Louis* gave him his second World Ton title and the moniker 'King of Speed' from Eric Simian and crew.[9] Significantly, the design deficiencies of *Resolute Salmon* hadn't been corrected by Britton Chance. His narrow-ended centreboarder, *North Star*, spent too much time broaching to be any sort of threat in that La Rochelle regatta. Two decades later Simian's delight in his boat and its designer remain undiminished.

Joe Louis could not be defeated in any event. After having won 'La Semaine de La Rochelle' and the 'Spid'Or Ouest France', the 'World Championship Selections', the [1977 'World Championship 3/4 Tonners' as well as the 'Semaine OE Cowes', it has been sold to Brits ... This man was just a genius, so much ahead of his time. Never we had such a talented naval architect in France. That's probably why Frenchies' performances are so bad in the 'America's Cup'.[10]

Though wood was still good for Bruce, *Joe Louis* had showed the construction way forward. But hi-tech composites were still uncharted waters for boats of this size, and in an esoteric field still in its infancy there were fewer experts in it than grog shops in Auckland's Mt Roskill. The Farr office would need another pair of eyes to see into the future of exotics.

Peter Walker had maintained regular contact with Russ and Lynda Bowler following their return from San Francisco. It was more than their shared past and that Gordon's Dry humour that drew Peter back to Russell. There had always been a tangible warmth and inclusiveness about Ces and Jean's home, something solid and secure now transplanted into Russ and Lynda's life. Unconsciously, perhaps, their friendship had carried a special compensation for the loss of his own childhood family.

As current national eighteen-footer champion it wasn't hard to get Bowler hooked on the idea. Promo glossies had been produced, and the Farr office's possible involvement in a professional eighteen-footer circuit sailing out of Long Beach, USA, the self-confessed sailing nut was soon twitching at the tiller. But although the idea would eventually sink into a West Coast sunset, it was lure enough to get him doing after-hours consulting work for Bruce Farr Yacht Design. He began by checking out the rudder stock in *Uin-na-mara*, the first Farr Two Tonner since *Gerontius*.[11] As other jobs followed, Russ's interest grew. Although his career with Kingston Reynolds was rocking along, he had been interviewing 'a lot of smaller partnerships ... and had come within a whisker of taking a couple of them up.'[12] Bruce's offer changed all that. Like the law, engineering was more about precedent than possibility and, at heart, Russ was an experimentalist. Here was a chance to indulge his creative passion in the sport he loved, get closer to the profit distribution than you could in a large organisation doing a lot of government aid work. Of course there was risk. But large firm or small, it all looked fragile. In fact Russ had reached the conclusion that a small consulting firm was arguably more secure than a larger counterpart which 'was just a bigger pot waiting to crash.'[13] There was also the country's crippling rate of taxation to consider. With his substantial bonuses now having 66 cents in the dollar ripped out of them, it wasn't long before the legitimate tax savings a small yacht design business could offer began filtering into the career equation. Even so, and despite Lynda's support of this shift in Russell's thinking, it would take the best part of a year of inward digestion and the orderly handing over of current projects before he was ready to make the move.

The sharp punches of prejudice from the local establishment had been absorbed by Bruce, thanks mainly to the talented dinghy sailors who had sailed his designs and proven them on the water. But these would be mere ripples compared with the tsunami about to be unleashed by yachting's international rulemakers. It was tough enough attracting commissions, let alone drawing yachts fast enough to win. And if each design advance could be dismissed with the stroke of a pompous pen, how certain was your future? The penalty for wide sterns had shown the consummate ease with which politics had bedded professional ethics in the ORC boardroom. At least Bruce now knew the score. So as the artist in him drove him on, the helmsman in him scanned the horizon for new sources of income.

Alpha Marine Ltd, in which he still held half shares, was the obvious builder to be offering his new ideas to the cruising yacht market. But there were complications. Despite Murray Crockett's best endeavours, Alpha wasn't living up to its phenomenal early promise. It was partly organisation, partly the speed of change to the IOR, and partly that their Canadian licensee was pumping out 727s without paying for the Alpha-supplied plug, let alone a royalty for each ex-factory boat. And when Murray and Bruce emptied their pockets trying to recover the burgeoning debt, the licensee convinced the notice-serving sheriff that the 727 design was theirs, not Alpha's. With Crockett and Farr inexperienced and hopelessly under-funded to fight a legal battle half a world away, the production arrangement that had promised much was about to sink out of sight without returning a cent. That was like New Zealand Yachts come back to haunt Graham Painter. Still providing consulting services to Bruce Farr Yacht Design, he was of the opinion that unless Bruce was involved with Alpha Marine — and not just as a silent partner — he was better off out or Crockett removed. It was an unsettling moment. On the one hand Graham had the Farr 11.04 up and running with Don Mosley's Glass Yachts and Compass Yachts across the Ditch, and was working on other production deals. On the other, Murray had successfully produced the first Farr production keel-yacht with a company Bruce part-owned. But Painter and Crockett didn't get on. That much was clear. It was also clear to Bruce that production, while important, was not his primary focus. So instead of stepping in, he stayed glued to his board. He was busy. It was where he felt safe. It was where he knew exactly what he was doing, and where the money was made to keep his business going.

***

When Doug Bremner made Kim McDell an offer he couldn't refuse, Kim abandoned his small accounting practice and part-time work for Gwyn Woodroffe's Power Marine and joined Sea Nymph Boats as Financial Controller. Sales Manager Peter Gribble was talking diversification the moment he walked through the door. For Kim the answer was clear.

'Doug, we should be doing a little Bruce Farr yacht.'

'Who's he? If we do our own production yacht it'll be designed by Sparkman & Stephens.'[14]

But that was not to be. With his restraint of trade in the carpet industry about to end, Bremner made another of those offers: vendor-financed purchase of a 36,000 square-foot factory with 50 employees, a substantial dealer network and a respected name: 'the predominant force in the power boat business'.[15] So with parental guarantors, bank guarantees, a sleeping partner and their hearts in their mouths, McDell and Gribble found themselves the proud owners of Sea Nymph Boats Ltd.

Having been a close friend of Peter Walker's since his early eighteen-footer days, and having had great success with Farr designs, there was only one port of call for Kim.

'Bruce, we want you to do a trailer yacht. Sea Nymph's got the manufacturing set-up and the sales network. But this is a very competitive market and we need something that will bring trailer-sailers into a whole new era.'[16]

The initial sketches weren't different enough for Kim. But when he spied 'some fairly way-out looking drawings to one side',[17] he knew he'd found the answer. With its wrap-around windows it not only looked different, it looked like another Farr breakthrough. Shorter and cheaper than the current crop of trailerables, it hit the market as the world fuel crisis broke. Suddenly wind was in. And so was the Farr 6000. From a standing start, Sea Nymph Boats were turning out two a week. They sold 30 in 1977, and would set up an Australian dealer network the following year. It helped that they had made it a strict one-design class. A clever device, it would provide competitive racing and maintain resale value. In fact, so successful would it become that Sea Nymph would eventually sell an astonishing four hundred Farr 6000s. And with three more versions they would eventually take total sales of Farr trailer-sailers through the eight hundred mark. Bruce Farr Yacht Design now had a hedge against rulemaker whims and another source of income besides couture commissions. As well, these were local royalties, readily collectable. It was a brilliant business move and a great commercial success.

***

Out on the Hauraki Gulf Auckland's design triumvirate was tearing up the racetracks. Laurie Davidson was now a light-displacement addict. With co-skippers Roy Dickson and Barry Thom at the helm, his *Black Fun* had won the 1977 National Quarter Ton Cup, pushing the Whiting and Farr designs into second and third overall. Significantly, the 727, world champion just two years before, couldn't foot it with the centreboarders.

Having pooled their *Bus* money, Murray Ross and Paul Whiting had been working hard at moving up a Ton. And fast. It may have looked like a 'join-the-dots' job,[18] but in just four weeks they had what they were after: a *Newspaper Taxi*. Chock-full of gear and rigging innovations, sporting Ross and Jones sails and with Murray Ross on the tiller, the daggerboarder was untouchable. Even with Helmer Pedersen steering *Cotton Blossom*, Peter Spencer's fixed-keel Farr could only win one race in the South Pacific Half Ton Cup.[19] Even Ian Gibbs' *Candu II*, the *Taxi's* fixed-keel prototype, was left a distant third.

No one had yet produced a centreboarder for the glamour One Ton Class, and, as expected, three Farrs from Design 51 — the wooden *Country Boy*, the fibreglass *Mardi Gras* and *Love Lace* — finished one-two-three in the 1977 One Ton Nationals.

Murray Ross:

> To this day I'm sorry about that decision [to sell *Newspaper Taxi* after just three months]. Financially, we should have stayed and done the Half Ton Cup and gone to Sydney [to contest the Half Ton Worlds]. But the One Ton Cup was to be held in Auckland and we wanted to do that instead.[20]

It was a decision that carried a hidden cost, one for which the ultimate price would be paid two years later when the lines of risk and competition met in the middle of the Tasman. A development of the *Bus* and *Taxi* themes, the big-bodied *Smackwater Jack* had its ballast moulded into the hull floor and just enough weight (90 kg) in the daggerboard to sink it. Rumour abounded that *Jack's* decks were light. And when Belgian Marcel Leeman, international juror, the ORC's representative and chairman of the measuring committee for the 1977 One Ton Worlds, later saw the multi-striped machine he shook his head: 'I think this is not intended by the rule!'

Leeman was right. Stability was not the issue. Structures were. But the dangerous trends emerging were doing so precisely because the ITC had failed to attend to issues of significance while busily making changes to the IOR in order to eliminate anything which deviated from their concept of 'the norm'. Penny Whiting:

> We'd be up all night gluing bits of foam on boats and glassing them over because of the changes to the IOR. So yes, the rule changes did change the designs. But we knew if we put a bump on here and another there the measurement was different. So we'd whip a boat out of the water and whack a bump on it [just to make it rate]![21]

Murray Ross adds:

> I mean, we got on a bit of a roll. And that's where I think Bruce is damned good; he doesn't let himself get too far out there. I find, actually, that he's quite a conservative designer.[22]

Not only had the ORC not closed the centreboard loophole. It had discouraged, by way of penalty, having weight in the centreboard. That led directly to designers saving weight in their hulls in order to get as much internal ballast on board and so reach the IOR's minimum centre of gravity factor. Only then could designers work below the stability for which they were being assessed. But while the IOR encouraged this trend it made no provision for scantling requirements.[23] This made it possible for designers, should they wish, to design hulls of insufficient strength to keep the water out. In short, in order to protect the design types it favoured, and in obeisance to the American cruising contingent's wish to preserve their traditional right to centreboards, the ITC and ORC only paid attention to those design features which threatened the heavy-displacement boats — the numerically dominant in the IOR fleet.

It would make no difference to the rulemakers that Bruce Farr Yacht Design had found ways of reducing weight without sacrificing strength and would pay minute attention to the construction and scantling details of their first retractable keel-boats. Or that they put extra lead in their drop-keels to boost stability and make their boats safer.[24] Despite the drum they banged, safety wasn't the issue for the ITC and ORC; politics was. And Farr, with his soon-to-be unbeatable centreboard designs was about to be propelled to the top of their Most Wanted list.

Chapter 16
Annus Mirabilis

*If as a designer you can increase performance only frac-
tionally and gradually, you will be welcomed to the club.
If, however, you are clever enough to make a quantum
jump and so leave all the others behind, there will be
cries of 'foul' and the red-faced referee will blow his whis-
tle. Bruce Farr and his school have now been honoured
in this way. It is a signal honour ... the Oscar of yacht
design.*

Jack Knights in *Sea Spray*[1]

*I have long been an admirer of Bruce Farr's work and he
has certainly paid his dues as the whipping boy of the
IOR.*

Robert Perry in *Sailing Magazine*[2]

It was one thing for Laurie Davidson and Paul Whiting to be experimenting
with daggerboards. But when the word got out that Bruce Farr was heading
down that track a clutch of New Zealand's yachting elite came knocking on
his door. Light-displacement Fever had entered the country's bloodstream.
Jim Young had even taken to chopping his One Tonner *Checkmate* in half,
swapping its original keel for a centreboard and its old name for a new one:
Heatwave.

Bruce Farr's One Ton retractable-keel design (Design 64) 'featured a
long, fine, easily driven hull with very little obvious rating distortion.'[3] As
with Design 65, the Half Ton version, wetted surface was reduced in
exchange for sail area gains. The first of the new generation of One Tonners
to be launched was Don Lidgard's *Smir-Noff-Agen*. Keen to try his hand at
offshore racing with his brother Jim, Don knew he'd have to bury the hatch-
et with Bruce if he was to be in with a chance.[4] Next in the drink was *The
Red Lion*; having come close to One Ton glory in *Jiminy Cricket,* Stu

Brentnall was going to try again. Ray Haslar had had enough of ploughing through oceans of water on *Barnacle Bill*, the heavy S&S design. Flying across the same blue stretches on the Farr One Tonner *Love Lace* had convinced him to take the light-displacement plunge and launch *Jenny H*. Last but not least was Graeme Woodroffe's *Mr JumpA*. The biggest sail carrier of the new Farrs, it had a varnished hull and cutaway transom. Piped into Westhaven by Pebo's boisterous bagpipes, Woodie had added an *A* to its name in deference to his major sponsor, the Mr Jump knitwear label, and the IYRU's rule 26 which prevented advertising on the hull. Having recently been beaten into fourth in the OK Dinghy Worlds, he also thought it a good idea to acquire the contest's winner, Peter Lester, as tactician. That decision confirmed what many already knew: the best dinghy sailors in the world were now sailing boats from the boards of Farr, Whiting and Davidson.

Paul Whiting's *Candu II* hadn't done the business for Ian Gibbs. And still hungry for that elusive world title he went back to Farr. Combining his daughters' pet names, Gibbs dressed his new boat in Ross and Jones sails and called it *Swuzzlebubble*. It would be the fastest Half Tonner out of the blocks and remain the season's frontrunner. Dr Peter Willcox, a Walker friend, also put up his hand for one of these hot Half Tonners. Knots quicker than its namesake from the New Zealand Land Wars, *Gunboat Rangiriri*, with its laminated Kevlar and glass hull, was the first Farr boat built by Marten Marine.[5] Sailed by Peter Walker with his hand-picked crew, this would be the first Walker campaign using a Farr design. Then there was the evergreen Roy Dickson, already one of the great contributors to New Zealand's yachting reputation. Unhappy at being caught on the wrong side of the latest rule change, he had Alpha Marine convert his Farr 920 into a centreboarder.

The Half and One Ton trials to select New Zealand teams to contest the Worlds were scheduled to be run together. But before they had got under way the New Zealand Yachting Federation (NZYF) was sending out disconcerting signals. Another low pressure system of opinion was heading in from overseas and they'd have to take a long hard look at this explosive trend to centreboarders. Not wanting to buy into the heavy versus light debate, the NZYF nevertheless insisted on self-righting tests for centreboard yachts before they would issue them with Category 2 certificates.[6, 7] This in spite of the Farr office's calculations showing that with drop-keels down their designs would recover from a knockdown quicker than their fixed-keel predecessors, and with boards half-retracted at the same rate. The debate was batted back and forth between owners and officials, the ground shots growing in intensity as both groups edged towards the net. Finally the NZYF dropped a point. Provided one centreboarder from each designer from each Ton class was hauled down and successfully self-righted, their offshore committee would accept independently audited calculations for those boats which didn't take the physical test. Singling out one design type and not another was discriminatory enough. But trying to sink the latest race yachts from the country's best designers was the next best thing to pillorying Farr, Davidson and Whiting outside the Town Hall.

***

The Epiglass five-race One Ton trials of 1977 produced some of the best offshore racing New Zealand had ever seen. The seventeen-boat fleet immediately split into two — the six centreboarders followed by the eleven fixed-keelers — with the competition fierce in both 'divisions'. And the weather threw everything at them — ten- to sixty-knot winds, flat calms to big seas. From crash-gybes to broken masts, from knockdowns to smashed spinnaker poles, all survived. The Farr boats had the better of it on the Olympic courses. But in the ocean races *Smackwater Jack* came out on top — by three minutes from *Jenny H* in the intermediate race and by six minutes from *Smir-Noff-Agen* in the 360-miler — giving it the series. That meant six centreboarders — one Whiting, four Farrs and a Young — would make up the Kiwi team to contest the Auckland One Ton Worlds in November.[8]

It seemed impossible, but the Half Tonners produced even closer racing than the One Tonners, with *Swuzzlebubble* finishing ahead of *Gunboat Rangiriri* and the Davidson-designed *Waverider*. In the intermediate race, *Waverider*, *Gunboat Rangiriri* and *Swuzzlebubble* ended a titanic struggle with only eleven seconds separating them after 135 miles. Roy Dickson's *Instinct* fell out of contention in the first race with a broken mast. Significantly, it would free him to sign on as tactician and co-helmsman for *The Red Lion* in the One Ton Worlds.

In the Ceramco-sponsored trials to select the New Zealand Southern Cross Cup team *Swuzzlebubble* finished ahead of One Tonners *Smir-Noff-Agen* and *Jenny H.* It would be an all-Farr team of retractable keelers for the biennial regatta in December. Gibbs' *Swuzzlebubble* remained the talking point. Its form had been so consistently outstanding all season that Design 65 was now being mooted as a one-design ocean racer.

However, the centreboarders' dominance in these trials only confirmed officialdom's worst fears. These boats were so fast they were not only odds-on to win the Worlds. They were likely to alter the face of offshore yachting and the composition of the IOR's fleet. Still, it was a shock when the Royal Akarana Yacht Club (RAYC), sponsors of the Southern Cross team, submitted their three Farrs for Category One certificates and the NZYF refused to say whether or not they would issue them.[9] The rumbling controversy had just become a major test of character for Bruce Farr Yacht Design. Suddenly it was fighting for its life: 'We are left in a position where we can't do any more design work for New Zealand clients because we don't know what the inspectors consider good or bad.'[10]

But no sooner had the NZYF finally agreed to treat lift-keelers as fixed-keel yachts and issue the certificates than the CYCA waded in with an even bigger stick. They would not allow centreboard yachts to compete in the events they controlled: the Southern Cross Cup and the Sydney–Hobart. This in spite of their having recently cleared the Farr retractable-keeler *Hecate* to cross the Tasman to Auckland to compete in the One Ton Worlds.

If cables had been flying intermittently between Newport, London and Sydney before this, now they hurtled through the ether at rapidly diminishing intervals. And those going from the NZYF's chairman, Harry Julian, to his CYCA counterpart had gone from hot to blistering.

> If the three Farr lift-keel yachts are excluded from the New Zealand Southern Cross Team we will withdraw Stop

Not that it takes much for an Ocker to ruffle a Kiwi's feathers. But in the process there had been another dramatic wind shift: Bruce Farr now had the unequivocal backing of his country's yachting establishment. And when Harry Julian's cables failed to elicit anything more than intransigence from Gordon Marshall —

> We have pressure on us local and international to come up with a way of assessing these boats for ocean racing proper Stop

— off went more sizzlers to the IYRU and the Australian Yachting Federation (AYF), strong supporters of the CYCA's position with their own concern about the seaworthiness of modern designs.[11]

> The NZYF will not be taking this latest piece of parochial rule-tampering by the CYCA lying down Stop[12]

Not that it was all one-way traffic. At least one Sydney commentator could see how the growing interest in the CYCA's perfidy was compounding into increasing embarrassment for Australian yachting.

> It seems that Gordon Marshall and the faceless men of the Cruising Yacht Club [of Australia], faced with the prospect of seeing the Southern Cross Cup retained by New Zealand again, have decided to eliminate the Northern New Zealand team by simply refusing their entry on the grounds that they are not seaworthy. The magic formula is, of course, not available. Just think what might happen if clever young Bruce Farr got hold of the formula. Why, he might prove that one of those heavy-displacement juggernauts might also be found to be unseaworthy.'[13]

November 1977 was a month when owners of these new retractable keelers must have been wondering what they'd bought into. For Farr, Walker, Hill and Cochran, it was like flying paper planes down tunnels of fire — there were no guarantees, just heat and so much smoke you couldn't see where you were going. But Bruce Farr Yacht Design wasn't about to be asphyxiated. Out came Bruce fighting (albeit politely as usual). 'If they hit centreboarders only, what they are doing could be viewed as an attempt to get rid of one type of boat. It may be coincidental that [New Zealand] has selected three of those new types to defend the cup.'[14]

When a copy of the CYCA's self-righting formula finally reached New Zealand it was immediately clear that the Farr boats could not comply. Nor, for that matter, would many of the more recent Half and One Ton fixed-keel-boats to which it would not be applied. The impossible formulation decreed that the Farr One Tonners were to have an extra 360 pounds added to the 'I' point of their masts and the Half Tonners an extra 250.[15] Off went more Julian cables. Still the AYF refused to get involved; the self-righting issue was a club issue and they were powerless to intervene. In a last-ditch effort to save things Bruce Farr and Dick Jones, Commodore of the RAYC, booked a flight to Sydney. But the day before they were due to leave the hard-fought-for cable arrived. At a full committee meeting the previous night the CYCA had passed *Swuzzlebubble, Smir-Noff-Agen* and *Jenny H* as fit to compete in the Southern Cross Cup and the Sydney–Hobart. Harry Julian was delighted.

'This whole affair has come to a sensible conclusion!'

In one sense, yes. But in another it was just more of the same. In reality it had been a display of politics so inept you could hear Machiavelli choking on grave dirt. The formula the CYCA had used to try to outlaw the Farr yachts had been prepared at the beginning of November by a meeting of the ITC chaired by Olin Stephens. Now Marshall's hubris could be understood; his mathematical formulation had come at a greater speed and from only a slightly lower altitude than those Mount Sinai tablets. But lost in their maze of political intrigue, the ORC and ITC now found themselves without a back door. How could they endorse the disqualification of a type of yacht which their rule specifically included? When they realised the size of the international incident they had been surreptitiously involved in there was only one way out — a hands-up. Thus it was that the race committee of the Royal Sydney Yacht Squadron (RSYS) announced, on November 18, that a cable had just been received by the AYF from the ORC.

> Centreboard rule comes into effect 1/1/78 Stop We also confirm that for administrative reasons RSYS cannot apply stability rule to Half Ton Cup Stop Therefore no new rule valid for Half Ton Cup Australia Stop[16]

It seemed that everyone except the Kiwis had egg all over their clocks. There was only one problem. So much time had elapsed since the beginning of this nefarious business that it wasn't just drying. It was starting to crack and peel.

***

[Smackwater Jack] *had some glaring weaknesses and was a bit of a mongrel tight-reaching. We spent five weeks modifying it [after the trials] and ended up making it slower! It was both a bitter disappointment and a lesson. But I got some relief because Stuey Brentnall dumped his Hood sails and bought a set of ours for* The Red Lion *and went out and won the Cup. If it hadn't have been for that I'd have had to get on the pills![17]*
Murray Ross, New Zealand yachtsman and yacht designer

Having finished fifth and last of the Farr lift-keelers in the national trials, *The Red Lion* was the surprise mover in the One Ton Cup, although its consistent 2/1/1/4/3 win was no more surprising than *Smackwater Jack's* spectacular failure. Pre-tournament favourite, the big-bodied Whiting could manage only seventh with a DNF/6/4/6/DNF. It was, though, a marvellous result for Brentnall and his crew and a stunning triumph for Farr.[18] Not only had his boats survived gale-force conditions and spectacular knockdowns. *The Red Lion*, *Mr JumpA* and *Smir-Noff-Agen* had filled the first three places in the world championship of the glamour One Ton class.

***

As guests politely sipped their gin and tonics during the RSYS cocktail reception for the Half Ton Worlds, Pebo strode into the dining room belting out notes on his bagpipes. Refusing to believe that the performance was anything other than planned, the guests sat bemused as the *Swuzzlebubble* crew's virtuoso performance ended with a rendition of 'When Irish Eyes Are Smiling' for defending world champion Harold Cudmore. And smiling they were as Harry steered his Holland-designed heavy-displacement centreboarder to a two-hour win in the light-aired fifth and final race of the series. But the smiles vanished along with the champagne when *Gunboat Rangiriri* crossed the line in second place; *Silver Shamrock III* had needed two boats between itself and *Gunboat* for Cudmore to keep his title. Having made a spectacular recovery during the early morning hours, Walker and Willcox, Rob Blackburn, Kim McDell and Gordon Fraser had blasted past

Tony Bouzaid's *Waverider* and Ian Gibbs' *Swuzzlebubble* to claim second
spot in the long ocean race and present Bruce Farr with his third World
Ton title for 1977.[19] It was a truly remarkable achievement, one without
historical precedent in offshore yachting. Sadly, given the convoluted
politics of the IOR, it was one destined never to be repeated.

***

With three wins between them in the first three races, *Jenny H* and *Smir-
Noff-Agen* had already proven themselves the outstanding boats of the
Southern Cross series. And as they strode through Sydney Heads, not only
were they racing boat for boat with the fleet's Two Tonners, *Swuzzlebubble*
was so far ahead of the other One Tonners that Gibbs and his crew looked
odds-on to win the Sydney–Hobart. Until, that is, two 60-knot fronts, one
after the other, slammed into the fleet. The seas they produced were so
massive and confused that by the time *Swuzzlebubble* entered Bass Strait a
safety-harnessed Dick Jones had twice been washed over the side. Ian
Gibbs:

> We had a quiet discussion after that — there was no panic, no drama. The
> decision was that we would be risking the boat and lives if we continued. To
> me that was the seamanlike decision, even though we hated to turn back.
> There were thirty-footers coming one after the other. The first was no problem
> and the second was no real problem either. But the third was coming in hard
> and fast and I was all too aware what would happen if the stern got washed
> around .... It was like sailing on a surf beach.[20]

So in spite of there being no damage to his Kevlar-glass hull, sails or ringing, Ian Gibbs and his crew retired. But things weren't looking so good on
Smir-Noff-Agen. Don Lidgard had become so dehydrated from vomiting that
he was coughing up blood by the time he called it quits.

> We were starting to launch off seas. We'd try to go along the top of the waves
> to dissipate some of the forward motion, and then slide down the other side.
> But when we went up and along a particular sea with breakers on the top we
> found ourselves in a hole and pitch-poled, falling down the front. Two of the
> crew were down below at the time. They staggered back up on deck, their
> voices like little girls': Do that again Don and the boat'll go down! We'd shat-
> tered all our forward stringers. We had some pipes in the cockpit which we
> used to put our feet against to brace ourselves. We took them off and man-
> aged to tie and pack them in up forward to brace the hull. We had the main
> off, just sailing under jib as we caught the seas into Bermagui.[21]

Ray Haslar went closer inshore and nursed *Jenny H* through the worst of the storm. His fifth on corrected time gave New Zealand its fourth Southern Cross win in five attempts. It did nothing, though, to ease the build-up of bile in the CYCA. That *Jenny H* was the only centreboarder to finish the race gave them the perfect excuse to vomit all over Farr. Ostensibly to write a report to make the Sydney–Hobart safer in the future, the CYCA issued a questionnaire to all entrants who had retired from the race. It didn't matter that from 133 starters there was a record of 58 retirements and that many of the heavy-displacement fleet had suffered more damage than the lift-keelers. Having fallen victim to the heavy-equals-safe philosophy, Marshall and the CYCA used the information against Bruce Farr and his light-displacement followers with all the political panache of a Pol Pot and his Khmer Rouge.

1. *'It was common knowledge that the one [light-displacement] survivor, who suffered broken ribs during the passage, had come to an arrangement of sale before the race that required her to get to the finish line. This may have had some bearing on whether she continued or not.'*[22] The fact that *Jenny H* finished top boat of the series and won the Southern Cross Cup for New Zealand was conveniently overlooked by the CYCA as valid reasons for finishing. Using Haslar's sale of *Jenny H* to American John Kilroy to try to discredit Bruce Farr's reputation was particularly unsavoury.

2. *'Half of the hull failures came from only 5% of the fleet. Had the ultra-lightweights not been with us, hull failures would have amounted to 4%.'*[23] In issuing these percentages the CYCA failed to mention that there had been no damage to *Swuzzlebubble*. The skipper of one of the smallest boats in the fleet had made his decision to retire on the basis of sound seamanship — a principle endorsed by the CYCA except, it seemed, when it came to Farr boats.

3. *'This race signalled the first occasion when "ultra-light construction" faced up to a Bass Strait crossing. They proved quite inadequate for storm conditions ... .'*[24] Given that two Farr 1104s — *Piccolo* and *Rockie* — had finished first and second in the previous 50-knot Sydney–Hobart and that the Auckland One Ton Worlds had experienced gale conditions the previous month, it seemed the CYCA was suffering from its use of selective memory. As well, evidence from previous offshore regattas in New Zealand had conclusively proved that light-displacement yachts not only out-performed their heavy-displacement rivals in wild conditions but were safer and had better sea-keeping qualities.

4. *'The fact that some yachtsmen need to be protected from their admitted inexperience is shown by the comments in questionnaire answers from "ultra-light construction" owners, who said that they had ... "never experienced seas like these before".'*[25] Not only did this wilfully

ignore the combined experience of Haslar and Gibbs in the Northern New Zealand team. The reality was that few, if any, in the fleet had seen conditions like these and wouldn't again until the horror Fastnet of 1979. 'There were no backs to the waves and you'd come out of them completely airborne.'[26]

5. '*Whilst the ITC and the ORC have made clear their intention to re-rate these types of yachts and so reduce their rating to speed advantage under the present rule, thus arresting the design trend, we have to face up to the safety hazards of those already built ...*'[27] Finally the truth was out, this sentence saying it all: Bruce Farr's designs were just too damned fast.

In the aftermath of the mayhem Farr's designs had, in fact, emerged better than most. Of the seventeen Farr boats in the fleet only five withdrew. From sixteen starters, S&S had five withdraw. Doug Peterson had six from ten retire; Peter Cole eleven from twenty-one; Michel Joubert three from six; Ron Swanson six from nine; and Scott Kaufman three from three.[28] With those percentages it wasn't hard to read between the lines. Nor did you need to be an English PhD to understand why the CYCA wasn't buying Ray Haslar's opinion.

The Farr centreboarders are quite definitely the fastest 37-footers in the world ... Why should an improvement be dragged back into line with something inferior? If more of the people who criticise these boats sailed them before opening their mouths, there would be nowhere near the amount of gum bashing that we've had.[29]

It was an extraordinary end to an extraordinary year for Bruce Farr and his Parnell team: three world titles and their three lift-keelers producing New Zealand's biggest-ever win in the Southern Cross Cup. And yet not extraordinary enough, it seemed, to stop the northern hemisphere makers of rules from dipping their dainty quills and drafting an end to the promise these extraordinary designs offered.

With tension mounting between Trade Consultants and Bruce Farr Yacht Design, Cathie Cochran had taken herself off to Fiji for a holiday. There had to be more to life than this. There was. On her return she handed in her notice and flew with Bruce to Sydney for the Half Ton Worlds, later introducing him to her family in Brisbane. She stayed on there for a time of reunion. But after six weeks of missing Bruce, she returned to Auckland, took up a position as a property manager for a real estate company and, at the beginning of 1978, moved into his large room in the rambling Remuera house. It wouldn't be until May of that year, after Bruce's return from

Newport, that they would set up a flat together. Faced with a complete collapse of his faith in the sport's ruling bodies, Cathie would be more than a lover and friend. She would become his emotional Rock of Gibraltar. And he hers.

PART II

The war against New Zealand's Plastic Fantastic was waged both on and off the water. In whispered words, in letters, in the public bars of Freo, in the hotel suites of lobbyists, and in the press. Yet even after the door was closed on Glassgate, members of the Sail America Foundation continued to lobby all and sundry to have fibreglass banned in Twelve-Metre yachts. The all-round ability of KZ-7 could not be matched. Which was why the San Diegans went searching for new advantage, technological and financial, and delayed setting a date for the next America's Cup. Greed was once more having its way with the One Hundred Guineas Cup.

When four men met in secret in Sardinia in June 1987 they had no idea that their decisions would soon lay bare the fragile psyche of the world's most powerful nation. Out-thought and out-designed by these New Zealanders as they mounted their Deed Boat challenge, the new holders of the Cup not only ignored the rules by which they held the famous trophy. They so wound up public opinion that they had the mayor of New York City spitting tacks and the Kiwis as Japanese heading for Pearl Harbour. Thus it was that the twenty-seventh challenge for the America's Cup dissolved into another war of words, machismo, money and patriotic pride. That New Zealand won the America's Cup at only its second attempt there is no doubt: it was awarded to the Mercury Bay Boating Club on 28 March 1989. That it was taken away six months later — in a decision presided over by a judge who would later be arrested by the FBI — is also indisputable. But although the decision stuck, what wouldn't go away was the 133 feet of sublime creativity. Conceived and engineered with breathtaking speed, KZ-1 would change the course of America's Cup history and so pave the way for New Zealand to win again and next time keep the famous silver ewer. But the giant yacht that emerged from an Auckland shed was more than just a triumph of design and technology, more than just a Spruce Goose to fly only once for a mile, or an immovable Eiffel Tower. Just as America had been a messenger of challenge to British imperial and industrial might, so was KZ-1/New Zealand to American technological pride and hegemony in the event that still remains the symbol of yachting supremacy.

Because the beginnings of the Deed Boat challenge were shrouded in such secrecy, what follows in the interlogues is a reconstruction of events and should not be read as fact.

Chapter 17
Omelettes and Eggs

You cannot make omelettes without breaking eggs.
David Edwards, Chairman Offshore Racing Council,
May 1978[1]

If one form of design ... proves it has a positive advan-
tage, it is the rulemakers' job to bring that design back
into line with the rest of the fleet so as to ... protect own-
ers' financial investment ... So, if anyone comes up with
something that is obviously better, the rulemakers feel it
their duty to bring in a penalty to make any development
equate with whatever is the norm.
Ron Holland in *DB Yachting Annual* [2]

It is hard to date this philosophical hotchpotch accurately, although it had
probably been floating around since 'Olin Stephens ... began to think aloud
about introducing special ratios of beam to depth. He believed that in this
way the trend towards dinghy and even scow-like shapes could be arrested.'[3]
Olin talking with the ITC about the pending light-displacement flood and the
couples they might entice on board — England and America, Italy and
Germany, Holland and Sweden — was a bit like Noah chatting with Shem,
Ham and Japheth about the benefits of arks and who they might coax into
their holds. Indeed, it signalled another big storm. If Bruce Farr Yacht Design
thought they had just been through a patch of rough water in Australia, it
would be nothing compared with the flood of retroactive legislation about to
come pouring out of England as the establishment moved to lock in the
value of their boats and the reputations of those who had designed them.

Snow was still on the ground when Bruce arrived in Newport. There in
a semi-official capacity for the NZYF's offshore committee, and with its
blessing and support for some of his proposals, he was hoping to bring
some sanity to the matters due for discussion at this April meeting of the

ITC. If he didn't and the already-circulated displacement-length (DSPL-L) and sail area-displacement (SA-D) ratios were ratified in London, they would become law that November and light-displacement yachts would be rated out of existence.[4]

He had gone well prepared. He had enlisted the help of his old Moth Class mate Vern Smith, who, at his behest, had developed a version of the IOR program in which the weighting of the various factors in the rating could be changed, allowing many options to be tested to try to find a combination that would give a fairer result for a wide range of boats. With Vern's version of the program, they were able to conduct a 'test' on 30 New Zealand yachts, a sample that represented a good cross-section of the world's IOR fleet. These comparative results formed the basis of Bruce's dissertation, in which he showed how the core IOR formulation might be altered to give some encouragement to heavier boats and penalties to lighter boats but in a fair, across-the-board manner. With the NZYF's approval he sent his paper by the due date to the ITC. But little did he know as Cathie drove him to Auckland International Airport that he was about to fly into a mile-high wall of reaction.

> This was my first glimpse of ORC/ITC politics — where somebody might simply neglect to distribute something to all the players if they didn't like it. I don't think my submission ever got distributed around the ITC.[5]

It was scary knowing that the future of his business rested in the hands of rulemakers on the other side of the world. It was even scarier learning that they had no interest in what he had to say. But having picked up the tab for the trip, Bruce wasn't about to sit in silence while delegates debated the future of his art and his business. When the opportunity presented itself, he rose to address the meeting and obliquely to challenge the chair.

'Mr Chairman, the rule changes the ITC has proposed and recommended for adoption are clearly unfair to a certain type of boat and knowingly so. Why don't you formulate a change that would be fair to all?'

'Because the ITC feels it has a mandate from the ORC to be unfair to the minority. It is our duty to protect the value of the existing fleet.'

'But what of the boats making up the remaining ten percent of that fleet? Are they no longer to be considered part of the IOR?'

'Thank you, Mr Farr. Next question ...'

It wasn't hard to spot the spinner at the centre of the web. It was just hard believing who it was. Hard to believe that the man who was the idol of many was playing the political game that hard; that the innovator had turned atavist in the twilight of his career in order to protect the competitiveness and value of his designs. But here in Newport Olin Stephens and his team — the makers of the rules — were openly spinning obstructionist threads to stop the sport developing along the lines on which it was naturally evolving. Not only did they have the ORC's support. They had supporters right around the world. Among them were Sir Peter Johnson from England, Hans Otto Schumann from Germany and Pat Haggerty from the USA. Even the Australian, Jim Robson-Scott, who held the one Australia-New Zealand vote on the ORC, was an opponent of light-displacement. Then there was American Gary Mull, the only other designer on the ITC, and England's Major Robin Glover, the IOR's Chief Measurer, who saw his Mark IIIa 'brainwave ... as emergency first aid for old boats'.[6] And as this groundswell of support for Stephens' concepts grew, it spread from officials to delegates, to owners and to leading designers who had as much to lose as anyone from the perceived Farr threat. The latter, with Ethnocentric Doug their leading light, met twice in England following the ITC's April meeting. Their communiqué supported the ORC's proposals and was openly critical of such things as bow-down trim which gave light-displacement designs an advantage under the old IOR Mark III. Their effective lobbying of the ORC's London meeting of 20–21 May would bring about wholesale changes to the IOR not envisaged the previous month in Newport.

Contained in the IOR Mark IIIa amendments were rule variations designed to improve the ratings of those boats which were heavier and had less sail area relative to their size than those now defeating them on the water. To do this the ORC adopted the ITC's DSPL-L and SA-D ratios and the mathematical formulae that went with them. Not that these ratios encouraged a design retrospective to an older style of yacht with spoon bows and tapered counters, or even further back to clipper bows and fantail counters. They merely engineered a return to the generation of yachts which currently dominated the IOR fleet, not in speed but in numbers. In so doing the ITC and ORC had not only wound the design clock back to the cumbersome style of yacht most Americans and English owned. They had set it ticking towards the IOR's demise.

Sounding dizzily perspicacious in his omelettes and eggs speech, Chairman Edwards quite missed the lapping of water at his ankles. While the IOR was sinking under the weight of intransigence and intrigue, he was blithely whisking up support among the party faithful. 'A rating rule can be more accurate if the boats are similar instead of various.'[7]

Fortunately for the chairman but unfortunately for the sport, Bruce Farr wasn't there to ask him similar to what. Nor to point out that his desire to see a limitation of yacht types ran contrary to the IOR's Rule 101. 'It is the spirit and intent of the rule to promote the racing of seaworthy offshore racing yachts of various designs, types and construction on a fair and equitable basis.'[8] It seemed inconceivable, but there it was. Lost in the fog of patriarchal politics, the ORC was about to take its noble tenet and run it right up on the rocks.

In the end the effect on light-displacement was to be more devastating than even Olin Stephens could have imagined. That wasn't surprising. Three world titles in one year to three Farr designs had added an unstoppable momentum to this growing rearguard action. Now it was clear to all and sundry just how easily that young Bruce Farr had stolen the offshore game out from under their heavy-displacement waterlines. There was only one answer, even if it was many-sided. They would introduce four Mark IIIa rules with 'boats built before 1973 ... entitled to pick the most favourable result produced by the four',[9] including an age allowance which would translate directly into time on the racetrack. The changes, expected to be rubber-stamped at the next ITC meeting and ratified by the ORC in November, would be binding from 1 January 1979.

The victory was emphatic. As Farr's had been on the water, so the rule-makers' was on paper. The Kiwi designers had been bundled out of play. And the Farr-style yacht, which had proven itself as having better sea-keeping qualities than its heavier displacement rivals, as being cheaper to build, and faster and more fun to sail, was now effectively excluded from life within the IOR. Under the new formulation *Jenny H*'s rating increased by 5.9 percent. So did *Country Boy*'s. *Joe Louis* went up by 5.8 percent, *Gunboat Rangiriri*'s by 4.7 percent, and *Cotton Blossom*'s by 6.3 percent. Paul Whiting's *Smackwater Jack* increased by 4.3 percent, his *Newspaper Taxi* by 7.4 percent. Laurie Davidson's *Waverider* received a rating bump of 4.4 percent. On the other hand, all heavy-displacement types had their ratings reduced. Among them was *Pathfinder*, with a rating decrease of 8.1 percent. Doug Peterson's *Streaker* went down by 2.1 percent, his *Hot Canary* by 5.9 percent. The Salthouse Cavalier 32s decreased by 9.6 percent, and Ben Lexcen's *Anticipation* by 4.9 percent. The rating difference between Bruce Farr's *Jenny H* and Olin Stephens' *Pathfinder* now stood at an unconscionable 14 percent. The ORC had just endorsed a rating difference between heavy- and light-displacement yachts that was big enough to steer a supertanker through.

> We had won all these world championships and were feeling pretty good about things. Then with a flick of the pen in November we'd been set up to be wiped off the map. In the space of a month we'd gone from having six months' worth of IOR work to nothing. I was angry and hurt. But the bottom line was we had a business to save.[10]

In fact, Bruce was gutted. How could they destroy the southern hemisphere fleet to protect the value of their own? It was quite beyond the artist in him to comprehend why the ITC had not striven to write a rule that would allow him to design the boats his clients wanted. And it was beyond the pragmatist in him to grasp why the ORC had set up 'a bus route to take people places they didn't want to go.'[11] But existing side by side with the artist and the pragmatist was the survivor. They were about to join forces and go looking for another way.

They'd have to. The 1979 Quarter Ton Cup sailed off Sajima, Japan, in which frantic seas had sunk one centreboarder and capsized others, seemed to endorse the ORC's stand on light-displacement. As would one of New Zealand's great yachting tragedies, the loss of Paul Whiting, his wife Alison and crew on *Smackwater Jack* in January 1980.[12] No matter that Maurice Dykes's stock Farr One Tonner *Chick Chack* came through the same gale unscathed, that the Fastnet Race of 1979 had claimed seventeen lives along with a number of heavy-displacement yachts, or that the Davidson-designed *Waverider* and the Farr *Swuzzlebubble* had survived gale-force conditions in the North Sea while racing for the 1979 Half Ton Cup.[13]

***

By this time Bruce and Cathie were happily settled into a charming villa in Sentinel Road, in Auckland's Herne Bay, a home Cathie was soon redecorating. Bruce not only jointly owned this dwelling with Cathie. The impoverished self-employed designer of a decade ago was now the proprietor of the country's leading yacht design business, a business now producing a handsome annual income that would enable him to buy his own investment property in eighteen months' time. Cathie Cross:

> It was wonderful to be on our own. The home was our sanctuary. We'd close the door on the world some weekends and forget the politics of yachting. By then the business had grown and Bruce's life was hectic. Yet we seemed to find a balance between work and play; we entertained a lot and Bruce found time to relax. Our two years there was a very happy time.
>
> My original impressions of Bruce never changed, although I learnt that he was also rather shy and sensitive. But he had a quiet strength and determination, and grew more confident and self-assured. He was warm and generous to his close friends, generous with his time and help; often there'd be a friend or someone in need using the flat downstairs. I don't remember any effort needed by either of us to adjust to living together. We shared the chores, and Bruce helped me with the restoration of the villa. He was also a very caring person. I remember the day my father died, the way Bruce guided me through the motions of normality. The way he drove me round all day listening to my emotional rambling, just being there to cling to when I needed to be held. Then there was my old china marmalade jar he accidentally sent flying one day. It was worthless except for its sentimental value, a memento from my former life with Steve. But instead of consigning the many pieces to the bin, Bruce painstakingly glued them back together, all without my knowing. It wasn't a wasted effort; I still use it every day.

Yes, I loved him — passionately, and dearly as a friend. We talked a lot about the future, about marriage and children: what they would be like and how we would bring them up. It seemed the natural outcome of our relationship. I was incredibly proud of him but not in the least bit overawed by his talents — which were very much secondary to the person he was. I remember consciously thinking how lucky I was to be so happy. I couldn't imagine myself ever loving anyone else.[14]

Robyn Curtain — English, elegant and with a sharp accounting mind — had replaced Cathie as Bruce's PA. For eighteen months she took his calls and arranged his schedules as the business mushroomed. With growing demand from overseas, Bruce juggled his time between his drawing board, his phone and aeroplanes. Increasingly, it seemed, his overseas clients wanted him in person, not just as a name on the bottom of their lines plans. His became a punishing schedule of short trips to the northern hemisphere and back. No sooner would he be over his outgoing jetlag than he'd be heading home again, preparing to adjust, through another bout of fatigue, to Auckland's time zone. Roger Hill continued to produce immaculate working drawings of Bruce's latest designs. Peter Walker was busy attending to local clients, designing interiors, testing the newly launched creations on the water and shouldering his share of drawing duties. For a creative team it worked like a well-oiled wheel. But as demand continued to grow it became increasingly clear that they would need another pair of hands — a need made more acute when Peter's life began unravelling in the latter part of 1979.

It was while he and and Annie were living with a friend at Auckland's Karaka Bay and building their dream home next door that their marriage came apart. Unsure, now, of where to call home, it was enough for Annie to leave with Frith and Iain. And when Bruce asked what he could do to help and Peter said he needed a place for Annie and the children to stay, Bruce and Cathie offered their downstairs flat. There were no secrets in the office. And as Peter's work rate dropped, his workmates picked up the slack, and Sally Gray, who had replaced Robyn Curtain, the emotional moraine.[15] Loyal till the end, Bruce left it up to Peter — he could stay on, employment conditions unchanged, until matters resolved themselves.

***

With only eighteen months between them, Bruce and his brother Alan had always been close. But even before he left Leigh on his High School scholarship, Alan was becoming emotionally distant. Student life in Auckland, his degree in mathematics and the esoteric world of computers were as far removed from country life as they were from his father's worlds of fishing and printing. A further distancing occurred when he and Pam married and moved to Wellington. Surviving a fatal car crash with punctured lungs and severe concussion at the end of that year was especially traumatic for Alan. But it wasn't until they moved back to Auckland in 1976 that Jim and Ileene understood just how serious Alan's condition was. In between times of treatment, he held positions with Burroughs and Auckland University. From there it wasn't far to Bruce's Parnell office, and the brothers would meet for lunch. It was good to have contact again, though sometimes hard for Bruce. Hard to accept that his systems analyst brother was now in and out of Ward Ten, the psychiatric unit at Auckland Hospital, diagnosed a manic-depressive.[16] It was even harder knowing that while Alan had been so highly regarded by Burroughs they had sent him to Detroit in 1974 to research a new computer system for the Auckland Savings Bank, it was increasingly difficult for him to hold down a job. The circumstances were not only sad. They were coloured with a cruel irony. The painfully shy sibling had emerged strong and successful and a darling of the media while the brilliant, progressive brother seemed determined to end his life.

For weeks, months even, Alan would be fine. But with changes to his medication and his inconsistency in taking it, his condition deteriorated to the point where Pam could no longer cope. Unable to sit through the long nights of talking then care for their two small children the next day while Alan retired to bed, she eventually asked him to leave. It was the hardest decision of her life. His continued attempts on his life, with Bruce at his side while his stomach was pumped, made it seem even harder. Then there were attempts at reconciliation that didn't work out.

By now Jim Farr was semi-retired, fishing part-time, and he and Ileene were able to care for Alan when he moved back home following a rehabilitation programme. And even when they took a much-needed three-month break on their boat, they had Peter and Annie Walker move into their Tudor Street home to help keep an eye on his progress. But that last year was a difficult one. At times Alan seemed increasingly lost and distant, at others closer to his parents than he'd been since childhood. And even the last time he left home he appeared to Jim and Ileene to be happier and more at peace with himself than he'd been for a long time.

It was on 13 December 1979 that Bruce received the call he feared but hoped he'd never get.

'Mr Farr, I'm afraid we have bad news. Your brother Alan — Could you please come in and identify the body.'

The day prior to his suicide, he had moved into a room in a house with others near us in Herne Bay. Bruce and I had gone over to have a look at it and he was due to come to us for dinner the next night. But in the middle of the afternoon, amidst all the Christmas shoppers, he threw himself off the top of a city car park building. It was numbing for us all, but poor Jim and Ileene were beside themselves with grief. Yet my overwhelming feeling was one of relief — relief for Alan that he was finally free of his torment. Bruce helped Pam with Alan's affairs. We also talked a lot, though more about Alan's difficulties before he died than about his actual death. Bruce coped strongly with the ordeal but must have buried a lot.[17]

He went straight back to work after the funeral.

**

The decision to move became an obvious one. Tension had been mounting with Graham Painter's Trade Consultants, and the Parnell offices were no longer big enough for the two organisations. So when Don St Clair Brown offered Bruce the loft in his downtown building, adjacent to leafy Albert Park, it seemed the perfect setting. And it was not long after the open-plan office was completed that Russ Bowler climbed the stairs to his first working day as a fully-fledged member of the team.

All of a sudden, from being in a worldwide professional organisation, I was in this tiny group with Bruce, Peter and Roger. They were doing a great job of their drawings. And as I became familiar with the opposition's products, it quickly became apparent ours were superior. I was head-down-tail-up for the first six months trying to figure out what the game was all about, trying to understand Bruce's boats, and trying to become a useful member of the team. I read anything I could lay my hands on — files, magazines — immersing myself in the whole technical and business side. This all took time because we had a heavy workload and demanding deadlines. I spent time on the water as well, learning how to sail Farr boats.[18]

To describe in a few short sentences what Russ Bowler brought to the organisation is to risk shrinking the range of gifts he carried through that Kitchener Street door. It would be misleading to see him as a Gil Wyland to Sparkman & Stephens, as a Rod to an Olin Stephens, or a Butch Dalrymple-Smith to a Ron Holland. He was not just a talented engineer with a growing reputation. He was the country's leading pioneer in small-boat exotics as well as one of its most gifted helmsmen and designers of racing dinghies. At a time of trauma for Bruce Farr Yacht Design, he brought, along with his lateral-thinking mind, a calmness of presence and a quiet but steely resolve. He also brought a sharp business sense. All would be vital to the business in the uncertain times ahead.

As he had in crises past, Bruce stayed belayed to his board. If the ITC had made broken poetry of his race-boat designs, he would turn to cruising yachts and the IOR could go rewrite the *Iliad*. There was plenty of encouragement for his new direction. The Farr 38 (11.6), a one-design cruiser–racer, was developed for Gary Hyde and Trevor Fell, Hyde having been keen to move up to Division Two racing after his success in *Mountain Dew*, a version of Bruce's early *Moonshine*. There was the Noelex 30, commissioned by Marten Marine.[19] And after the wild success with their first trailer-sailer, Sea Nymph's joint managing directors were back for more of the same.[20] They were also keen on something more substantial — like a compact cruiser-racer about 10.20 metres long. This in spite of the New Zealand government's boat tax of 1979. Kim McDell:

> We were having an evening for the Outboard Boating Club that particular night. There were about fifty of their members doing a tour of the factory. At about 8 o'clock the phone went. It was Jan, my wife. I've just heard the news, she said. Muldoon has put a twenty percent Sales Tax on boats and caravans as from midnight tonight! Before we could catch our breath the phone was red hot; twenty-odd dealers around the country were all wanting their boats before midnight! So there we were with a factory full of boats to get out before the witching hour. But we did it. In the morning there wasn't a boat left inside. They were all sitting out in the street with invoices attached![21]

It is difficult to understand where one of the country's most brilliant political minds could have dredged the idea up from. Whether it was out of his heart for the little man, or from watching Auckland's wealthy swinging at anchor around him as he tucked away gins on Harry Julian's bridgedeck, history has yet to say. But there was no doubting the effect of Rob Muldoon's tax. Overnight the boat-building industry crumpled. Having an Australian network and trailer-sailer orders from the USA helped Sea Nymph survive. Many didn't. And for Bruce Farr Yacht Design it was yet another crevasse to cross.

Chapter 18
Diaspora

Freedom is what you do with what's been done to you.
Jean-Paul Sartre, French philosopher,
novelist and dramatist

It was during Russ's first week at work that the design team jumped in a car and drove to McMullen & Wing's where they inspected the $68\frac{1}{2}$-foot *Ceramco*. With its deck being attached the aluminium monster looked magnificent. It was the biggest Farr yacht to date and a big step up for New Zealand. As Farr, Walker, Bowler and Hill walked around its multi-framed hull with its 'subtle clipper bow and rakish retroussé stern',[1] it wasn't hard to picture it flying their flag as it surfed the Southern Ocean. Keith Chapman recalled:

> The first time I saw *Ceramco New Zealand*, the eighteen-strong team had all been brought together at 21 Gabador Place, Panmure. I'll never forget standing on a platform at the bow and looking down at this sinister-looking shape. The lines and its low coach roof were just awesome. I think we all came out of McMullen & Wing's shed fired up about what we were going to do.[2]

Tom Clark was too. While the challenge committee went about selling $500 debentures to anyone who would sign on for the dream, his Ceramco Industries was underwriting construction.[3]

> [*Ceramco*] was significant for two reasons. One ... it was an absolutely fantastic boat and showed [out] against all the others ... Two ... we got some corporate dollars in. I had a bit of a how-do-you-do on that one. I went off on a Sunday and committed the company to the whole bloody deal. I called a directors' meeting and on the Tuesday I told my directors what I'd done! I was certain that if I'd asked them there was no way I would have got it. So there were two options then: one was to go with it, the other was to sack me.[4]

One wonders why, with all that support, Peter Blake should try to minimise the design team's significance. Was it his lack of talent as a sailor that produced this need to garner glory for himself at the expense of others?[5] Or was it something darker and deeper? Whatever the reasons, in the Farr-Bowler story it was a pattern that would begin but not end with this campaign. In reading *Blake's Odyssey: The Round the World Race with Lion New Zealand*, one could be forgiven for thinking that the principal concepts that made *Ceramco* fast for its rated size — 'less emphasis on the rating rule', '[lack of] hull-line distortion', '[absence] of control problems', 'a fair, fast hull' and 'a boat which could maintain high speeds readily'[6] — were more the result of 'Blake's own thinking'[7] and instruction to the designer than their being foundational to the light displacement theories Bruce Farr had developed a decade before. The reality is, all that Blake brought to BFA was a DNF and last in his first two Whitbreads, the need for a fast boat, the knowledge that Farr produced them, and the lightly held conviction that 'heavy displacement and masthead rig [made for] the wrong sort of boat'[8] — a notion he'd soon dispense with when he ordered *Lion New Zealand*. Of course there was discussion — the constraints of the IOR, Blake hoping for more boat for less money (an attraction of light displacement), and Farr suggesting another three feet without adding much to displacement or to cost. But to make great play of his ideas and 'the considerable input ... [he] had to offer'[9] is fanciful and hubristic. You didn't go to BFA with light displacement theory. It's what you got when you went to Farr and why you went there in the first place.[10]

Blake had not only cast the relationship die; he would use it to form vicious rumour.

'Bruce, would you care to join us after the Sydney–Hobart? We'll be dipping down into the Southern Ocean, calling in at Milford Sound, then making promotional calls on our way up the coast.'

'You bet.'

'Without your designer's fee?'

'Sure.'

But by December 1980 Bruce had still not received official notification of the trip.

'Peter, can you confirm I'm on the crew for the promo trip? If not, that's fine. I just want to know so I can plan a summer cruise with Cathie if I'm not going with you.'

Blake was curiously noncommittal, and would remain so until the eleventh hour.

'Sorry Bruce, we'll be taking Peter Montgomery instead of you. We want him to do live broadcasts as we cross the Tasman and travel up the coast. I'm sure you'll understand.'

'Of course. That's fine. Good luck, and I hope you win the Big One. We reckon you've got the boat to do it!'

They did. Only the fourth yacht to win line and handicap honours in the Sydney–Hobart's history, *Ceramco New Zealand* took just two days and eighteen hours to complete the 630-miler.[11] It was an emphatic first-up win for the downwind flyer.

All was going like clockwork for *KZ4400*. 'If Blake's XI had any doubts about what they were involved in, they were gone now. *Ceramco* and her crew were celebrities.'[12] There was not only that marvellous double win. The country was turning out in its thousands to cheer the 'Porcelain Rocketship' on its journey back to Auckland.[13]

Also on a journey of its own was the whispered news that Blake had refused to take *Ceramco*'s money-grubbing designer because he'd insisted on trans-Tasman airfares for his girlfriend and himself before he'd make the trip. When Bruce eventually heard that news he could only shake his head in disbelief. There was only one person who selected crew. Only one person who had the authority to make decisions on airfares had Bruce asked, which he hadn't. And there was only one person among *Ceramco*'s crew, as history would later show, who could have had an interest in creating a rumour of such maliciousness. That person, Bruce and Russell would learn, would one day be prepared to go to such lengths to discredit others for the purpose of self-gain that he would reach beyond the bounds of rumour and publish his deceit in books and videos and repeat it in the press.'

Fortunately Bruce and Russell's relationship with a Morges-born sailor was decidedly more enlightened. Perhaps that was because Pierre Fehlmann Jnr was Swiss and his first offshore contest was nearly his last. Following his dramatic rescue in the 1976 *Observer* Single-handed Transatlantic Race (OSTAR)[14] he took to heart the advice that 'it was bad to stay on a bad experience'.[15] The following year he formed the Swiss Ocean Racing Club with family members and seven friends. With the money raised by their non-aligned club they rented an S&S design, named it *Disque d'Or* after its cigar-making sponsor and sailed it to a creditable fourth in the 1977–78 round-the-world race.[16]

Believe me, sailing a Swan 65 in the Roaring Forties is not very interesting. I had heard of *45° South* and *The Red Lion* and knew we needed a fast down-wind boat like these if we were ever going to be successful in the Whitbread. Bruce was the first with light-displacement. There was no one else to turn to for this type of boat. Designers like German Frers and Olin Stephens were still designing heavy-displacement boats. Also, light-displacement is less expensive and that was very important to us. So during the Whitbread stopover in Auckland I had a meeting with Bruce. At the time my English is not good but I managed to put to him three projects: a boat for the 1981–82 Whitbread, a lake boat for my brother Claude, and a boat for the single-handed racing.[17]

***

'This is Italy calling. This is Mr Farr?'

'Yes it is.'

'Please you hold for Mrs Lievi.'

'Mr Farr?'

'Yes it is.'

'My name is Silvana, Mr Farr.'

'Please call me Bruce.'

'You don't know me, Bruce, but my husband and I meet you through *The Red Lion*. It was sold to Riccardo Lenzi? You know?'

'Sure.'

'Luciano, my husband — he was Italian Finn rep and European champion — crewed on *The Red Lion* when it sailed for Italy in the 1978 One Ton Cup. Luciano he says you are revolutionary designer. Bruce, about my call — you know the Centomiglia, the hundred-mile yacht race on Lake Garda?'

'I do.'

'And do you also know the Millemiglia?'

'Yes.'

'Ah good! Then you know this is the famous motor race, founded near Brescia where our yacht race begins. Ours is a hundred miles long, the motor race a thousand. So to make for public interest we think to give them similar names. But to tell you quickly, Bruce, my father he is the founding member of the Circolo Vela Gargano Yacht Club which runs the Centomiglia beginning from September 1951. And Luciano and I make spars for yachts which race the Centomiglia.[18] This includes Classe Libera, the biggest racing yachts on the European circuit.'[19]

'Okay.'

'Luciano he tells our friends if they want to win the Centomiglia they need a Bruce boat. And they say they would be honoured if you would design one for them.'

Long an inspiration for poets and painters, Lake Garda stretches its narrow arm through three north Italian regions. Metaphors abound in its enchanting landscape, though not always for the sailor. Its character can change in seconds as the local wind, the Peler, charges down the Alps at 30 knots to turn the normally placid waterway into an angry sea. Yet no matter the conditions, there has always been excitement for those competing in the hundred-mile race and for those watching from its amphitheatrical shores.

From the Centomiglia's inception there were no rules save those common to yachting worldwide. Although a variety of designs raced throughout its first two decades, it wasn't until the first yacht designed specifically for the Centomiglia won a wind-blown race in 1974 that an open class was discussed. The idea gripped the imaginations of contenders and officials alike. And with the birth of the Libera Class a few years later, Centomiglia lovers were pouring passion into purpose-built boats to win the supreme prize in Europe's lake boat circuit.[20] Silvana Lievi:

> With the arrival of Bruce's first two designs, *Grifo* and *Farrneticante*, it seemed all the other boats built for the Centomiglia were fifty years out of date. They were so exciting to watch as they flew down the lake at twenty knots with ten of their crew on trapezes. In 1981 they were first and second. There was no comparison with the others. It was a race only between these two. Sadly for my husband, *Grifo* was the winner. Although he'd won in 1970 with a boat called *Champagne*, and despite his hard work and undying passion for *Farrneticante*, Luciano can never win the Centomiglia as helmsman of our Farr![21, 22]

But although Design 97 had proved an outstanding success, it hadn't been a cakewalk. In fact the Lievis had had to pass through a period of embarrassment as *Grifo*'s original purchaser walked from the contract and their team. At the time they weren't to know the *Grifo* group would soon reform with a different sailing crew. But when asked if they wished to proceed in the face of this uncertainty, Bruce and Russ had only one reply. Mutual esteem and friendship began from that moment and continued to grow after the Lievis took over the contract. During their long international calls, Luciano and friends would sit around the design table at night, relaying through Silvana comments on wind and water conditions, opposition yachts, and the various types of wood available in Italy from which the intricately stringered and framed Farr hulls could be built. They also asked Bruce to give their boat an Italian name. Using the shut-eye/index finger method, he chose the word for 'incoherent'.[23] For those who watched *Farrneticante* finish second to *Grifo* in the 1981 Centomiglia, the name made even less sense. The two 46-footers had come from a designer half a world away who had never even seen Lake Garda.[24] Silvana Lievi:

> *Grifo* and *Farrneticante* have left an imprint on almost every new boat on the European lakes. Even the cruising boats that now sail the Centomiglia look like Farr boats. And even though these first two boats were wooden, the influence of their hull shapes continues in those made of composites. Luciano says that Bruce and Russ have given to yachting the passion of sailing. And even though it is many years since we owned *Farrneticante*, this beautiful boat still races on Lake Garda and still does very well. To sail a Bruce boat is always a pleasure. We think that's why it has a new name, one we like very much. It is now called *Pleasure*.[25]

***

Within weeks of Russ arriving at the Kitchener Street offices the Farr team was talking seriously about leaving the country. But this, they knew, wouldn't be the usual Overseas Experience for a group of young New Zealanders. Not only was a permanent return unlikely. They also had a complex set of questions to answer before they left the country. While somewhere closer to the world's major markets had been considered previously, frenetic activity in 1979 and Alan's tragic death had precluded further study. But Alpha's misadventure with the Canadian production of the 727 had taught Bruce that to capture a mass market you had to be there. He knew Americans didn't like doing business unless they could meet you face to face. And Europeans were increasingly wanting him to visit them, to inspect and test their boats on the water. Then there was the Laser 28, an international project which had Bruce regularly flying a triangular route between Auckland, Canada and England as he tried to get it off the ground.[26]

> Do we want to stay and be a big market force in New Zealand and be a small part of the world scene? Or do we move to the northern hemisphere and try to become a big part of the world scene? At the time we had quite a bit of our focus on production boats because we saw that as being the way you could turn yacht design into a viable business. In terms of production designs we saw the bigger markets of Europe and the USA as being essential to make it work. So part of the impetus to leave was to get into a much larger market place with the idea that we might be able to hook up with some big production companies, do some good work, and get well rewarded for it.[27]

Cathie researched a lot of locations. And to help them with their decision they engaged the services of John Vandersyp, an Auckland management consultant who Bruce had met through sailing. He chaired their team meetings and skilfully guided them through their options:

1. England,
2. Europe,
3. West Coast USA,
4. East Coast USA.

1. England: After countless hours sailing the incomparable Hauraki Gulf and revelling in Auckland's subtropical climate, none of the team could get excited about English weather or the Solent.

2. Europe: The success of *45° South* and *Joe Louis* ensured that Farr designs already had a high profile in France. French wine and the Côte d'Azur sounded wholly inviting except for one thing: no one, not even Bruce with a French grandmother, could speak the language.

3. West Coast USA: With Farr designs already doing well on the Bay, San Francisco looked appealing. But why move to a market serviceable from Auckland and not significantly closer to Europe?

4. East Coast USA: The sale of Woodie's *Mr JumpA* to Bostonian surgeon Dr John Wysocki and the winning of its class in the 1978 Southern Ocean Racing Conference (SORC) made Florida another likely location. It seemed, though, too removed from key yacht production areas and also promised stifling heat in summer. Picturesque Newport, home of the America's Cup, workplace of several yacht designers and summer home of America's wealthy, looked more likely. But although, like Florida, it had an excellent boat-building infrastructure, it lacked that all-important facility: a nearby international airport. It was also bitterly cold in winter.

Ian Bruce, a friend and associate through the Laser 28 programme, suggested Annapolis, Maryland. A quick check revealed the most likely location yet. It was situated on beautiful Chesapeake Bay. It was mid-point East Coast USA. Its climate was more equitable than either Florida's or Rhode Island's. It was a yachting centre with a service infrastructure. As a one-time fishing port, as the site of the US Naval Academy — one of four US military universities — as the last resting place of the famous schooner *America*, as host to numerous yachting regattas and home of several sailing schools, it had a long boating and naval tradition.[28] And at Washington DC and Baltimore, it had international airports less than an hour away. Suddenly Europe, the Americas, even Australasia, seemed within easy reach.

That was more than enough for Bruce to fly off to the USA in May of 1981 to conduct a reconnoitre. Dinny White recalls the first time he met Bruce:

> I was sitting in my Annapolis office up on Church Circle trying to figure out how to get the next lot of business in when there was a knock on the door. This chap pokes his head in and asks: 'Do you know where I can find Jobson Sailing?' I immediately recognised the accent. 'Are you from New Zealand?' His face lit up. 'How did you know?' 'Easy,' I said, 'I'm married to one!' We introduced ourselves. 'Bruce Farr, the yacht designer?' I asked. He nodded. 'I've seen a lot of Farr yachts in New Zealand but none over here.' 'Well, we're hoping to change all that,' he said with that smile of his. With that I told him Gary Jobson was just around the corner and asked him if he'd like to come to dinner and meet a couple of fellow Kiwis.[29]

Having settled on Annapolis, there were still two major hurdles to overcome. With Russell about to become a partner, the existing business would need to be restructured. The second and rather more vexing problem was how to fund a move halfway around the world.

It was left to lawyer Craig Dunbar to form the new company: Bruce Farr & Associates, Inc (BFA). While Bruce held most of the new company's shares, that, in time, would reduce to an almost equal holding shared with Russell. Payment of a percentage of the fees for any work done for the old company by the new would enable a continuity of business between the two and make for an easy transition for clients.

When Bruce arrived home with his glowing report of Eastport, with anecdotes of the friendly people he'd met — Jean and Dinny White, Jim Allsopp of North Sails, Jack Lynch of Chesapeake Cutters, Gary Jobson — the famous blue crabs and details of his meeting with immigration lawyers, the decision was immediate. For those committed to the move, there was now a lot to do and only weeks in which to do it.

Still the principal owner of the business, Bruce took it upon himself to finance the entire move. For the 32-year-old it was not a small undertaking; it meant, among other things, selling his investment property. There were seven adults, one child and household possessions as well as office equipment to transport to Annapolis. Then there were the first month's accommodation costs to meet as well as keeping the new business liquid during its establishment phase.

> Yacht design business is a business which requires subsequent sales and multiple construction of boats to get compensated for the original cost of design. So there can be quite a period in its early life when you are actually losing money. Fortunately, 1978 to 1980 had been good years and I was able to manage the move as well as cover the US start-up period. But it did mean I had gone from being comfortable in New Zealand to having all my assets invested as loans in the American company. As well, there were no guarantees of success. It was a venture which might or might not have succeeded. It wasn't a position I enjoyed, though it did make me committed![30]

Swamped by the chaos in his personal life, and saying that he wanted to be no further removed from his already estranged family, Peter Walker could not commit to the move. It was a sad though inevitable decision for the three close friends who had shared so much of their personal and working lives over nearly two decades of competitive sailing.

Roger Hill and his wife Christine, the youngest of the team, were redecorating their house and building a boat when the decision was made. They quickly finished both, sold them, then piled their possessions into a Farr 740 BFA had bought for shipping to Baltimore. Roger's chief responsibility in the office was the drawings. He had these microfilmed then packed along with the drawing machines and boards for freighting by air. There was, however, one set of lines plans he wouldn't let out of his sight. Design 121, the Farr 10^{20} for Sea Nymph, was tucked firmly under his arm as he and Christine waved goodbye to family and friends and turned towards Customs.

The decision set Russ and Lynda on a mad scramble to finish their Northcote Point home. There was the painting to finish, a driveway to go in and fences to be put up. They had purchased the property in 1975 even though the site's splendid harbour views demanded a home beyond their budget at the time. They had removed the old original dwelling, had a large shell built, and spent the following years labouring on it themselves. Having decided to rent rather than sell, Russ suggested it to former corporate sponsors of his eighteen-footers as an ideal property for their relocated executives. He then charged into Auckland's largest appliance and furniture store and bought an entire household inventory on tick. The time it took to get their home into a rentable condition gave Russ the space to transfer remaining business matters into the hands of Bob and Sue Farrell and to close down the Kitchener Street office in an orderly manner. It was like leaving Lake Rotoma for the very last time. The end of an era. Although they'd moved fast, it was six weeks after the last of the Farr team had left for Annapolis before Russ and Lynda, with five-year-old Gareth, were able to climb on board a plane, put their seats back and give themselves over to thoughts of the life that lay ahead.

Interlogue 1

INT. RUSSELL MCVEAGH MCKENZIE BARTLEET & CO, AUCKLAND. 11 JUNE 1987.

A fresh-faced lawyer puts down the phone, gathers up papers from his desk, heads down the corridor. He pauses at the hallowed door. As a past commodore of the Royal New Zealand Yacht Squadron, the firm's senior partner will be the perfect person to listen to this idea.

McKenzie: Yes Andrew?

Johns: John, what would you think of advising our clients to challenge for the America's Cup under the Deed of Gift?

McKenzie: Fay and Richwhite?

Johns: Yes.

McKenzie: I think you'd better explain.

Johns: Well, by not announcing in Perth when or where the next regatta would be, the San Diego Yacht Club has broken with America's Cup protocol and left itself wide open to a challenge under the Deed of Gift.

McKenzie: Oh?

Johns: It's here in the first Deed. I quote. Any yacht club of a foreign country, dot-dot-dot, shall always be entitled to the right of a sailing match for this Cup. It's up to the challengers to initiate a challenge, John, not the defenders. What do you think?

McKenzie: What do I think? What do I think, Andrew? I think you're bloody mad!

Johns: Perhaps we should keep it quiet then, at least for a while.

McKenzie: Andrew, run with it if you wish. But don't worry about me keeping it quiet. If I repeated it to anyone else they'd think I'd gone insane!

INT. OPERATIONS ROOM, FAY RICHWHITE AND COMPANY LIMITED, AUCKLAND. AFTERNOON. JUNE 12, 1987.

Fay: Andrew. Welcome.

Johns: Michael, I've been doing some reading and uncovered something that might be of interest to you. The deed by which the America's Cup was gifted is a challenger's not a defender's deed.

Fay: Really?

Johns: The way the deed is worded the initiative rests with the challenger. That is, the defender only defends after the challenger has challenged. And it's always been that way. All you have to do is to advise the Cup holder of your vessel's name, rig and dimensions when you submit your Notice of Challenge.

Fay: It's really that simple?

Johns: Yes. It's there in the first Deed of Gift. Look. Any organised Yacht Club of any foreign country shall always be entitled, dot-dot-dot, to claim the right of a sailing match for this Cup.

Fay: Are you sure about this, Andrew?

Johns: As sure as I can be without further research. Michael, it's the same wording used in each of the three Deeds of Gift. Furthermore, it was the donor's express intention that it should be, and I quote, perpetually a Challenge Cup for friendly competition between foreign countries. Even the December 1956 amendment to the Deed, reducing the waterline length from ninety to forty-four feet to permit the use of Twelve-Metre yachts, doesn't alter that. So I think it's safe to say you could bypass the multi-challenge format and go straight to the Deed of Gift as the basis of your challenge. Well, what do you think?

Fay: You mean we wouldn't have to consult the Challenger of Record?

Johns: Not if there isn't one. It isn't part of the Deed, and has only been America's Cup protocol since 1974. The Challenger of Record is merely a convenient way of organising a challenger's regatta if there is more than one challenger. So if there's only one challenger —

Fay: It's an interesting proposition, Andrew. Very. Yes, Andrew, why not? Why bloody not? Let me get David in.

In just ninety seconds the merchant-banking partners agree to commit their resources to determining the legal viability of a Deed Boat challenge. The opportunity is too much to resist. As is completing the job begun by KZ-7, and the thought of restoring 'the biggest show in town' to its pre-War glory. It might also be a fast track to the centre of the world's financial stage.

Chapter 19
Free Fall

Earth in beauty dressed
Awaits returning spring.
All true love must die,
Alter at the best
Into some lesser thing.
'Prove that I lie.'
> W.B. Yeats in *The Winding Stair and other Poems*[1]

If Annapolis were Casablanca, Marmaduke's Pub would
be Rick's.
> John Alden in *Baltimore Magazine*[2]

Like a miniature Manhattan, Eastport sits with its tidy grid of avenues and streets between two waterways. Joined to Annapolis by a metal drawbridge since 1949, the area was originally patented by Quaker Robert Clarkson in 1665. It has served variously as farmland, fishing village and garrison as well as a dormitory suburb for the state capital's workers, watermen and trades-folk. Today refurbished Queen Anne and Four Square homes rise grandly between modest clapboard houses, trendy eateries and the maritime busi-nesses of the quiet neighbourhood. To the south, Back Creek packs a forest of masts into its marinas, while the sites of the boatyards which once built the skipjacks, bugeyes and pungies,[3] schooners and luxury launches and the PT boats for World War II still sit facing Annapolis along its northern fore-shore: Spa Creek.

> As dusk falls purple and pink over Spa Creek, there is no better place to catch up on sailing gossip from all over town — and from as far away as Australia ... The sailors are easy enough to recognise from their ruddy faces, sun-bleached hair, and huge, waterproof techno watches ... But increasingly at Marmaduke's you'll see men with paler faces, wearing wire-rim spectacles and plain flannel

shirts instead of Goretex. Their talk is filled with terms like 'digi-edit, gnomen, geosims', and 'ASCII bytes'. They are engineers and programmers and they are speaking the 'new' language of sailing, with all its computer-aided design work and velocity projection test programs ... In the process they are turning Maryland's capital city into the East Coast centre of yachting technology.'[4]

That wasn't the mix when the Farr team first pitched their design tent at 121 Eastern Ave. Marmaduke's, on the corner of Third Street and Severn Avenue, may have been the preferred pub for sailors, where they could watch home-made videos of the day's yachting events played back by Bill Heim, the owner.[5] And George Hazen may have just arrived in town with a head full of 'Fast Yacht', his performance prediction programme.[6] But in 1981 yachting was still more black art than science, more amateur than corporate-driven. And although the local community was welcoming to the designers and their families, there was a cold indifference to light-displacement. In fact the only Farr boat they could find in Annapolis was a half model of *Why Why*, the 1976 North American Quarter Ton champion, screwed to one of Marmaduke's walls. The 727's place among a host of heavy-displacement winners left Roger, Russ and Bruce with a feeling of disquiet.

The Maryland summer of 1981 welcomed the self-exiled Kiwis with fifteen days of temperatures over one hundred degrees Fahrenheit. Having flown out of an Auckland winter, it was like landing in a Bessemer converter. Roger Hill:

> I'll never forget my first night there — I thought I was going to die. Chris and I had been put up in an old hotel on a corner in downtown Annapolis. For much of the night I sat next to the open window desperately trying to breathe. It felt like a massive asthma attack! Finally, in the early hours, Chris called Cathie and Bruce. They rushed over, collected us, and we stayed glued to their air-conditioning unit until we found a place of our own two weeks later.[7]

They were a tight quintet as they set up their modest Horn Point office with its wall-to-ceiling windows and panoramic view of Chesapeake Bay. There they fell about at the quaint Americanisms they daily encountered, sought personal and business credit, and prepared to take on the world. Sally Gray stayed with the Hills until she found her own apartment. Jack Lynch, a pioneer in computerised sailmaking and the person who would later make the Laser 28 sails, offered his South River home to Bruce and Cathie while he was setting up a plant in Ireland. And Russ and Lynda rented in the pretty riverside suburb of Severna Park. Soon the drawing boards were up and the Noelex 30 being completed for Marten Marine. Then it was all hands to the Farr 10[20]. There was still a lot of work to be done on Design 121, a key project for Sea Nymph Boats and a promising royalty earner for the new Farr office.

As well, the Laser 28 programme was taking on a life of its own. Ian Bruce, 'the father of the Laser', had been undaunted by the demise of his original business. As he once had with Bruce Kirby's lines, so his creative engineering mind was now fully occupied with turning Design 91, a small cruiser-racer, into the phenomenon of the eighties that the Laser skiff had been for the seventies.[8] Bruce remained involved in the detail of design while Russell worked on DuPont's Kevlar[9] laminates and the two-mould injection moulding system with which they hoped some thousands would be produced.[10]

Yet despite their bustling beginning the goss at Marmaduke's was unsettling — the industry was slipping into recession. Not only was the rising cost of IOR boats having an adverse effect on trade. As the depression dug deeper, American production companies were only wanting new product if it was safe and cheap. That meant fat-turkey types, not sleek light hulls with designer names attached. In fact, East Coast companies were showing as much interest in Farr designs as Bostonians had in East India Company tea. If the expatriate team were going to survive in their country of adoption they would either have to change its yachting psyche or hunt elsewhere for business.

Being battered and bruised by the ORC had had a profound effect on Bruce. The designs encouraged under the IOR since 1978 were wide, heavy, with oversized keels, undersized rudders, too much sail and were mostly handling disasters. This was anathema to a designer of light-displacement flyers, and Bruce had turned to non-rating designs for sanity and survival. At his lowest, before they had left New Zealand, he and Cathie had talked about retreating to an island in the Hauraki Gulf where he could draw his boats at leisure and try his hand at painting. He was not one for outward displays of emotion, but this much was evident: his heart was no longer in it and his boats were off the pace. His *Anchor Challenge* could only manage third place behind the Fauroux-designed *Bullit* and the Davidson-designed *Hellaby* in the Quarter Ton Worlds sailed in Auckland in January 1981. And his Admiral's Cupper for the Lidgard-headed syndicate couldn't even make the New Zealand team. True, there had been construction problems beyond his control with Design 82.[11] But it was now abundantly clear that the draconian IOR was more than suited to the derivatives of Ron Holland's pre-Mark IIIa *Imp*. As the Cork-based Kiwi climbed back to the top of the tree he took Farr clients — Stu Brentnall, Ian Gibbs — with him. And when the Holland-designed *Swuzzlebubble III* became the top points scorer in that year's Admiral's Cup, it seemed the once brilliant Farr star was definitely in decline.

***

*Saturday, 29 August, dawned damp and overcast, battle
flags indicating a gentle south-easterly ... The dock was
awash with ... onlookers and I recall an intense moment
of pride as we manhandled our charge so that her bow
was facing the channel.* Ceramco *was a sleek, almost sin-
ister thing and I knew that all were admiring her beautifully fair lines.*

Keith Chapman, New Zealand yachtsman[12]

Like its crew and most of New Zealand, the hopes of Bruce Farr &
Associates were riding high above the hounds as *Ceramco New Zealand*
reached down the Solent on Saturday 29 August 1981.[13] And with just
cause. *Ceramco* had not only won the Sydney–Hobart with blistering
downwind speed. With Bruce and sailmaker Jim Lidgard on board during
the Maxi Seahorse series, Peter Blake and his crew had learned how to
improve its balance and windward speed in their lead-up to the seven-
month race.

Design 90 was the first race boat Bruce had drawn since the IOR
changes of 1978. 'To reduce [rating] penalties [he had] ... extended beam
and depth measurement points then ... [drawn] a sharply sloping transom.
The result was something more than a yacht, a beautifully shaped, elongated big dinghy, proportionally the narrowest of any of his designs.'[14] He
had also given it a triple spreader rig with a heavily tapered effect above
the hounds. That reduced the weight of the mast's top section, although the
supporting diamond system gave the rig an extreme appearance. The mast,
extruded by Allspar Australia, was supported by standing rigging supplied
through Terry Gillespie's Navtec. The rigging was rod, not the usual wire,
and was detailed to be continuous and bend inside tubes over the spreader
ends; this rod had been used on the race-boat circuit and was the latest
technologically.

But questions were fired like bullets after it exploded a hundred miles
north of Ascension Island. Was the metal faulty? Was it sturdy enough for a
round-the-world racer? Was it overloaded, overtensioned? Had it been
weakened in the Chinese gybe which flattened *Ceramco* en route to
Australia? Whatever the cause, the race was over for the New Zealand crew
24 days from the start. Their dreams, their country's and those of a tiny
Annapolis office lay tangled in the rigging on the deck. And although
Ceramco proved it was the flyer of the fleet by winning the downwind leg
to Auckland and being the only boat in the fleet to win two of the four legs
on handicap, after its mid-Atlantic dismasting it could never do enough.
Third in the race for line honours, *Ceramco* finished eleventh on corrected
time behind *Disque d'Or III*, Pierre Fehlmann's fourth-placed Farr, and
Digby Taylor's fifth-placed Davidson, *Outward Bound*. Conny van

Rietschoten, the first skipper to win line and handicap honours in the Whitbread, remarked to his sail coordinator, Grant Dalton — a *Ceramco* crew reject, along with Chris Dickson — that had Blake asked Farr for a 73-footer instead of a 68-footer, *Flyer* could not have kept up with *Ceramco*.[15, 16] And Pierre Fehlmann remained convinced that had the IOR not been changed after *Disque d'Or III* was built, he would have won on handicap.[17]

These were the 'ifs' and 'maybes' that only added colour to the Whitbread. In the end it was the 76-foot *Flyer* which confirmed German Frers' winning way with large racing yachts and reinforced his blue-chip connection with the USA's North-East. Despite his attention to detail, Bruce Farr still appeared to be the *enfant terrible* of offshore yachting, a brilliant but dangerous designer of lightweight 'collapsibles'. It didn't matter that *Ceramco's* fissure of fatigue was beyond his control. A fractional rig failure was reason enough for his opponents to continue their tut-tutting and for most American yachties to keep a watchful distance from the avant-garde design team. It would take someone big of heart and vision to bet on the Farr Big Dinghy type in the Maxi-racing world before the tiny Eastport office could consider itself secure.

Ceramco's dismasting could not have happened at a worse time for Bruce and Russell. Commissions were lean, the incoming cash more a trickle than a flood and insufficient to cover the outflow. The Laser 28 was demanding more and more of Bruce's time and was not yet ready to spin its money-making wheels. There had been a small trailer-sailer for G. Meschini of Italy and a 48-foot cruiser-racer, *Sangvind*, for Dr Jerald Jensen of San Francisco. *OPNI* and *Azzardissimo* had been exciting projects. And the Fehlmann brothers were back talking a Lake boat and a Maxi for their company Decision SA. But not only were the funds in Bruce's loan account diminishing fast. Russ was on the wire too. Mortgaged to the hilt in New Zealand, he loaned the business whatever cash he could to help keep things afloat. That left him just enough pennies to pay his and Lynda's rent and buy a $200 Ford.

As their circumstances became more and more constrained Bruce began thinking about race boats again. With the production boat market run out of wind, maybe the racetrack was where you survived until things came good again.

The Southern Ocean Racing Conference (SORC) was the biggest racing circuit in the USA. With more wind and waves and offshore events than other East Coast regatta, an all-round boat designed for the Florida venue might be the way to go. The only problem for Bruce was that SORC boats raced with IOR ratings — and if he were to recommit himself to the race boat market, it would mean trying to find a way through the crazy 'house of mirrors'.[18] *Feltex Roperunner* had taught him that his light-displacement theories did not easily translate into the heavy type of boat now being demanded by the rule. But if he could make incremental changes to it — if he could make it marginally wider in the stern and a little lighter than the weight the rule now favoured — maybe he could regain some of the reaching and running qualities of his former world champions. It helped that Hazen and Stern had set up shop just down the road and were keen to hook up with the expatriate designers.[19] In fact, George Hazen's VPP was especially useful in testing how much a keel's wetted surface could be reduced before it adversely affected a hull's performance. The process quickly became a two-way street: Hazen offering his program and consulting services, with Bruce and Russell doing a lot of new writing as they created their next design.

Ted Simpkins, a Florida-based liquor distributor, had owned a Farr Half Tonner, *Farr Out*, for several seasons and had offered Bruce advice before the Farr team left Auckland for Annapolis. He fell for Design 124, and his fractionally rigged *Free Fall* was the first of a new generation of Farr IORs. Not only would it become the boat to beat in the events leading up to the 1983 SORC. From this generic design Bruce and Russell would develop the potent One Tonners that would dominate the eighties. Design 124 was the chute to end the free fall.

***

They created a new standard; a new dimension in the 'factory team' approach that is used in motor racing and some other sports so that everything is aimed towards the success of the product ...

Duncan Spencer, American yachting writer[20]

A chronic asthma victim, Geoff Stagg had hundreds of days off school and suffered academically. It wasn't until he was pushed out from Muritai Beach in a P Class that he found where he belonged. He graduated to Cherubs and, with his father Ray as forward hand, became three-time Wellington champion. At eighteen he began building a 45-foot John Spencer plywood keeler in his family's backyard. Three years later *Whispers II* was 'terrorizing the local Wellington scene'[21] and won the 1973 Auckland–Suva Race on handicap. In 1975 Stagg convinced a group of 30 Wellington businessmen to part with $1000 each to build another Spencer design. Although *Whispers of Wellington* failed to make the New Zealand Admiral's Cup team, Stagg's relationship with John Spencer brought him into contact with Tom Clark and Peter Blake, and eventually with *Ceramco*'s designer. He would stay with Cathie and Bruce while he was pitching for 'Blake's Eleven' and after being selected, while *Ceramco* was having trials.[22] Bruce liked his charging energy and the pair got on well. But it wasn't until after the Whitbread that they talked seriously about working together. In Hawaii for the Kenwood Cup, Geoff was on a Japanese entrant and Bruce on *Sangvind*. Having declined an offer from Chris Bouzaid to manage American sales for Hoods, the Ford Motor Company rep was now ready to cut a deal.

> It all came about because we couldn't figure out how to meet Geoffrey's expectations from our modest design receipts! It made us look closely at what we were doing, and the numbers just didn't stack up. We'd design a boat for as little as five percent of its value, but the person who sold it got ten percent and the person who brokered it three years later got a further ten percent. So it was in answer to this conundrum that we set up the marketing arm: Farr International. That, we hoped, would create the cash to pay Geoffrey as well as getting us into new boat sales, brokerage and project management.[23]

Farr International was formed in September 1982.[24] Right from the start they worked as a team, Geoff chasing production contracts for Farr-designed boats, presenting his *Ceramco* slide show, campaigning boats, working the boat shows. And although they saw the SORC and international regattas as their focus of race-boat promotion, as sailors with a product to sell they also flew their flag in local Annapolis races. Dinny White recalls:

> We saw this when we first raced with Bruce and Geoff on Bert Jabin's *Ramrod* in the Governor's Cup race to St Mary's City in 1983. This is the biggest event in the area — a hundred miles long and three hundred boats in the fleet — with major accolades going to the first boat over the line. For casual sailors like Jean and me the race was a revelation. In a hundred and five degrees and nearly a hundred percent humidity they were totally focused, from two in the afternoon until when we finished at five o'clock in the morning. I was just amazed at their concentration — how, during the night, they were busy picking up the currents on the edge of the channel just south of the Patuxent River. They understood the dynamics of tide and flow and where the currents were just by reading the contours of the bottom. I thought, God Almighty, where do they get this from? They know far more about the area than someone like me who's been sailing here for years! It was a lethal combination. Bruce was navigating, including the tidal planning. Geoff was steering and running the crew, with Bruce and a friend, Fred Potts, assisting on the helm. And Stagg was merciless. No one's feelings got spared. I've never seen such intensity. Most of us got no more than two hours' sleep! But we did it. In their first big Annapolis yacht race, on a brand-new Farr 37, we were first across the line.[25]

The win was sweet but sales stayed sluggish. Geoff was barely scratching out a living selling Farr 37s for an East Coast builder, Dickerson's.[26] But Bruce was travelling more — checking on production, testing clients' boats and sailing them in regattas — and custom work was growing. Russ had lifted a double burden from him — the engineering of designs and the running of the businesses — and was casting around for help for Roger in the lines plans department. They found it one night at Marmaduke's. He was a peripatetic Australian with a distinguished career designing and building boats. Graham Williams and his wife Jenny had just settled in Annapolis with their young son Trekker, and yes, he'd be delighted to do some consulting work for the young New Zealand designers.

From the moment they walked through the Horn Point doors, Cathie had worked alongside Bruce, helping to get the office up and running, then turning her hand to promotions. Under the special visa arrangement she could be paid by BFA. But after six months she felt the need of an income not derived through Bruce, and took an office job with the Australian Embassy. Without a Green Card this was the only other legitimate employment available, which meant travelling daily to Washington DC with Christine Hill, herself employed by the New Zealand Embassy. Cathie understood the pressures of a new business in a new country and, to start with, hadn't minded the time she'd spent on her own. But as Bruce's trips increased in number and duration, a new pattern began emerging. In Auckland their home had been a haven, their weekends a time of restoration. Not only had this quality time diminished. Cathie sensed a shift in Bruce: the sensitive, caring person she had fallen in love with now seemed oblivious of her deeper needs. She thought she could cope with a temporary submergence of her identity as the business demands grew. But what she couldn't find room for was this feeling of not belonging. As she felt their relationship slipping away, the allowances she had been making for Bruce's eternal absences began turning into resentment.

My one regret was that we never resolved our feelings. I guess I'd become too cross to be reasonable, Bruce too withdrawn to talk. He was under tremendous pressure. But I felt more and more left out, felt I was just another person relying on him and that he'd lost track of why I was really there. I badly needed a little TLC but he didn't seem to see things from my perspective any more.

So I took a long hard look at my situation. I was living in a home which belonged to someone else, with a man I saw less and less often and who seemed unprepared to offer any sort of commitment for the future. I finally had to admit to myself that Bruce was an extraordinary talent who was born to design boats. That his genius and the normal life I longed for would never go hand-in-hand. I had virtually made up my mind to go back to Australia. But when the opportunity to take part in Antigua Race Week came up, I jumped at it. For me it was a final chance for us to get our relationship back on track.[27]

Built in 1725 as a dockyard for the repair and maintenance of the Royal Navy's fighting ships, English Harbour was abandoned in 1889. The advent of steam-powered vessels had rendered its strategic location redundant.

Using traditional methods and materials and the original building plans, work began in 1951 to restore it to its former glory. Charter skipper Steven Cross had always loved the way 'the Georgian buildings of English brick somehow blend[ed] perfectly with the exotic, tropical landscape'.[28] For him there was no more romantic setting than English Harbour, Antigua.

Bruce and Cathie were crewing on Ian Bruce's *Early Byrd* when the Farr 38 crossed tacks with a Swan 48 at the beginning of the first race of the five-race regatta. Cathie did a double-take: could the person on the helm of *Scaramouche of Warwick* be her ex-husband? Her second moment of dis-belief came when Bruce flew off halfway through Race Week. And although he would return before the regatta ended, it would do little to assuage her feeling of abandonment.

'Hello Steve!'

The bronzed sailor went weak at the knees when he heard the familiar voice. Turning from the notice boards with their interim results, he saw her standing on the balcony above him. He felt more angry than pleased as he climbed the stairs, their meeting more formal than pleasant. He'd fought long and hard to erase her from his mind and now it seemed she was about to invade his world again. In fact, so unsettling did he find it that he spent the rest of the week hoping not to see his ex-wife again.

The prizegiving was held at the Admiral's Inn, 'a glorious setting over-looking the bay'.[29] The *Scaramouche* crew spent the early part of the evening drinking a lethal brew from the silverware they'd won. But it would take the steel band, swaying palms and an obliging moon — plus a few more rum punches — before their skipper would gather up courage enough to cross the expanse of lawn to the woman in the long white dress standing all alone. Bruce had gone to get Cathie a drink and hadn't yet returned, way-laid yet again by someone wanting to talk boats. It was not a matter of if but when Steve would ask her to dance. And as they walked to the fort at the head of the bay and sat among the ruins, it was as if everything crumbled that night — the walls between Cathie and Steve, Bruce and Cathie's relation-ship.[30]

Love arrives and dies in all disguises.[31] Months later Bruce was still unable to fully understand what had happened. Had he changed as he'd fought to save the businesses? Was his need to draw boats greater than his need for love? Cathie had handed him her life and trusted him to hold it. He had loved her in return. But had his desire not to fail overwhelmed her fragile gift?

There wasn't time to ponder let alone answer these questions when Sally Gray buzzed him. She had that man from Morges on the line wanting answers of his own.

'Bruce. You have the test results on the Kevlar panels from DuPont, no?'

'Sorry, not yet, Pierre. And Russell is still doing ours in his garage.'

'But Bruce, we must have this lightweight Maxi. You know why we must have this lightweight Maxi?'

'To win the Whitbread? To make us happy with a contract?'

'Because, Bruce, the Maxi is ten days faster than the small boat. And if you are not on a Maxi then no more girls when you arrive in port!'

Chapter 20
A Piece of Work

The trouble with being as dependable as a rock is that you get stood upon.

Roger Highfield and Paul Carter,
English editors and biographers of Albert Einstein[1]

A predominant characteristic ... of the behaviour of those I call evil is scapegoating ... They sacrifice others to preserve their self-image of perfection ... What possesses them, drives them? Basically, it is fear. They are terrified that the pretense will break down and they will be exposed to the world and themselves.

M. Scott Peck, American psychiatrist and author[2, 3]

Having been told by Peter Blake that he was considering an entry in the next Whitbread race, Bruce wrote to Tom Clark on 21 April 1982, indicating his office's keen interest in designing for them again. Tom replied that, yes, they had 'unfinished business [but wanted] to keep an open mind in regard to a new entry for the next race'. He added: 'I do not imagine any concentrated effort in regard to planning will take place until early 1983'.[4] On 9 August Bruce replied that they had 'built up a mathematical model sufficient for hydrostatic and rating calculations, for a boat of very similar general style to *Ceramco*'.[5] He enclosed a preliminary sketch of their Maxi, a list of dimensions and comparisons with *Ceramco*, and an estimate of research and construction costs.[6] 'We plan to begin some preliminary developments immediately ... drawing up a lines drawing ... and doing [a] computer velocity prediction analysis ... We are also currently setting up a joint research programme with DuPont in Wilmington, Delaware, to do extensive test work for high-tech Kevlar and other exotic reinforced plastic structures.[7] He concluded: 'I believe that we will be able to produce boats with a rare combination of extreme high speed and more than acceptable safety and reliability.'

It wasn't until the following January that Bruce again wrote to Tom Clark. He copied Peter Blake with the same text after congratulating him on being nominated for New Zealand Yachtsman of the Year. Both letters contained details of a Maxi design proposal, including appendixes on research, drawings, support services, principal characteristics and dimensions of the new design. The fee structure with its upper limit of US$83,000 was broken down and unequivocal offers of exclusivity and confidentiality made. 'These estimates are based on the assumption that the design would be for your exclusive use until 30 June 1986 ... You should be assured that if we are to accept more than one commission for the race, we are absolutely committed to the principle that each client's personal input, ideas, requirements and design detail must be kept absolutely confidential to that client.'[8]

> When *Ceramco* sold ... the decision as to which design office to use was narrowed down to a choice between Farr and Holland ... I had, by now, done a lot more work on what I wanted for a Whitbread Maxi and detailed my thinking to Farr and his design team. *Ceramco* was an extremely good vessel but there were certain points that we wanted corrected. Farr was really keen to do the boat and we made solid progress in our negotiations until we struck the question of exclusivity.[9]

In fact, *Ceramco* had not been sold either before the issue of exclusivity was first raised or when Peter Blake shifted the goal posts by demanding not just an exclusive design but the exclusive use of the Farr office's services for the next Whitbread race — an important distinction he failed to make in *Lion: The Round the World Race With Lion New Zealand*. Nor had he 'detailed [his] thinking' to Farr. On the contrary, Blake, it seems, knew so little about exotic construction and had done so little 'thinking' that he hadn't even made up his mind between hi-tech or alloy construction.

3.2.83 Ref Carol 499

Attn Bruce Farr

Queries Re Round The World Maxi

1) How will construction recommended stand up to long periods of intence cold (minus 10 deg. C.)?

2) Wld yacht be a viable proposition if built in alloy or wld it become too heavy?

3) We remain very insistant that if a decision is made to have a new yacht by 31 March, it must be a totally exclusive design with no other yacht from the same designer of over 65 ft rating being commissioned for the same race. unless this is the case a commission from us will not be forthcoming. What wld be the charge if applicable?

Regards

Peter Blake[10]

The exclusive purchase of one's time is to designers what censorship is to artists — it inhibits freedom of choice and expression, and can severely limit income. So when Blake's first written message clattered through their telex machine, Bruce and Russell knew they had a problem. How could protection be traded fairly for both client and designer? Sally Gray copied Geoff with the message, and the partners sat down to talk. After a harrowing year and a half, sales were picking up and they were finally re-emerging as designers of fast race boats. But there was a downside to demand: with growing annual overheads, they could not devote themselves to a single Whitbread project for their normal one-off fee and stay afloat as a business. As the question of exclusivity careered around the room, Bowler, Farr and Stagg tried to nail the issues it raised. Finally they were agreed. Their sympathies lay with a Kiwi challenge and especially with Peter Blake. And he, it seemed, by raising the matter first, had accepted that there would be a cost for the exclusive use of their services. Tom Clark, the trio surmised, would surely understand. No captain of industry would begrudge reimbursement for lost opportunity. Nevertheless, they would give a detailed answer to obviate misunderstanding.

3 Feb 83
Attn Peter Blake — Urgent
Thks yr tlx Ref. Carol 499. I had a letter ready to go re exclusive design but will summarise here.
1. Minus 10C sounds like normal winter here. Seriously tho, we have no information to suggest freezing any problem. we will be conducting freezing tests soon for kevlar foam laminate for another project and will ask DuPont for comments on Nomex in meantime. Obviously no problem in aerospace applications at much lower temps. We believe resin system wud be primary concern.
2. Alloy is probably viable in that overall weight and stability cud probably still be achieved but it wud suffer significantly in performance, particularly if not very sophisticated. Firstly, removal of weight from ends gives exotic fibre advantage because of reduced energy loss when pitching and rolling and therefore greater speed thru seas, and better handling downwind as bow will lift more quickly when pushed into back of waves. Secondly, the substantial weight savings — perhaps 40 percent — in hull and deck mean keel can be thinner with much less drag and improved profile shape for more efficient lift because less ballast req'd external to reach req'd stability due to increase in ballast ratio.
3. I am very anxious to settle exclusivity question quickly so that we can proceed with you with confidence and deal with other people fairly. As you are aware I am concerned that we may not be able to reach agreement as I feel our losses may be potentially greater than your gain.

We cud only guarantee not to accept commissions to design boats over 65 ft rating for the race after the date that exclusivity agmnt is signed and money paid (or a date set for payment to secure exclusivity). You shud be aware that we are currently discussing 3 other serious commissions for Maxi yachts for the Whitbread race which we are currently confident cud produce at least 2 confirmed commissions.

We feel that we must measure compensation from 5 points of view:

A. Reduced cashflow/income to support worldwide efforts.

B. Lost worldwide promotion value in countries where boats are being built.

C. Lost promotional value during buildup and race due to only one Maxi.

D. Risk of unforseen accidents removing all chances of successful result.

E. The time effort that we anticipate putting into your particular project cannot be supported by your design fee alone.

I am sure you can appreciate that this is a difficult value for us to assess but here goes.

As an opening proposal I suggest a payment of US dlrs 220,000 (two hundred and twenty thousand)[11] for above 65 ft rating (to rule in force at design time) for the 1985 Whitbread race. Total or most would have to be apid upfront to secure protection, and wud be subject to our seeking opinion from counsel in relation to possibility this protection being in violation of US law and other proposals and your holding us harmless in event of any action arising therefrom.

This proposal expires 17 February 1983 and is not binding on us until it has been agreed and monies paid.

I wud appreciate your earliest reaction to this and any counter proposal that you might wish to make.

I will be away from Fri afternoon (4th) till 14th Feb.

Kind regards to all,

Bruce Farr[12]

The Farr office, however, was already committed to designing Whitbread Maxis for two other customers and all they were prepared to offer was the same hull with small modifications towards our design criteria.[13] We had a problem. We didn't want anyone else getting the advantage of our thinking. We asked what it would cost to buy exclusivity.[14] The answer was a figure which was prohibitive. Tom and I saw red.[15]

Clearly, when he asked 'what it would cost to buy exclusivity', Peter Blake must have known that Bruce Farr & Associates (BFA) had not by then 'committed to designing Whitbread Maxis for two other customers' or were only 'prepared to offer ... the same hull with small modifications'. What makes this falsity in *Lion: The Round the World Race with Lion New Zealand* even more outrageous is the fact that BFA would not sign the first of their Maxi commissions until July of that year, a full five months after Blake had been

offered the exclusive use of BFA's staff, resources, services and design philosophy for what amounted to a four-year period for that race. And it would
be nine months after he supposedly 'saw red' at the reasonable figure Farr
put up for discussion (not as a demand) at his request that they would sign
their last two. Further, there is nothing to show Blake's thinking held any
'advantage' that would have materially affected the design of a Maxi round-
the-world racer, or indeed for anyone except that mythical version of himself he pitched to his sponsors and supporters.

> Faced with this situation, we decided to approach the Holland office with a
> strict set of design criteria to see if they could produce what we had in mind.[16]

If Blake was 'seeing red' at a figure he had requested and Farr had put up
for discussion, he certainly wasn't saying so — or at least not to BFA.

> 14.2.83 Ref: Carol 775
> Many thanks for Maxi information. We will contact you again as soon as a
> decision to proceed has been reached.
> Regards
> Peter Blake[17]

> Some people also thought we contradicted ourselves by commissioning
> Holland who had already done a boat for Rob James which wasn't that far
> from our thinking. The difference was that the Farr office would not even give
> us an assurance that they would work on no other Whitbread boat once they'd
> received our submissions and ideas and produced a design for us.[18] This was
> the least we could expect if we were to feel sure that nobody else would ben-
> efit from what we had to offer.[19]

As is his way, Peter Blake blamed Bruce Farr for not offering 'an assurance'
when, in fact, it was Blake who had failed to respond to the offers of assurance already on the table: the offer of design exclusivity and confidentiality
contained in Bruce's letter of 5 January 1983, and the exclusive use of BFA's
design services in Bruce's telex of 3 February 1983.

> 16 Feb 83
> Attn Peter Blake
> Thks yr reply re exclusivity. Is it possible for U to give any indication of reac-
> tion to our proposal. I am not sure we can hold off other possibilities until 31
> March without some measure of agmt between you and us on what terms and
> form of exclusivity wud take when and if project is agreed to by Ceramco.
> If we are to seriously consider an exclusivity proposal I believe it wud be ethi-
> cally wrong for us to carry on discussions with others, but I don't see how we

can make this decision without some proposal from you or at least a reaction to our proposal.

I appreciate that goahead cannot now be given before 31 March and I am only asking for some feeling of what Ceramco's position might be, as we must make our decision now.

I hope you can help us out. Pse call collect any day this week if discussion wud help.

Kind regards to all

Bruce Farr[20]

By now the Farr team was forging ahead with their third generation round-the-world racer, producing a prototype they hoped could be used as a basis from which to develop individual designs to best meet the wind and sea conditions each client expected in the race. Although an innovative concept, there was a risk to conducting expensive research begun before client contracts had been signed. However, if clients agreed to share the expense it could prove cost effective for all who partook of its benefits. To increase rated speed for the Maxi, Bruce was now taking more account of the IOR than he had with *Ceramco*, further refining the shape that had proved ideal for the downwind work on the old clipper route, the rugged Roaring Forties. But Russell was entering uncharted waters; no one before him had engineered a boat of this size out of Nomex-cored Kevlar or devised a way of holding it all together. Most maxis were built of alloy. Or, in the case of *Kialoa*, the substantial laminations Ron Holland had used over its end-grain balsa core had produced a relatively heavy hull. Bruce Farr:

> There is a big risk and we take it very seriously. We worry about the people and the boats and we do lose sleep. It's a tough balance between safety and speed. It is not about designing a boat that gets around the world. That's easy. The hard part is designing a boat that will be strong enough to get it around the world safely but at the same time being light enough to be the fastest.[21]

As their minds roamed the boundaries of lightness, speed and strength, the designers looked for a new place to start. They knew if they used exotic panelling they would be unsure if the internal structures made of composites would be strong enough to hold the hull together.[22] Without something more substantive, the massive loads from rig and keel could tear the hull apart. The answer lay in memory. If Russ could translate *Jennifer Julian's* rudimentary space frame into a giant internal structure, that might just do the job. The concept was brilliantly simple and originally shared with Blake: if the body is built around a fabricated frame, much of the weight removed from the hull could then go into the keel. That would not only generate more 'horsepower'. The surplus weight needed to meet the IOR's stability rule could be centrally positioned to reduce pitching moment. It all added up to a faster, lighter, more stable boat. Even at this conceptual stage their Maxi was chalking up firsts. It would be the first Farr Maxi and the first of light-displacement. It would be the first Maxi made of exotics and the first with an internal load-bearing frame. They didn't know it then, but it would also be the first Maxi to take to the skies.

While their pioneering work was generating interest from Fehlmann, Padda Kuttel and Digby Taylor, nothing had been heard from Blake for over four months. But when he finally called from England, he certainly sounded chipper. *Ceramco* had been tentatively sold. And although Ceramco Ltd would not be involved again, Tom Clark was still enthusiastic. Blake had only one question: Could an exotic plastic boat be built that is strong and reliable? We have absolutely no doubt, Russ Bowler replied.[23]

In the meantime Bob Farrell, Bruce Farr & Associates' New Zealand representative, continued dialogue with project leader Clark. Things seemed to be hotting up.

13 6 83
Attn Bruce
After lengthy talks with Tom ... I think we have contract coming up. Have had
to offer Tom NZ exclusivity ...
Regards, Bob[24]

Farrell followed this up with a phone call. 'It's good news, Bruce. They have the money and we are preferred designer.[25] So we have the job subject to you and Russell agreeing to New Zealand exclusivity.'

'Confirmed, Bob, and for that there'll be no charge. And tell Tom, in terms of the building programme, we'd be looking at ordering hull and deck materials in September which would give us a January start.'[26]

Further negotiations followed between Farrell and Clark concerning Geoff Stagg. Tom Clark was adamant; he wanted Stagg's involvement:

(a) to represent them in the Farr office;
(b) to work only on their boat;
(c) to be available for tuning and racing;
(d) to sail all or part of the Whitbread.

This was agreed by phone and the rates for Geoff's involvement set on behalf of Farr International.[27] So near at hand did a contract appear that Bruce wrote a critical path analysis for the complete design programme. It looked as if things were settled, the deal almost done.

24 Jun 83
Attn Bruce
Spoke with Clark who said Blake will contact U re design contract problems

still current. He still says he wants some sort of design priority or exclusivity to
go ahead. Also said Ron Holland has just offered the total Colt Cars design to
him ... Confirmation of designer will be 30 June ...
Regards, Bob[28]

On 19 August, under a covering letter, Bruce sent six sets of drawings to
Peter Blake for promotional purposes use. He needn't have bothered.

August 25, 1983
Attention Bruce Farr and Geoff Stagg:
Design decision reached today. Regret you are not successful this time but
many thanks for all your efforts.
Regards Tom Clark[29]

Russ and Bruce sat back in their chairs, the wind knocked out of their sails.
As Peter Blake would later say: '*Ceramco* had turned out to be everything
that we'd asked for'.[30] And Holland's maxis 'in strong winds from behind ...
were totally out of control and you couldn't even contemplate a yacht with
that sort of handling problem for the Southern Ocean, which dictates that
you have a boat which is very controllable and runs and surfs well'.[31] The
leaps in logic were monumental. And after making a mountain out of the
exclusivity molehill, not only had Blake commissioned a boat based on an
existing Holland design, but the former *Colt Cars* package had a half-built
near-twin sitting waiting in Moody's Solent yard to challenge it. The next
day Bruce sent a letter to Bob Farrell asking him to recover the design
information and drawings, as well as establishing from where Alan Sefton
had acquired the copy he had used in his magazine. In the end, however,
it wasn't the loss of control over their information or even the failure to
secure a hoped-for commission that shocked Bruce and Russell — that was
commercial life. It was the sickening sense of *déjà vu*. Not only had Bruce
not asked for, let alone insisted on, airfares for Cathie and himself before
he'd agree to join *Ceramco* at the end of the 1980 Sydney–Hobart. But
here, out of the blue, was another denigrating rumour about his demand
for money.

21 Sept 1983
Attn: Tom Clark
Although we were obviously surprised and very disappointed with your decision to go with Holland, I appreciate your letting us know promptly. Have
been expecting some sort of official correspondence from Peter Blake but
nothing yet.
There is a story around Auckland which has come from people close to your
camp that the reason for the Holland decision was that we were going to

'I was in love with the 727. It was a beautiful boat, easy to handle and a joy to sail.' Roy Dickson.

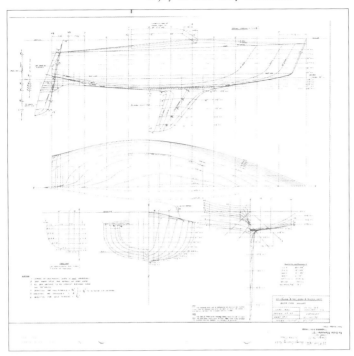

Lines plans of the Farr 727 (Design 37). *(Bruce Farr Yacht Design.)*

Vind-Too on a plane. The Farr 727, with its physics-defying ability, disproved the theory that a displacement keel-boat in flat water cannot move a fraction faster than the product of 1.3 times the square root of its length. It introduced to the international keel-boat community the principal Farr philosophies still used today around the world in racing keel-boat design: fractional rig, light displacement and weight concentration. *(Bruce Farr Yacht Design.)*

Turning the 727 plug at the Crockett family home, Mt Roskill, Auckland, 1973. *(Murray Crockett.)*

Genie, the first 727 out of the mould at Alpha Marine, Devonport, Auckland, 1974. *(Murray Crockett.)*

New Zealand's first keel-boat world championship win in a yacht designed, built and crewed by Kiwis: *45° South* (formerly *727*).

he soon-to-be world champions putting away a
w on the trip across the English Channel to
eauville. *(Bruce Farr Yacht Design.)*

French fishermen blockading the
entrance to Deauville Harbour during
the 1975 Quarter Ton World championship. *(Murray Crockett.)*

The new Quarter Ton world champions, Deauville, 1975. L-R: Rob Martin,
Graeme Woodroffe, Roy Dickson and Murray Crockett. *(Murray Crockett.)*

The breakthrough Farr 1104 (Design 51) conclusively proved that Bruce Farr's light displacement theories worked just as well at the bigger One Ton level.

Noel Angus's *Prospect of Ponsonby.*
(Bruce Farr Yacht Design.)

Stu Brentnall's *Jiminy Cricket.*
(Bruce Farr Yacht Design.)

Graeme Woodroffe's *The Number* (later *45° South II*).
(Bruce Farr Yacht Design.)

The three Farr world champion designs of 1977

Eric Simian's Three-quarter Tonner *Joe Louis* (Design 56), La Rochelle. *(Bruce Farr Yacht Design.)*

Stu Brentnall's One Tonner *Red Lion* (Design 64), Auckland. *(Bruce Farr Yacht Design.)*

Peter Willcox's Half Tonner *Gunboat Rangiriri* (Design 65), Sydney. *(Bruce Farr Yacht Design.)*

Ceramco New Zealand (Design 90).
(Bruce Farr Yacht Design.)

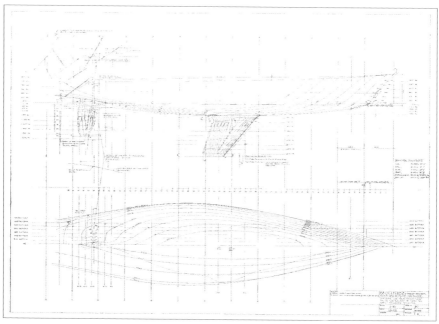

Lines plans.

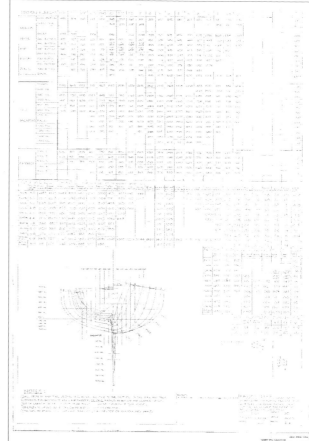

Offsets.

Ceramco New Zealand. (Bruce Farr Yacht Design.)

The first Farr European Lake Boat design

Thrills and spills on *Farrneticante* (Design 97), Lake Garda, Italy.
(All photos courtesy of Bruce Farr Yacht Design.)

The Farr 10²⁰ and Farr MRX

The first design completed in Annapolis, the beautifully balanced Farr 10²⁰ (Design 121) was a hit with New Zealand's cruiser-racer addicts and a great commercial success for Sea Nymph Boats. The hull was later modified by BFA into the match-racing Farr MRX built by McDell Marine.

Sea Nymph's 100th Farr 10²⁰. Joint Managing Directors, Peter Gribble and Kim McDell are third and fourth from right. *(Bruce Farr Yacht Design.)*

Farr MRXs in the heat of battle. *(Ivor Wilkins.)*

The first business premises of Bruce Farr & Associates Inc: Horn Point, Eastport, Annapolis, Maryland. *(Jim and Ileene Farr.)*

'If Annapolis were Casablanca, Marmaduke's Pub would be Rick's.' Marmaduke's, Eastport, Annapolis, 1995. *(John Bevan-Smith.)*

Nice work if you can get it. L-R: Russ Bowler, Geoff Stagg, Nick Danese, Bruce Farr, Bob Munoz. *(Bruce Farr Yacht Design.)*

Design 124
The design that made the breakthrough into the tough US market.

Freefall. (Bruce Farr Yacht Design.)

Migizi, with Russ Bowler on board, winning its division of the 1983 SORC. *(Bruce Farr Yacht Design.)*

The Farr 40

Design 136 and its progeny carried Bruce Farr & Associates and New Zealand to the top of international offshore yachting.

The 1983 New Zealand team, seen here in Sydney on their three Farr 40s, produced the largest-ever winning margin in the history of the Southern Cross Cup. L-R: Owen Champtaloup's *Geronimo* skippered by Stu Brentnall, Tom McCall's *Exador* skippered by Ray Haslar and Del Hogg's *Pacific Sundance* skippered jointly by Geoff Stagg and Peter Walker. *(Bruce Farr Yacht Design.)*

Bruce Farr and Peter Walker on board *Kiwi* (Design 181), 'big boat' member of the winning Kiwi Admiral's Cup team in 1987. *(Ann McGlashan.)*

Design 131
The breakthrough Farr design that changed the shape of the Maxi.

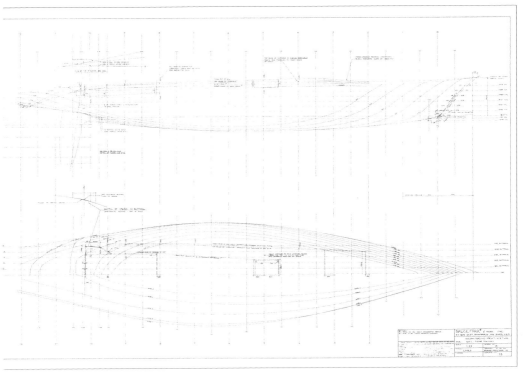

Design 131 lines plans. *(Bruce Farr Yacht Design.)*

Pierre Fehlmann's *UBS Switzerland*, line honours winner of the 1985-86 Whitbread Round the World Race. *(Bruce Farr Yacht Design.)*

The Bowler breakthrough

The breakthrough Bowler drawings that answer Conner's question: 'The last seventy-eight 12-metres built around the world have been built in aluminium so why would you build one in fibreglass unless you wanted to cheat?' (*Bruce Farr Yacht Design. Used with permission of the New Zealand Challenge Design Group: Russ Bowler, Laurie Davidson, Bruce Farr and Ron Holland.*)

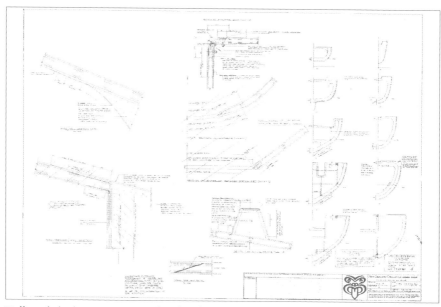

Hull and deck construction details of *KZ-5*.

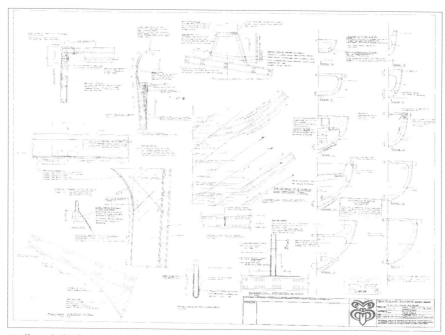

Hull and deck construction details of *KZ-7*.

The powerful fibreglass *KZ-7*, twice world champion (1987 and 1999).

The delivery and launch of *Fram X* (Design 185).
(Bruce Farr Yacht Design.)

A proud moment for the Norwegians as they collect the One Ton Cup. L-R:
HRH Crown Prince Harald, Odd Roar Lofterod, Stein Foyen, Kjell Arne Myrann,
Petter Hagelund and Einar Koefoed.

charge a 300,000 dlr deisgn fee. This is incorrect and very bad taste.

I dont see any value in this sort of thing for you, particularly if it continues and we are forced to make public statement to correct facts. I would appreciate anything you can do to stop the stories continuing and repair the situation.

Best of luck with project and I look forward to following your progress.

Regards,

Bruce Farr[32]

Oct 13 1983

Bruce Farr and Associates

121 Eastern Avenue PO Box 3457

Annapolis MD 21403

I have let the response to your telex of 21st September languish on my desk whilst I cooled down. I know nothing of the source of the rumour mentioned by you and amongst many rumours have not heard that one. I would respectfully suggest that the tone of your telex to meleft a lot to be desired. Regards to Hoeff and yourself.

Tom.[33]

14 Oct. 83

Attn: Tom Clark

I am surprised at your reaction to my telex as I thought you would be as concerned as we were about this rumour.

For your information it has come back to us through 3 different sources, one of whom said he heard it discussed in a group at RNZYS which included Tim Gurr and had immediately beforehand included Peter Blake. These people passed it onto us because of disbelief and because the statement seemed to be offered as a reason for going with Holland.

I passed it back to you out of genuine concern for the welfare of your project as well as looking for the quietest way to kill it. I see no value or reason to be at loggerheads with your group and there was no adverse tone meant in my telex.

I hope that we can all get through our involvement in our respective projects as friendly rivals. This is certainly the desire of all of us here.

I look forward to seeing you in NZ soon.

Bruce Farr[34]

On October 10, Peter Blake's handwritten letter arrived in Annapolis.

31-8-83

Bruce,

Very many thanks for all your information. I have nothing but the highest praise for your designs but for the next project have decided on a slightly different type of approach — for better or worse.

I look forward to some great racing against your new maxis, and wish you the best of luck with them.

Regards,

Peter[35]

Bruce replied by letter.

November 7, 1983.

Dear Peter,

Thank you for your recent letter and I too look forward to seeing you at the starting line.

Thank you also for the return of our drawings. We believe you also have confidential information regarding the boat and feel that it would be best if you were to return this also, thus reducing any possibility of it falling into the wrong hands.

Best regards

Bruce Farr[36]

By any yardstick Blake's deceit was staggering, as were his lies and rewriting of the time line concerning the choice of designer in *Lion: The Round the World Race With Lion New Zealand*. But prior to its publication Blake had suffered another embarrassing loss and had had to face up to his sponsors yet again. Was there for this failed sailor[37] only one fail-safe method by which to save face successfully: scapegoat others — in this case Farr and Holland, who were more concerned with designing yachts than fighting flagrant lies? [38]

Interlogue 2

INT. DINING ROOM, LUXURY HOTEL. PORTO CERVO, SARDINIA. EVENING. JUNE 21, 1988.

Four men sit at a table. Not a grey hair among them. They study the elaborate spread of food, the butter statues, the guests ordering steak, watching it cut and cooked. The one with the lopsided grin casts his eyes around the room. There are too many Americans. Too many syndicate heads doubling as Twelve-Metre delegates from the meetings begun that morning. But what the hell. Leaning forward he adjusts his frames and begins sotto voce.

Fay: Bruce. Russ. What would you say to a Deed Boat challenge?

Bowler: ... !

Fay: Yes. In a ninety-foot waterline yacht.

Farr: You've got to be joking!

Fay: No joke Bruce. Explain it to them Andrew.

Johns: Conner's claims that we were cheating were bad enough. But I think all potential America's Cup challengers have had a gutsful of Sail America's arrogance and San Diego's intransigence. I mean, we still don't have a date let alone a venue for the next regatta and no indication of when these might be announced. But what really started me thinking was a call I got from Tom Ehman last Easter. The Twelve Metre Association's technical committee would, he said, be recommending that fibreglass be banned from the Twelve-Metre Class. And in a stunning leap of logic he assured me that that was in the best interests of yachting. So, with that, I took a closer look at the three Deeds of Gift.

Bowler: And?

Johns: And it's there plain as day. In fact it's been there since July 1857, in the first Deed of Gift: the letter from the ex-owners of *America* to the New York Yacht Club. Any organized Yacht Club of any foreign country shall always be entitled dot-dot-dot to claim the right of a sailing match for this Cup. That's what it says.

Farr: You mean — ?

Johns: Yes, it's a challenger's not a defender's deed, Bruce.

Bowler: That's beautiful. Beautiful.

Farr: Why has no one thought of this before?

Johns: Because in the America's Cup's recent history the New York Yacht Club's efficiency has, in effect, become a tradition. They planned the next match before the current one was over. And challenged as soon as Bondy won in '83. It seems with the hiatus of World War II and the introduction of multi-challenges, the Deed of Gift has all but been forgotten.

Bowler: Forget the rift between Sail America and the San Diego Yacht Club. This'll produce one bigger than the San Andreas Fault!

Laughter.

Fay: OK — let's keep it down. There are too many ears around here, especially American ones. I think we should leave it at that, finish our meals and get an early night. Let's meet at the marina first thing in the morning and continue this then ...

Chapter 21
Back On Top

A Farr-Bowler design is like a Lennon-McCartney song; it is tangibly more than the sum of its structure and lines.
Richard Gillard, New Zealand singer-songwriter

New Zealand ocean racing yachtsmen have swept back to international stature with the greatest of their four Southern Cross Cup victories in Australia. The five-race series ... saw the three Farr 40s dominate and finish ... with ... the largest winning margin in the history of the Cup. Peter Campbell, yachting writer[1]

1983 was a watershed year. Karli was born to Lynda and Russ on August 16. And it was the year that everything changed for the small Annapolis office. The year in which two of their designs — one of 40 feet, the other of 80 — would alter so materially what a keel-boat could be and do that they would emerge three years later as the world's most potent force in racing-yacht design. But with business survival still high on the agenda as the new year dawned, that was only a distant hope.

Almost by osmosis, the ideas that would make *Migizi* a winner in the 1983 SORC began working their way through drawing board and keyboard into Design 136.[2] Could this 40-footer be for the struggling business what *Titus* had been for Bruce at the beginning of his keel-boat career? Perhaps, for there was an ocean of difference now. Not only were they in the USA with DuPont just up the road. Now Russ could explore the far corners of that technology and deliver lighter and lighter structures to complement Bruce's shapes. Although for the last five years they had kept a tetchy distance from the ocean racing scene, the IOR's new Centre of Gravity (cg) rule had pushed the door ajar.[3] Enough, at least, for Bruce to begin kneading the parameters of his shapes and Russ to aggressively pursue exotic construction methods. Bruce Farr:

After the changes to the rule in 1978, we tried to adapt our general style to a heavier type of boat and it didn't work that well. What we learned — though not quickly enough — was that in order to be successful under the new IOR we had to design a totally different type of boat; we had to go where others were within the rule and try to do better. We were encouraged by the performances of *Free Fall* and *Migizi* and from there we invented a new type of boat. Fortunately it worked, and that became the turning point.[4]

Designed for the rough and windy conditions of New Zealand and Australia, the yachts from Design 136 were given less sail area and more length than those conceived for the lighter airs encountered in northern hemisphere racing. Carrying small trapezoidal keels, their lightweight hulls were very stiff. All-round performers, they would handle with great efficiency a wide variety of conditions in many different locations. Not only would the 136s and their derivatives put the Farr name back in lights. They would become the keelboats of the decade, sailing off out of the eighties with almost more silverware than their 40-foot hulls could carry.

There was a bitter-sweet irony to this happy turn of events. Bruce, Russ, their staff and families had moved half a world away from the genesis of light-displacement to a place they thought would be closer to the beat of yachting's heart. Yet those who were quick to take a risk on this new generation Farr were precisely those who had taken the risks before: the gung ho Kiwi sailors and those prepared to back them. Simon Gundry, former *Ceramco* crewman, might have auctioned the 136 lines plans for a mere ten dollars when Geoff Stagg arrived from the USA and the boys had a big night out. But those who put up their hands clutching serious money in the coming weeks were those who would change New Zealand's yachting fortunes.

The first three boats built to this design were constructed in Divinycell core by Cookson Boats of Auckland and launched in 1983. All were 40 feet overall but rated close to 30 feet, the minimum length required to make the One Ton class.[5] Like *Titus*, *Prospect* and the revolutionary lift-keelers, these three Farrs were about to enter Kiwi yachting lore. *Pacific Sundance*, owned by Del Hogg and Brian Morris of Wellington, was to be co-skippered by Peter Walker and Geoff Stagg. Stu Brentnall was to campaign Owen Champtaloup's *Geronimo*. And Tom McCall's *Exador* was to be under Ray Haslar's control.

In an emphatic display, all three swept past the bigger Frers-designed *Shockwave* and *Solara* in New Zealand's 1983 Southern Cross Cup trials. It was another all-Farr team that produced a record 101-point winning margin during that Sydney summer. With Walker and Stagg making it sing, the perfectly balanced *Sundance* won three of the five Cup races. In an international fleet of 27, it finished top boat of the series with *Geronimo* second and *Exador* ninth. It was as if *Prospect of Ponsonby* and *Jenny H* had come back to haunt the regatta. By the time it was over and the historic records checked, Farr designs had won eight of the thirteen races the last three times New Zealand had won the Southern Cross Cup. There was no other designer in sight. Tom Schnackenberg:

> I remember cruising back into Sydney Harbour on *Geronimo* after one of the races. We'd cleated the sails and were chatting away eating sandwiches while the bow man steered. And without paying much attention to what we were doing, we sailed right past this Australian boat which rated two feet higher. One of their crew told me later how depressing it had been. They'd been fighting to keep up with us, timing us as we passed each mark on the way back in, but we quite effortlessly pulled away. That's how dominant those boats were. You'd just point them at the mark and off they went. They went upwind, downwind and across the wind. They were just excellent boats and really were that quick.[6]

Seven months later *Pacific Sundance* and *Exador*, together with Neville Crichton's *Shockwave*, went hunting for the Pan Am Clipper Cup in Hawaii. The New Zealand A team had the trophy in its keeping as they neared the end of the last event, the triple-point Round-the-State race. Until, that is, *Exador*, with sheets just started, stopped dead at the southern tip of the Big Island of Hawaii. Sailing near the surf line, 'suddenly two enormous walls of water reared up dead ahead ... They hit the face of the first one, which took the mast out, and the second swamped the boat ... The Farr 40s were so far up in the fleet that Kiwi Terry Gillespie on *Winterhawk* (ex *Ceramco*) in the Maxi division saw ... [the] giant wave with spray blowing off the top like smoke.'[7] Ahead on points going into the 725-mile race, *Sundance* and *Exador* had been first and second on corrected time, with *Shockwave* lying sixth overall. But while the freak waves cost the Kiwis the Clipper Cup, they were still ahead in the Mumm world championship for best national team. As well, *Sundance* finished top boat in Hawaii.

And when the Spanish Navy sailed *Sirius II* into third place in the 1984 One Ton Cup, Farr and Bowler were now no longer just players in the Pacific and Caribbean. *Sirius II*, with its slightly shorter waterline and increased sail area, had shown the northern hemisphere that their break-through design was a potential winner even in their conditions.

***

It was eighteen months after Cathie had sailed out of Bruce's life that he received a call from his flatmate's girlfriend.

'Bruce. Connie Warren. Geoffrey tells me you're a lover of art.'

'Yes.'

'Well then, that could make this perfect.'

'Really?'

'A fellow realtor of mine just happens to have a spare ticket to a local art gallery auction. Looks like you could both be in luck.'

'Connie, I have another date.'

'Bruce, this lady is drop-dead gorgeous. Cancel!'

He did. And on the fourteenth of September 1984, Bruce and Gail Hartman exchanged marriage vows on the beach at the Bay Ridge Club on Chesapeake Bay. Jim and Ileene, having flown in via Europe, were on their first trip out of New Zealand, a thank you from Bruce for their love and unstinting support. They sat through the opening round of toasts, listening to best man Dinny White describing the miracle it was for Bruce even to be among them. Gail, elegant in her lace overlay and simple sheath gown, blushed, remembering only too well those feelings of indignation she had had during her bridal luncheon. The more stoic she had become as she registered support for the man who was about to become her husband, the greater the difficulty her girlfriends had in holding back their laughter. Which really surprised no one given the previous night's events. For in spite of Bowler, Stagg and White supplying plenty of food, before the stag do had reached its zenith Bruce was seriously wobbly from the expertly spiked drinks.

> Bruce had jumped right into it. He'd decided that if 'it' was going to happen to him — whatever 'it' was — he'd just have to go with the flow. We'd been having a great time and Bruce was pretty far gone when ten of them grabbed him and pinned him to the floor. That's when surgeon Eddie Holt appeared with his rubber gloves and a bucket.[8]

Within minutes the waist-to-ankle cast was hard and Bruce passed out on the Eastport Yacht Club floor. In not much better shape were the bearers of this palanquin of plaster as they stumbled down three flights of outside stairs. The patient was eventually deposited at Gail's Bay Ridge home, where Gail and Bruce's dad placed Bruce in the bath and removed the cast. It said much for the groom's resilience that not a trace of the night's operation showed as he beamed at his best man's stories and filled in the bits he'd missed.

While the honeymooners cruised the coast of Virginia on a borrowed Farr 38, Jim and Ileene toured America. Two weeks later, after seeing them off at the airport, Bruce and Gail dined in Eastport. As they walked back across the drawbridge, a VW Beetle careered across the sidewalk, the drunk-driver ripping Bruce from Gail's side. Wheelchair-bound and with a rod to support his smashed tibia and fibula, it took more than a month before Bruce could return to work, and many more of rehabilitation before he could properly walk again. Pam Farr:

I remember him ringing me from his hospital bed asking me to make sure I met Jim and Ileene at the airport. I was to tell them he was fine — not that he was in traction! I mean, he could have been killed or lost his leg. 'Are you really all right?' I asked him. 'Yes,' he said, 'I'm really all right. Just make sure Mum and Dad hear from you first. I really don't want them to worry.'[9]

***

Seven weeks before the wedding, Bruce had received a letter from PR consultant Cedric Allan, an old yachting friend. Dated 17 July, it had been written just three days after the New Zealand National Party had been swept from power in a snap election.

> The campaign committee has asked me to write to you with our views on the design to produce New Zealand's first 12-metre.
>
> As you know we have given the responsibility for this to the Ron Holland Design Office ... [10]
>
> Nevertheless, it is the wish of both Ron and the campaign committee to make the fullest possible use of the design and technical skills available from all New Zealanders. Ron's approach is very much one of coordinating an input from a team.
>
> It is the committee's hope that you will feel able to make a contribution to the New Zealand challenge and this should be on a proper commercial basis. Financial arrangements should be made with the Holland Design Office and Ron will be contacting you to discuss the role which you could play in assisting the process of putting together the best possible challenger for New Zealand.'[11]

While it was good to hear from Cedric again, the tenor of the letter was about as realistic as a waterlogged *trompe l'oeil*. Bruce and Russell wondered how the arrangement could be on 'a proper commercial basis' if they were dependent on an opposition designer for payment and if the official challenger, the Royal New Zealand Yacht Squadron (RNZYS), wasn't prepared to top the project up with tink. It wasn't that they were 'being difficult ... [or] paranoid about secrecy'.[12] The number eight wire approach simply didn't work where they now lived — in a commerce where lawyers waved writs like weapons. It would have to be business as usual when it came to an inquiry for their services. Besides, they had offered to meet Don Brooke on his Fachler-funded research trip, and it was Brooke who had decided not to stop off at Annapolis on his way back to Auckland.[13] If their professional approach offended their compatriots then so be it. Having barely survived their first three years in America, they weren't about to risk it all, no matter how worthy the cause. Despite their feelings of patriotism, they knew if they fell over in the USA there'd be no one from New Zealand offering to pick them up. Besides, there was a lot they didn't know.

They didn't know, for instance, that the steering committee's Project Manager, Aussie Malcolm, had decided to ditch most of Don Brooke's six million dollar game plan, including the US$400,000 purchase of the Aga Khan's *Enterprise*.[14, 15] Nor did they know that for a mere six thousand greenbacks, winged via Monte Carlo to a bank in New South Wales, Malcolm and Holland had acquired 'the full set of plans and computer calculations for *Australia II*'s hull and keel'.[16]

Nor did they know that Michael Fay, a member of a syndicate building a Ron Holland luxury yacht for export, would soon be showing interest by having a beer with the RNZYS's designated designer while he was in New Zealand in November for the launch of *Lion New Zealand*.

In fact few yachties knew much about Fay, apart from his company's sponsorship of the three-Farr-boat Southern Cross team and *Lion New Zealand*. Few knew that he had made his early money out of property, or that his now blossoming career was built on tax reduction deals that made the nouveaux riches smile and the Commissioner of Revenue frown. Of course no one knew just how well Fay and David Richwhite would do out of Labour's volte-face and its decade of deregulation. But in the meantime, while Richwhite targeted tax, Fay was about to settle his still-rising star over another high-risk deal: the quest for yachting's Holy Grail. If Richwhite used the war cry — 'There is no morality in tax law!'[17, 18] — Fay had a more pragmatic maxim with which to harness his troops: 'Around here we apply the golden rule; I've got the gold so I make the rules.'[19] It would become an indictment on Fay and Richwhite, in their headlong rush for glory, that neither they nor the BNZ would tell the RNZYS whose gold it really was.

It was while he was searching for the much-needed shortfall to mount a successful America's Cup challenge that Project Manager Malcolm found himself in the mirrored and black-marbled foyer of Fay Richwhite and Co. Fay declared his interest. But it would take many more visits and protracted discussions before the boutique merchant banker was ready to sign his name on the dotted line. 'On the one hand, he wanted the addictive rush of adrenalin that came from being involved in the toughest deal in town. On the other, he was chilled by the fear of failure of such an awesome and challenging project. [He] could not come in as just another sponsor. If he was to be in, he had to be seen to be in, and to take the lion's share of the credit when the deal worked.'[20] After committing overdraft to help keep the show on the road, Fay finally threw down the gauntlet.

'You say this project is capable of uniting all New Zealanders behind our cause ... If the chemistry is that strong I want you to prove it by uniting Laurie Davidson and Bruce Farr with Ron Holland.'[21]

Unoriginal, but a masterstroke nevertheless. And the way Fay engineered it meant that, with time running out, if he didn't get what he wanted he'd turn off the taps and leave the challenge high and dry. It was a strategy so compelling that in the two pivotal meetings that followed it would change America's Cup history.

The first, a laid-back affair, took place when Russ Bowler dropped in on Aussie Malcolm in his Westhaven office in January 1985. His was a 'determinedly expressed view that if the Farr office was to be involved in a coordinated design exercise, there would have to be an understanding that the design team should seriously look at the possibility of [fibre]glass construction.'[22] Malcolm agreed, although he had no idea at the time of the significance of his concession.

The second meeting, on 18 February 1985, was arranged and chaired by Aussie Malcolm and took place in the premises of George Levine's Classic Car Company, Florida. Having Ron Holland and Butch Dalrymple-Smith, Bruce Farr and Russ Bowler, Tom Schnackenberg and Aussie Malcolm all in the same room talking a single design programme was as unlikely as the product being processed nearby. 'Outside, in acres of tar-sealed, barbed-wire-surrounded storage yards, were hundreds of brand-new American sports cars, straight from the assembly line. They were brought inside the vast factory complex where teams of men were attacking them with gas axes and blow torches. At the other end was flowing an extraordinary collection of perfect replica MGTDs, vintage Mercedes sports cars and 1950 vintage Porsche coupes.'[23]

It was during this historic meeting that Tom Schnackenberg, sail designer for *Australia II*'s successful America's Cup challenge two years before, suggested a design blanket. While it wasn't new — Starling Burgess and Olin Stephens had employed a similar arrangement when designing the post-war J-boat *Ranger* — it was an important and comforting concept.

'The world will know that there are four of you in there. They'll see the heaving and the wriggling, but won't know whose arms or elbows or knees they are. All that matters is that the plans get poked out from under the blanket at the end of the day.'[24]

***

Either [they were] telling no stories, or they had none to tell. In what was like a broken record, they simply proclaimed they had no serious problems ... They reported a staggering 370 miles in a 24-hour period. They claimed they could carry spinnakers easily in 40 knots of true wind, in control, with the spinnaker sheet cleated. Conceptually, UBS was a breakthrough Whitbread design.
Skip Novak, American yachtsman and skipper of
Drum England[25]

The technical experts were just as impressed. 'The "wedding" of aluminium to Kevlar aramids, selected by Bruce Farr [& Associates], has made possible an optimum ratio between load-carrying capacity and speed. It establishes a new benchmark in lightweight construction and technology.'[26] So said Alusuisse, Europe's biggest aluminium maker, after seeing their Bowler-designed aluminium fabricated frame fitted into the 45-millimetre-thick 80-foot hull. And of the mast, 'fabricated from six different extrusions comprising half-shell shapes in lengths varying from eight to 24 metres',[27] its maker waxed just as lyrical. 'Bruce Farr [& Associates'] concept is new in the field of highly stressed aluminium extrusions and promises further application ... in other sectors such as aircraft and vehicle construction.'[28]

Named after its principal sponsor, Union Bank of Switzerland, *UBS* could have been *Lion New Zealand*. It could have been Peter Blake's. That it wasn't wasn't surprising. Pierre Fehlmann was a straight shooter. Nor did the Farr/Bowler intelligence quotient send him into a spin. With his double degrees and army experience he simply ran his next campaign from Morges with Swiss-watch efficiency, his vision still untarnished after two tough Whitbreads. You didn't get bent out of shape with Pierre. You just learned how to cope with his punishing skeds or the next new idea that tumbled out of his briefcase.

And in the end each got what they wanted: Fehlmann a composite light-weight Maxi, fractionally rigged but stable and easily handled and driven; Blake a boat built to carry the 22 crew he'd decided were necessary to drive his boat at optimum speed 24 hours a day. Sadly for his sponsors, his idea was as bright as a one watt light bulb. To cater for that many crew, their victuals and equipment, it would have to be built, in the words of his mentor, 'like sixteen brick shit-houses'.[29] Although he'd ignored the Farr offer of exclusivity, Blake still had the option of commissioning a lightweight custom design as Digby Taylor had done.[30] Instead he launched a masthead yacht which, at approximately 36 tons (36,500 kg), was by far the heaviest in the fleet.[31] Displacing six to seven tons more than *UBS*, the race was over for *Lion* before the starter had fired his gun.

It was a different story for Fehlmann. Not only was his the lightest Maxi of the seven. His research had been so thorough that *UBS* had the Whitbread won on the third of its one hundred and seventeen race days. Basing his strategy on a computer study of four years of weather patterns, he tacked 'to the west and away from the light winds that the fleet [was] experiencing. He sailed that way for more than twelve hours ... and was into twenty knots before making the tack south. It was a devastating move that put *UBS* [into] Cape Town ... sixteen hours ahead of the fleet'[32] and over two days clear of *Flyer*'s record for the leg. And in spite of the South Atlantic gales which produced delaminations, wobbly keels, broken ribs and ripped sails among the other competitors, *UBS* came through that weather virtually unscathed. It was testimony not only to Fehlmann, his crew of dinghy sailors and Australian Chris O'Nial who had overseen construction.[33] Leg One had conclusively proved that the Farr-Bowler concept could work for the largest racing yachts in the toughest race of all.

Leg Two produced the extraordinary sight of *Atlantic Privateer*[34] and *NZI Enterprise* finishing within twenty seconds of each other after racing seven thousand miles across the Southern Ocean and down New Zealand's north-east coast. And although *UBS* could only manage third into Auckland, it would win the third leg by over thirty-two hours from *Lion*. 'For the Farr boats' masts, it's two down and one to go,' scoffed Peter Blake in Punta del Este.[35] What he hadn't counted on was that the only mast fully designed and engineered by the Farr office was on the Swiss boat and doing just fine.[36]

Changing spinnakers for each of his sponsors as he sailed up the Solent over two days ahead of *Lion* on the last leg, Pierre Fehlmann accomplished his three out of four leg win by eclipsing *Flyer*'s round-the-world record by sixty-four hours.[37] And as he stepped forward on 12 June 1986 to receive his trophy from the Princess of Wales, his smile said it all. So did Bruce Farr's.

'As a result of that [victory Bruce Farr & Associates] became known worldwide as name-brand designers.'[38] The yacht, flown by Fehlmann to Monaco for its launching in a military Super Guppy, had sent the Eastport office planing past the established big-boat names of Holland and Frers and given its European profile a major lift. Design 131 had not only changed the Maxi-boat world. It had propelled the Kiwi designers, from 40 to 80 feet, to the very forefront of ocean racing.

But they weren't the only beneficiaries. Their designs had built the platform from which New Zealand had won the Champagne Mumm World Ocean Racing Championship in 1985.[39] 'New Zealand had finished first in the 1983 Southern Cross Cup, second in the Clipper Cup [of 1984] and third in the Admiral's Cup [of 1985, with a team of *Exador* and *Epic* and the Davidson-designed *Canterbury*] — brilliant consistency using seven Farr boats out of nine in three completely different oceans and conditions'.[40] But Farr-Bowler boats weren't just burning up the racetracks. During that same year, Sea Nymph Boats were proud to announce that they had just sold their hundredth Farr 10[20], including one to Bruce.

Chapter 22
Lifting the Blanket

*If anything, I think that the GRP boats are more legal
than any other 12-Metre ever built.*
James Course, Lloyd's surveyor[1]

It is easier to believe than think.
James Harvey Robinson, American educationalist and
historian

While New Zealand's second winning of the America's Cup was another
remarkable feat for a tiny nation, it is a fatuous notion to think it had anything to do with red socks. And the minimal contribution (as opposed to
glory taken) of the person wearing them should barely rate a mention when
compared with his country's decade of endeavour to achieve that moment.
From designers to boat-builders, from sparmakers to sailmakers, from computer buffs to managers, from ground crew to sailors, a great many New
Zealanders had given unstintingly of themselves, their time and their talent
to bring about that win — and the first — in far less time than it had taken
their trans-Tasman neighbours.[2]

In the euphoria of the moment — May 1995 — and the subsequent
rewriting of events, it is too easily forgotten where the beginning of that
winning lay: with Marcel Fachler's entry fee and Russ Bowler's urging of
Aussie Malcolm to build the challenge Twelves in glass-reinforced plastic
(GRP). Englishman Peter Bateman was not 'the major proponent of using
fibreglass' as Russell Coutts solecistically claims.[3, 4] Not only were the
Holland-Bateman plans to build an aluminium Twelve-Metre already well
advanced before Farr and Bowler joined the New Zealand Challenge Design
Group, a letter of intent had already been given to aluminium boat-builders
McMullen & Wing.

> Without that letter [we] would not have booked space for the project. And it
> was only after all the designers got together that the decision to switch to

fibreglass was made, something I recall Bruce Farr pushing for. That decision took place in Michael Fay's office, a meeting at which I was present.[5].

It is a point not idly made — because that fibreglass decision is the cusp on which New Zealand's America's Cup history turns. In fact this Bowler-Farr initiative was to have an even greater impact on the history of the Cup than either Olin Stephens' trim tab and keel-rudder split or Ben Lexcen's keel.[6, 7] Subject to a process of approval from Lloyd's Register of Shipping that successive designers had considered too complex to confront, the glass-fibre concept would so outflank American thinking that the winning San Diegans, in their search for new advantage, would play both dirty and for time. 'I'm determined to do everything I can to make sure the 12-metre class association outlaws glass,' wrote Dennis Conner in Comeback.[8] But in their flouting of convention, Sail America would find themselves outmanoeuvred yet again by that same Kiwi ingenuity. And even as the SDYC resorted to a yacht ruled illegal to rebuff the Deed Boat challenge, the designers of KZ-1 were setting up a meeting of the world's best yacht designers, with American Bill Ficker in the chair, to provide a way forward.[9] Not only would this next Farr-Bowler move be the catalyst for a new rule and class to replace the out-dated Twelve Metres for future America's Cup regattas. It would create the most level playing field in the Auld Mug's history — and the opportunity for Russell Coutts to helm the winning boat in 1995.

***

'It shall be short, awfully heavy and shall be deeper than a canoe hull.'[10] That was the problem confronting Bruce and Russell when they first considered Cedric Allan's letter. The Twelve-Metre Rule[11] — framed by its makers to 'not allow dramatic developments ... [that would make] their expensive craft obsolete overnight'[12] — produced heavy-displacement 'lead mines'[13] which were the antithesis of their thinking.

Given its longevity[14] and given that it had been introduced into America's Cup competition almost thirty years before,[15] there was little room for design development in a Twelve-Metre yacht. Even the winged keel might not have been allowed had the New York Yacht Club (NYYC) not diverted attention from its measurement protest to the country of origin rule.[16]

But Bruce and Russell knew there was another area of design which remained largely unexplored: the use of fibreglass in Twelve-Metre construction. This had been approved by the IYRU Keelboat Technical Committee at a meeting in Knightsbridge, London on 4 November 1980. It had, though, a codicillary problem. Anxious to avoid the weight advantage gains aluminium had made over wood in 1974, the IYRU had stipulated in its minutes that 'GRP yachts shall not be of less weight, nor have a more beneficial weight distribution nor be less strong than an aluminium yacht built strictly in accordance with the requirements laid down in the design study carried out by Lloyd's Register of Shipping for aluminium yachts (dated June 1971).'[17] Being just one side of the legal coin — Lloyd's had not yet established the scantling requirements by which this rule would be controlled — it had proved an effective deterrent. So, convinced in 1982 that Lloyd's conservative approach to construction would produce a heavier boat of no greater stiffness, Englishman Ian Howlett had abandoned the glass-fibre option for *Victory 83*. Others had done likewise — the French; American Gary Mull, designer for Tom Blackaller's St Francis Golden Gate Challenge syndicate; even Australian Ben Lexcen: '[Fibreglass] is not as stiff as aluminium, it is harder to alter and repair. I know, I've checked it out twice for Twelves.'[18]

Farr and Bowler saw it differently. While a theoretical weight advantage was denied by the Twelve-Metre Rule, they knew a hull made of GRP would have significant secondary advantages. Aluminium's tendency to 'oil can' under heavy loads would become even more pronounced in the rugged seascape of Gage Roads. There would be none of that with fibre-glass — its greater structural strength and local panel stiffness would cause less hull drag and produce a better set of sail through higher rig-loading capabilities. In fact, the more they looked at fibreglass the more obvious it became. A glass hull could be built more quickly. An identical second could be taken off the same mould, a process faster and cheaper than building two aluminium boats. And being fairer in finish, it would easily comply with the no-hollows-above-the-water-line rule.

When Bruce and Russell arrived in Auckland for the first official meeting of the New Zealand Challenge Design Group, they were both impressed and taken aback. Under the blanket were three sets of lines plans: *Anzac Day*, a long heavy Twelve, designed with Perth in mind, that Laurie Davidson had begun for himself on Anzac Day 1984; *Victory 83*, bought from Ian Howlett; and the famous *Oz II*, replete with keel geometry. Bruce and Russ had heard of espi-onage, of course. But Lexcen's lines — nicked? To focus on, however, were the model-size drawings Ron Holland had faired up in his CAD-CAM computer from the three designs. There were also the test results of his early work at the Netherlands' Ship Model Basin in Wageningen and Peter Bateman's notion of two-boat testing to consider. Then Farr asked the question.

'If we're thinking of building two boats to the same design, isn't fibre-glass the best and quickest way of doing that?'

Holland and Davidson looked at Farr, looked at Bowler and smiled. All four knew what GRP could do. Chris McMullen was all too aware of the growing trend to glass. But no one else had been there in the Twelve-Metre business, as Bateman quickly pointed out.

'Without the hull deflection you get with aluminium, we'll get much better results from two-boat testing.'

'But the pressure on the programme ...'

'Think about the gains.'

'Can we — have we got time to get it through Lloyd's?'

'We won't know till we try.'

'No.'

Settled without debate, the complex issue of construction was handed to Russ Bowler. With this decision the New Zealand Challenge Design Group had also placed the loci of greatest risk and greatest gain in an office in Southampton. It could take months of negotiations, with the likelihood of failure far greater than success, before the result was known. In the meantime the materials and construction decisions would have to go on hold and the already truncated time-frame be even more compressed. There were no guarantees. But if Lloyd's approved Bowler's fibreglass plans, Boat Two could be in the water within three weeks of Boat One's launching, with all the attendant gains waiting to be unlocked.

'Who's doing the lines?'

For designers with many world titles to their credit, it seemed astonishing that this question should hang in silence. That it did spoke as much for their mutual respect as it did of their collective humility. Finally, and with an almost shy reluctance, Bruce agreed to take them on. He would work with Laurie on a first-draft concept of a set of lines before returning to Annapolis to complete the task. Ron would pick up sail plan, deck layout, keel geometry and share the rig with Russ. Laurie would be the team's rep in Auckland. After budgets had been set and billing arrangements established, the three-day meeting was adjourned. It was March. Time was short. Already it would need more than a minor miracle to get their first two boats in the water by December if they were to contest the Twelve-Metre Worlds in February. Bruce Farr:

> Our next group meeting was in Newport at a Howard Johnson's Inn. We met in the restaurant, figuring it was the one place where none of the locals would see us. We spread out the new lines drawings and basically kicked off from there. From these real things would have to happen — the building and testing of models at the Wolfson Unit,[19] real rigs ordered, real sails made, real boats built.[20]

It wasn't long, however, before cracks in the tracks appeared: as the freight from the early programme impacted on the new; as rigs and winches designed for aluminium Twelves arrived out of the blue; as decks floated down fax lines before the final lines plans were drawn. It would be enough to bend lesser minds and wills, this gestating of a Twelve-Metre yacht — their country's first — in just nine months.

To complicate matters, things were also changing in Annapolis. Thanks to *UBS* and the Farr 40, where Bruce and Russell had once gone looking for work now work came looking for them. Not only was their office under increasing pressure from commissions. As the only Kiwi nationals on their staff the two partners were unable to delegate any of their New Zealand Challenge tasks. Roger and Christine Hill had moved to Lymington, where Roger now worked for English yacht designer Rob Humphreys. Sally Gray had returned to Auckland. Her replacement, Bobbi Hobson, was American. So too was Bryan Fishback, Scott Gensemer, Jim Donovan and part-timer Bob Munoz. Graham Williams was Australian and Nick Daneze Italian.

The task for Russ Bowler was both intensive and convoluted. First he would have to produce a fully engineered set of drawings of an aluminium Twelve in order to extrapolate the weight calculations of its hull and deck. Only after these had been approved by Lloyd's could they become the benchmark for the fibreglass structure he would engineer from Bruce's lines. These would have to show that the fibreglass boat was not lighter, did not have a lower cg or a more beneficial weight distribution than his aluminium plans. But early in his negotiations he made a startling discovery. The Twelve-Metre Rule required that Lloyd's needed to know only a boat's total weight[21] and, as a consequence, they held no detailed calculations of a Twelve-Metre yacht in their Small Boat Registry. Thus the data needed for comparison with an equivalent fibreglass hull and deck remained unknown and it would be up to Bowler to provide it and to convince Lloyd's of its veracity. Steve Marten:

> I'm sure the hardest thing was to design an aluminium Twelve that Lloyd's would agree was acceptable in terms of strength, weight and weight distribu-tion. Russell could have designed a heavy one and Lloyd's would have said, 'That's fine.' But he had to design the lightest possible aluminium hull and deck otherwise he would have penalised the composite ones.[22]

It was problematic — too light and he would fail with Lloyd's; too heavy and he would sink the New Zealand Challenge. But to the designer who had revolutionised small boat construction in New Zealand and had recently engineered the world's first Maxi from exotics, it was also a challenge that offered delicious possibilities.

> Aluminium boats traditionally had an enormous amount of fairing and paint — 200–300 lbs, or possibly more in some extreme cases — on the outside to make them smooth. That was needed to cover the spots and joins but mostly because their hulls distort during the welding process[23] and also buckle when racing. So I took the position that you could build one without any filler, thus reducing its theoretical weight considerably. I even debated with the inspectors

such things as the weight of the weld. On the other hand, by the time you got enough glass into the structure to make it the same weight as aluminium, it became incredibly strong. So without the filler we had a much lighter boat. But we also had a much stronger one. Which meant not only could it carry greater rig loads with the improvement to performance that that brings. Unlike aluminium, our hull and deck required no reinforcement — and therefore no added weight — to those areas where deck fittings were added.[24]

To produce what no one had before him — a fully engineered set of drawings of an aluminium Twelve — and to convince Lloyd's it could be built as lightly as he said, required a freedom fighter's will dressed in the diplomacy of humour. Of the four designers Russ Bowler alone had the requisite mix of formal qualification, technical knowledge, experience and charm to succeed. As points of conflict and debate arose during his visits to Southampton, instead of voicing objections he would repair to Annapolis to think through Lloyd's position before returning with his own carefully constructed counter. If he needed help he sought it out — enlisting the services of Richard Honey, an Auckland University graduate working for High Modulus,[25] to assist with the engineering of the fibreglass structure. From beginning to end the process was snail-paced and taxing. But in spite of the mounting pressure from the Challenge management Russ patiently worked at gaining Lloyd's approval of every hull-and-deck inch until, on 23 September 1985, he walked, smiling, out of their Yacht and Small Craft Department in Southampton with their stamp on his fibreglass construction plans.[26] That Bowler had succeeded where others had failed was not just a triumph of intelligence and patience. Above all it had provided the breakthrough by which the Zealand Challenge could produce the strongest, fastest Twelve-Metre yacht in the world.

Which is exactly what KZ-5 hinted at in February 1986 when its talented young helmsman and multi-national crew won the opening race of the fleet-raced Worlds. With only hours of preparation and a basic rig and wardrobe, that KZ-5 went on to second overall in the fourteen-boat regatta was both a stunning and an unexpected result.[27] Seated before the media, Chris Dickson's china blues seemed to beam a cosmic warning to the ten other syndicates with their mega-dollar budgets and months of preparation.

It was not just the booster needed in Perth. It also sent a wave of support barrelling north from the Bluff and Michael Fay riding along in its wake. By the time he hit dry land, the now publicly declared leader of the New Zealand Challenge had morphed from merchant banking Wild Child to impresario. He whipped up support from the corporates, who identified with the buccaneer in his smile and the education in his vowels. He charmed the media with his accessibility, the public with his vision. A party animal, he held public parties and sold America's Cup replicas for silly sums of money. So infectious were his ways that the once conservative 'People's Bank' threw five million dollars at the campaign and opened their counters to queues of Kiwis lining up to do likewise. Perhaps it was just as well that the flightless ones were ignorant of the sub rosa lives of those leading the charge. If they had known that the Bank of New Zealand and Fay Richwhite and Co., together with another corporate icon, Brierley Investments Ltd, were about to do to them what the Wakefields had done to Maori, the New Zealand Challenge might never have docked in Freo. Still, it had. And for the moment commercial ignorance was patriotic bliss as the country awaited the launching of the Plastic Trinity's third member — the one that might just bring home yachting's Holy Grail and send those already-rising share and property prices soaring into the heavenlies.

***

The notions of 'team' and 'teamwork' are embedded deeply in New Zealand's psyche and culture. Without it the young colony would not have survived or flourished — nor, over a century and a half, developed its enviable prowess in team sports. In terms of New Zealand yacht design, the concept of a cooperating team approach first took root commercially in the Farr office environment of the 1970s. It grew under the auspices of the New Zealand Challenge Design Group in the mid-1980s and continued to develop among various groups both on and off the water and in various campaigns, including the America's Cup. 'Team' was not a 'recipe' developed by Coutts and Blake, as journalist Richard Becht opines.[28] The team idea — and a coordinator to tie the programme together — was, in fact, first suggested to Peter Blake by Russell Bowler in September 1992 and by Bruce Farr to the still-fledgling Team New Zealand (TNZ) group in January 1993 when they asked for his advice on the best way to go about winning the America's Cup. TNZ's 'masterstroke'[29] was to package New Zealand tradition and the Bowler-Farr ideas and trade them for sponsorship dollars. 'No part of [the Blake-Coutts] campaign was original; it was all imported ... and assembled',[30] the ideas and experience, even the sponsorship packages, inherited from previous challenges, including the design team approach first used in a New Zealand America's Cup campaign in 1985.

Ron Holland:

> After observing the first two boats we said: Okay, what will we do that's different for the third? That was the meeting during which we made the decisions to make it fractionally finer forward, a bit fuller aft, to give the bow more rake.

The direction we decided on was pretty evenly contributed to ... as we barn-stormed around a table. The drawing that produced the tank-test model, Bruce did, [although] it could have been done by either Laurie or me.[31]

The changes, however, were not just aesthetic. Having gained rigid power-ful hulls with minimum drag from Russ Bowler's work with Lloyd's, the aim was now to produce 'high stability and a very efficient keel'.[32] If that diffi-cult combination could be achieved then Boat Three would be able to carry its maximum sail area in the windy Fremantle conditions and have the side forces to resist the lateral forces of its sails.

Even as they continued to make joint design decisions — as their fax machines burned with ideas from three different points on the globe — more and more of the work was devolving to Annapolis. Having realised there was a gap in the spar end of the programme, Russell had approached Professor Peter Jackson, head of Auckland University's Department of Mechanical Engineering. He had a young man, Chris Mitchell, who was completing his thesis on spar design; he would, he said, be perfect for the job. He also suggested another of his talented students: Mike Drummond. With Bruce drawing the lines and Russell detailing construction, it made sense that their office should coordinate all the drawings for the final boat. This convention became an imperative after the first two keels were found to be considerably overweight; their checking of the volume and area calcu-lations of the third, as well as the weight estimates of the rest of the boat, would be absolutely essential. But that was not before Russell, as the design team's coordinator, had engaged the services of Richard Karn to help with keel design and the interpreting of test results. Karn, another graduate of Auckland University, lived on the beach at Napier where he designed and made windmills for farmers. While the Holland office had done the early keel work and continued to do the drawings of the more conventional types, there were concepts still to be chased down tunnels of thought before they could be discounted. Karn's was the perfect mind to sit across the table from Farr, as they floated a whole range of ideas from rudimentary winged types to a tri-element thin-winged keel with a huge inverse-tapered bulb. Bruce Farr:

Because of a girth station measurement halfway along the hull, Twelves tradi-tionally had excessive volume where the keel was attached. My thinking was: if we put aft sweep on the keel we could push quite a bit of the keel volume forward of the measurement point. This would not only spread the volume between the keel and hull but also make it easier to meet the girth difference measurement because it was no longer coming into the keel. Which was what we ended up with at the end of our keel research: a well-developed stretch-bulb keel. Most of the other syndicates had forward not aft sweep on their keels. And most didn't have bulbs. So ours — which was quite thin at the top and fat at the bottom and had small high-aspect ratio winglets — reduced drag and got the vertical centre of gravity (vcg) down low. That was an important part of the mix for *KZ-7*.[33]

As was its minimal girth penalty. That saving in rating loss enabled *KZ-7* to carry a bigger rig — a boon in light to moderate airs, though a hindrance in over twenty knots if the area were not reduced.

Another important design feature came directly from the fractionally rigged IOR racing yachts pioneered by Farr. The 'Kiwi yachts were the widest boats by over a foot [and] the extra beam ... was applied in the area of the sheet tracks, well aft. This allowed the sheeting angle of the genoas to be eased outboard by as much as four degrees in heavy weather ... [making for] more efficient sail setting as well as reducing the heeling forces in strong winds.'[34] The wider-than-normal decks gave a better working platform for the Kiwi crews. Changes to the original deck and cockpit layout on *KZ-3* and *KZ-5* were suggested by the sailors and relayed to Annapolis by Roy Dickson.[35]

'Team' now meant more than just the design group. Its net was cast wide to include all those engaged in the creative, management and manufacturing processes associated with the challenge. It included the sailors. It included Richard Fogarty, Project Manager at McMullen & Wing. Working stupendous hours, he had overseen the lofting and building of the wooden male plug for the three fibreglass Twelves before transporting it to Marten Marine in nearby Pakuranga. His company also built the keel moulds and keels and the load-bearing aluminium frames to be fitted inside the hulls once they received the shells back for assembling and finishing.

Nowhere was the team culture more alive than between designers and builders. Such was the compression of time that the team at Marten Marine would not know the details of their working brief until they arrived at work and pasted together the plans and instructions that had arrived overnight by fax from Annapolis. Every day, during the hours their work days overlapped, Steve Marten would speak with either Russ or Bruce, reporting on progress and checking facts.

I recall the day we finished laying up the first boat. It was a mammoth, crazy thing we did. We started off working normally, with morning and afternoon teas. Evening came and we stopped for a meal. We had a coffee break at 10 pm, thinking we were only an hour away from finishing. But that hour just kept on growing and we ended up doing a full eight-hour stint at the end of this sixteen-hour day. Which was when I rang Annapolis. WE'VE DONE IT! WE'VE FINISHED! Sounds good — I'll ring you back, said Russell in that laid-back way of his. What I didn't know until he did was that he'd had a composite expert in his office when I called telling him why it was impossible to build a fibreglass Twelve-Metre![36]

After a distinguished sailing career and fifteen years in banking, Steve Marten had faced his mid-life crisis early by plunging headlong into the fickle world of building racing yachts.[37] With a passion for both the product and the sport, he had quickly made a name for himself. He now had Farr boats like *Gunboat Rangiriri* and the Noelex 30 to his credit. At a time when fibreglass construction was little more than a cottage industry, he had built a company noted for its discipline and structured approach, essential when it came to working with Lloyd's and the exacting process of laying up. First the shop's temperature and humidity had to be set by computer, then every length of dry fibre measured and recorded against the Lloyd's-approved specification. The same had to be done with the resin in which every sheet was soaked. After the glass-fibre sheets had been applied to the hull, the scraps were collected, measured and weighed, and this amount was subtracted from the original weight. Also deducted was the net surplus resin in buckets, on rollers, on the plastic tarpaulins under the mould and on the laminators' gloves, boots, overalls and hats. After all that, if minor variations occurred in the lamination process, the resin-to-glass mix could be quickly adjusted and built back up to the Lloyd's approved weight in James Course's tables.

Course, Lloyd's surveyor, was present at all times, carefully checking figures and procedures, signing his certificates as he worked the same long hours as Marten Marine's staff. The entire process of measuring, weighing and checking was so rigorous that he would later comment: 'We are talking about a degree of weight control far in excess of any GRP marine construction ever built before. We were more than 100 percent satisfied that the requirements were fulfilled ... [that] the laminate was uniform throughout the length of the yacht with no "lightened" ends.'[38]

In fact, the manufacturing process had been so carefully controlled that all three hulls ended up weighing within 15 kg of each other. Given that each hull weighed in excess of 2000 kg, this minuscule difference would have knocked cold Conner's cheating challenge had the facts been revealed in the fiasco that would become known as Glassgate.

But that still lay ahead. For the moment it was enough for 'The Weapon' to receive its champagne kiss from Dame Naomi James before being shipped out of Auckland on the *Jebsen Southland* the following day.[39, 40]

It was July 1986. The calm before the storm.

Interlogue 3

EXT. PORTO CERVO. EARLY MORNING. JUNE 22, 1987.

Four men are walking down a marina arm. The historic town behind them has barely begun to yawn. A haze of silence hangs over the harbour, the seagulls more floating than flying. The sun warms their shoulders, throws shadows on the buildings near the canal. The waterfront is hushed, its bars and cafés closed. Cypress trees stand sentinel in the surrounding hills where walls of bougainvillaea guard colour-washed villas. The Stella Maris Church glows in the early morning light.

The quartet slows, comes to a halt at the bow of a luxury motor yacht. A circle forms. The comic of the group holds court. Some names are mentioned. Burnham. Conner. Ehman. Marshall. Then Burnham again. Laughter erupts.

INT. PORTO CERVO. MAIN CABIN, *SOUTHERLY.*

A brunette pours coffee and tea, fills tumblers with freshly squeezed OJ. The four help themselves to croissants and jam, to the bowls of dates and plums. The syndicate head gets an over-the-shoulder hug. A blush as she kisses his cheek. He smiles his lopsided smile, and begins with those rounded vowels.

Fay: Well, Russ, Bruce, what would you say to working on a Deed Boat challenge?

Bowler looks at Farr, Farr at Bowler.

Bowler/Farr: Yes!

Fay: Good. Good. What Andrew and I especially like is that the process seems a simple one.

Johns: If single-masted, our vessel can be no more than ninety feet on the load waterline or, if two-masted, one hundred and fifteen feet. All we have to do is to provide the name, rig and dimensions — length on load waterline, beam at load waterline, extreme beam and draught — when we submit our Notice of Challenge. Both parties then have ten months to make it to the start line.

Bowler: It's really that simple?

Johns: It's really that simple. It's there in black and white. Here, read it for yourself. Now here's the bare bones of what I've uncovered ...

In less than twenty minutes the lateral-thinking lawyer sums up his weeks of thought and reading.

Fay: It means we can go to today's meetings and it won't matter a toss which way we vote. On anything! On fibreglass, the budget, the venue for the next Twelve-Metre Worlds —

Farr: On fleet or match-racing. On gennakers —

Bowler: On moveable wings.

Fay: We can forget Burnham's stone-walling tactics.

Bowler: And Ehman's technical manoeuvring.

Fay: Suddenly all their B-S is irrelevant. Nothing we've been fighting for at these meetings matters any more. In corporate speak, gentlemen, we're talking takeover time! We've just lined up the Sail America Foundation for International Understanding in our sights. But we'll need something from which to fire our broadsides. And quick.

Johns: There's only one thing we don't know and won't until we've issued our Notice of Challenge. Does the San Diego Yacht Club already have a challenger? Which means we've got to keep this thing tighter than a Swiss timepiece.

Bowler: How, when you have to use a yacht club to issue a challenge?

Farr: Yes. The Royal New Zealand Yacht Squadron has something like two and a half thousand members.

Bowler: Tough keeping that lot quiet!

Johns: Good point, guys. But Michael and I have already worked that through. Toby Morcom runs the Mercury Bay Boating Club out of his clapped-out Zephyr Six!

Fay: It might not be prestigious but it's small.

Johns: And Toby's utterly dependable.

Farr: Sounds good to me.

Bowler: And me.

Fay: So let's have a ball at the Association's meeting today. Let's go and stir the bloody thing up!

Johns: First let me end with a funny story. I'm at Leonardo da Vinci, en route from Kennedy after my America's Cup research in the New York Library and my meetings with our New York attorney, George Tompkins. Suddenly I'm thinking: What if someone finds out I've been in the Big Apple and not in London as I've said?

Farr: So you buy a London daily?

Johns: Right, and read up on the weather. And sure enough, the Kiwi crew who collects me from Albia Airport also picks up a friend of his who's just flown in from London. Question numero uno: How was business in the city, Andrew? I didn't see you on the flight, pipes up his mate. Nor I you, said I acting surprised. But wasn't it unbelievably muggy and warm for London? Oath, he said, never known it so bloody hot!

Laughter.

INT. COSTA SMERALDA YACHT CLUB. MORNING. JUNE 22, 1987.

Four men are walking across the lobby. Thoughts from their early morning meeting sit deliciously on their minds. Inside the conference room they file towards their seats. Gianfranco Alberini, Commodore of Yacht Club Costa Smeralda, looks tidy in his double-breasted gold-buttoned blazer. He extends another welcome to the contestants in the Twelve-Metre Worlds, observers and Association members. Big of smile and big on camaraderie, there's the commerce of hauteur about the Comte. Yesterday, Chairman Henry Anderson had welcomed and addressed those participating in the two days of meetings, offering a special word of thanks to Tom Ehman and the Technical Committee for their extensive work redrafting the Rating Rule.

Ehman and John Marshall, leading lights on this small but influential committee, are also in attendance. Respectively, they are also design coordinator for Stars & Stripes *and Chief Executive Officer of The Sail America Foundation. They have been working hard to have the Twelve-Metre Association ban the use of fibreglass. Their work is not yet done.*

Bowler (aside to Farr): The San Diegans still can't accept we achieved more stiffness for the same number of kilos in our fibreglass hull.

Farr (aside to Bowler): I know. Very untidy logic.

Johns (aside to Bowler): Just who are they trying to fool? Everyone knows the weight of a fibreglass Twelve-Metre is the same as an aluminium one. Our Lloyd's certificate says so!

Bowler (aside to Fay): They need to compare apples with apples. There are huge cost variations in both forms of construction. A carefully built aluminium yacht can cost twice that of a nasty job. And they're still flannelling on about the cost of a hi-tech fibreglass option which I know they're not familiar with.

Fay (aside to Farr and Johns): It's their aspersions that get up my nose.

Farr (aside to Fay and Johns): If we could check the thickness of *Stars & Stripes* plates I think we'd know who's been doing the cheating. The number of times I saw it floating below its legal limit in Perth —

Johns: Point of order, Mr Chairman!

Anderson: Mr Johns.

Johns: Mr Chairman, can we please have clarification as to the costs being discussed here. If you'll excuse the pun, they don't seem to add up. Are Messrs Ehman and Marshall referring to the construction of a Twelve-Metre yacht? Or are they referring to Mr Conner's budget for the Twelve-Metre Worlds?

Laughter.

Johns: And his retinue of seventy!

Anderson: Mr Ehman.

Ehman: We don't like to see our Kiwi friends getting all hot and bothered over the fibreglass issue, Mr Chairman. I for one would be happy to see *KZ-3, 5* and *7* remain in the Association provided, of course, there are no more built in the same material.

Bowler: Gee thanks, Tom!

Johns: I personally consider the banning of fibreglass would be akin to replacing computers with typewriters. Is that the way the Association wants to go, Mr Chairman? Backwards?

Anderson: Thank you Mr Johns. Mr Howlett.

Howlett: Quite frankly, Mr Chairman, I've become rather tired of all this ... this Kiwi-bashing. I recommend a change of heart. Let us as an Association be grateful for the breakthrough the New Zealand designers, engineers and boat-builders have made. Their efforts have done this Class an enormous amount of good, to say nothing of the wonderful competition they provided in Perth. But to the point. As some of you will be aware, fibreglass has been allowed in the Twelve-Metre class since 1980. That would seem to make a nonsense of this motion. And if my memory serves me correctly, every Six Metre built this year has been built not out of aluminium but out of glass-fibre. We should not turn the clock back without overwhelming justification for doing so. Practically every race boat in the world is built of this wonderful material. So why not Twelve-Metre yachts?

Anderson: I'll put the motion to the meeting. Those in favour? Against? I declare the motion lost.

Hurrahs.

Anderson: Mr Marshall.

Marshall: Mr Chairman, I would like to move a motion that the Twelve-Metre Association accepts the introduction of moveable wings into the Twelve-Metre Class.

Gilmour: Don't want to be wing-keeled again, eh John?
Laughter.

Petterson: Mr Chairman, as the information supporting this motion has already been tabled, I would like to —
Anderson: Yes, Mr Petterson, you may.
Petterson: As the meeting is aware, you need a vast amount of money and a shipload of science to develop moveable wings. If this motion is passed it will effectively prevent yacht clubs from all those nations without an aerospace industry from competing for the America's Cup.
Bowler: A what-load of science, Pelle?

Laughter.

Marshall: You guys can't have it both ways.
Ehman: Aren't you Kiwis the ones who advocate progress? Well, this is progress.
Marqueze: Ah oui! This is progress. This motion I support.
Petterson: That's because France has an aerospace industry! For the sake of the future of the Twelve-Metre Class and the America's Cup I ask the meeting not to pass this motion. This is not about sailing, Mr Chairman. This is Star Wars!
Schnackenberg: Pelle's quite right about the industry infrastructure required to build these things. Moveable wings are complicated structures and will be enormously expensive. I'm sitting here with my Bengal Bay hat on wondering who in Australia, Japan or New Zealand will be able to write the programs we'll need to design these things. Frankly I can't think of anyone. What do you think, Michael?
Fay: The wilder the better as far as I'm concerned. The America's Cup has always been the biggest show in town and I don't see anything wrong with that. I think the rest of the world will agree. People don't want to watch safe little boats sailing safe little races. And no one has to play if they don't want to.
Anderson: In which case I will put it to the meeting. Those in favour —?
Farr (aside to Fay): Michael, don't vote for this!
Fay (aside to Farr): Bruce, I can raise more money than all these bastards put together. If some of them drop off as a result of increased costs so much the better. Quick, tell Russ and Andrew to stick their hands up ...

Chapter 23
A Glassgate in a Winebox

We have seen the birth of a new generation of 12 Metres.
I can foresee the day when aluminium is obsolete.
James Course, Lloyd's surveyor[1]

Complicity is not sudden, though it occurs in an instant.
Anne Michaels, Canadian poet and novelist[2]

You put two old rivals on a stage, place a mike between them, and if you've already filled their heads with words like 'tricky' and 'prohibit', and if one is pumped up by a win,[3] the other by a tipple, expect the unexpected: 'The last seventy-eight twelve-metres built around the world have been built in aluminium so why would you build one in fibreglass unless you wanted to cheat?'[4]

That there was less veracity in the question than yeast in a chapatti mattered not a toss. The mixture had been cooking since September and now, on 8 December 1986 as the sparks arced between Conner and Blackaller and the press, it ignited.

But what was it about, this mana for the media, this thing called Glassgate, this bad-assed headache for the Kiwis? Many things, as it turned out.

Fear of obsolescence

Russ Bowler had taken an American invention and used it to do what no one had done before him: engineer a fibreglass Twelve-Metre.[5] That the great American technology dream had been hijacked by some laid-back blokes from Down Under was about as comprehensible to these US sailors as dialectical materialism — and even less palatable. Bruce Farr:

> They simply didn't want to believe that these guys who had never designed an America's Cup boat before sat down one day in 1985 and in six months developed a faster Twelve-Metre than anyone else in the world.[6]

Complicity

Prior to the beginning of the challengers' elimination contest a letter dated 16 September 1986 had been sent by the Sail America Foundation for International Understanding (SAF) to the Challenger of Record.[7] Both that letter and its reply were circulated to all challenge syndicates. The former read: 'We would like to mention two procedures that should be included in the measurement process to ensure that all yachts are legal 12-Metres ... First, we request that the measurer or the Lloyd's surveyor take samples of the keel at various points to determine its specific gravity.'[8] '[Second,] we request that the Lloyd's surveyor take core samples of all composite-construction yachts to ensure that the laminate meets the "as-built" Lloyd's specification.'[9] Being the only composite yacht in the contest, that meant *KZ-7*.

Sitting in BFA's temporary Eastport offices on Fourth Street, Russ Bowler fumed. Being accused of cheating by two Americans who had themselves been censured by the SORC for cheating was almost beyond belief. 'Kill it dead,' he advised the BNZ Challenge. 'If Alberini wants to core *KZ-7*, fine, as long as the aluminium boats get cored too. Then we'll know who's really been doing the cheating!'

Untidy logic

The two main issues covered by SAF's inflammatory letter — a Twelve-Metre's vcg and lengthwise weight distribution — were verified by Lloyd's at the plan-approval and construction-inspection stages. 'By the time the boats got to Perth that part of the measurement had been completed in the boats' documentation'.[10] Knowing this, the Challenger of Record should simply have rejected SAF's requests. Instead, Gianfranco Alberini replied for Yacht Club Costa Smeralda with an obsequiousness that was ominous: 'The importance of the matter is such that ... I am therefore asking the measurement committee to study a system to control the type of material and the specific gravity of the keel at various points during the measurement which will be carried out ... on yachts entering the semi-finals. A similar procedure may be applied for the plastic hull. Should any irregularity be found, the yacht in question will be disqualified and substituted by the runner-up.'[11]

Ignorance

Tom Blackaller: 'The fibreglass construction and the Lloyd's survey is a very shaky thing and we'd like to know if that [*KZ-7*] was built right too ... These are the first [Twelve-Metre] boats that have been raced that are fibreglass and the Lloyd's rules are very, very cloudy. All Dennis wants to know, and I'd like to know too, is whether that boat is lighter in the ends than it is in the middle'.[12]

Not only was this statement ignorant of procedure. It was also reflective of the way Twelve-Metre hulls were usually inspected — at the plan-approval and post-construction stages — and the known tricks of the trade: switching thin for thick plates and grinding others down below the legal thickness after a Lloyd's inspection. Either out of convenience or choice, the Americans had ignored two key facts: (a) that, unlike the building of their own Twelve-Metres, a Lloyd's inspector had been present at all times during *KZ-7*'s construction, and (b) unlike an aluminium hull, one made of fibreglass could not be materially altered after it had been built.

Disbelief

'My own engineers have told me consistently since I've been in this game that you can't build a boat in fibreglass that's light enough and strong enough to compete, because the Lloyd's scantlings prohibit making the boat light enough yet strong enough'.[13] Given that *USA*'s designer, the upbeat Gary Mull[14] — Chairman of the ORC's ITC and designer of successful fibreglass Six-Metres — had failed to get his GRP Twelve-Metre plans approved by Lloyd's, perhaps Blackaller's disbelief was understandable. But therein lay the difference. What those before Russ Bowler had failed to understand was that inattention to detail was a fatal disease. Lloyd's were so scrupulous in their scantling requirements that they were concerned not just with a boat's overall weight and weight distribution but with its spread of panel weight.

Even while penning *Comeback* after he'd won the America's Cup, Dennis Conner's thinking remained borderline histrionic. '[*KZ-7*] was very, very fast and ... was around 800 pounds lighter in the hull than *Stars & Stripes* because she didn't have to be faired off with all the filler we needed to use.'[15] In reality, the fairness of *KZ-7*'s hull might have saved a maximum 100–200 lb of filler with another 50 lb, perhaps, saved through innovative detailing of secondary structures which supported major fittings such as winches and the forestay.

Cultural disconnection

'America is a land of lawyers and lobbyists, and it shows in yachting. New Zealand isn't. Kiwis, particularly the guys building and sailing boats, don't want to know about hearings. They don't want to get dressed up and go to committee meetings with Commodore Alberini. They want to go out and win sailboat races.'[16]

Nor were they interested in riding the Twelve-Metre Rule like bounty hunters. At least not like Dennis Conner when he used multiple rating certificates in the 1983 America's Cup defenders' series. Or as he did in Perth during the post-race weigh-ins. There, if his boat was overweight, he would exchange dry for wet sails or illegally move his crew to the side of the boat opposite the measurer's viewing position to prove that it was floating to the level of its rating.[17]

But cultural disconnection was not just about the mind and muscle arts of lobbying and intimidation. It was about approach: the use of spies during the Worlds and in the pubs and wine bars of Freo and Perth, the extensive use of photogrammetry and, above all, the use of science. Now that Scientific Applications International Corporation (SAIC) and Grumman had made the latest in technology available to SAF, their campaign had leapfrogged from big business and boys sailing boats to war.

Clay Oliver, a graduate and teacher from the US Naval Academy in Annapolis, and latterly a SAIC employee, had been seconded by John Marshall as the performance analyst and VPP specialist for the *Stars & Stripes* programme. He was also responsible for much of SAF's gathering of intelligence.

> I think it was John Marshall who made the mistake of getting Dennis into the plastic boat controversy. A lot of things were initiated by John or through ideas that came from the design technology team he headed. Unfortunately, by the time they got to Dennis they would sometimes come out the wrong way ... The word cheating was unfortunate. It was pretty much invented at the PR table ... and once it took on a life of its own they ran with it. No one thought it was cheating. In Dennis's mind it may have evolved into that. But the rest of us were only interested in why the Kiwi boat was doing so well.[18]

And it was precisely the reason for *KZ-7*'s superiority — established by the SAF design team from their photogrammetric evidence and on-the-water observation — that turned into Glassgate: it had less longitudinal moment of inertia in a seaway than *Stars & Stripes*. It was a known fact that less weight in a yacht's ends improved performance in rough seas. And this was true for *KZ-7*, though not because it was illegally light in the ends as Blackaller and Conner implied. Rather, it was because their aluminium boats required as much as eighty pounds of filler at the stem and stern to make them fair that they had a greater pitching moment at the radius of gyration than *KZ-7*. The Americans, who had totally overlooked the inherent advantage of their own invention, now used it to discredit the New Zealanders, two of whom had realised its potential for a Twelve-Metre yacht long before anyone else.

If the Americans had believed Conner's diatribe they would have issued a formal protest instead of running it in the press. But the more *KZ-7* proved superior on the water, the more convenient Glassgate became. Even the French joined in, complaining that the ultrasonic equipment the Lloyd's surveyor used to measure the four LVC semi-finalists worked only for aluminium and not for GRP.[19] But by then even Alberini had had enough and refused to re-open the Glassgate doors. Fay, though, wasn't finished. He'd asked Lloyd's to drill the hull, and the results matched exactly the recent ultrasonic tests.[20] He offered these — and the name of his New York lawyer — to Admiral Marqueze, the *French Kiss* syndicate head. That might have stopped Glassgate swinging on its brittle hinges. But as it had from the beginning, the obvious continued to be ignored. Steve Marten:

> Drilling holes in *KZ-7*'s hull would not have proved a thing to anyone because in order to make sense of a core sample you had to have the first part of the equation: the thickness it was approved to be. Only Lloyd's knew that. And only someone who had done all the work Lloyd's had could have cleared it as a legal Twelve-Metre. So in the end it came back to the fact that Lloyd's were the only people who could approve *KZ-7*, and they already had.[21]

What Glassgate did show, however, was the lengths to which SAF would go to regain the Cup Conner had lost in 1983. At the end of the LVC semi-finals it seemed inconceivable that New Zealand would not defeat the Americans and become the Royal Perth Yacht Club's official challenger. Before they began their best-of-seven final, *KZ-7* had set an America's Cup record by winning 37 of 38 round-robin races — a phenomenal performance for a first-time challenger. *Stars & Stripes* had won 31.

What, then, went wrong?

There was no 'well-kept secret that, at one time or another, each of the designers walked out'.[22] Nor did 'strikes'[23] or 'bickering'[24] occur among the design team, except in the mind of myth-maker Coutts who was involved neither in the design process nor in the campaign. Dissatisfaction did arise after the contest when claims made for work done that hadn't been done were addressed in the press by Bruce Farr. But the reality was that the four designers had delivered to the BNZ Challenge the fastest, fairest, strongest, best all-round Twelve-Metre in the world. Their reputations as the world's foremost designers of offshore racing yachts, their acuity, creativity and their commitment to the project and each other had seen to that. *KZ-7*'s record-breaking run was the proof.

Laurent Esquier had almost single-handedly turned a group of independent, sometimes wild, small-boat sailors into a sophisticated team that consistently beat the best in the world. Michael Fay had provided direction for the campaign, his charisma its energy. But 'within days of the crew selections being finalised ... [he] decided the time was right to give [Chris] Dickson his head. He withdrew from the dock, and shifted his campaign office to the BNZ Challenge House in Fremantle's High Street, leaving the dock area to the sailors and the technicians ... It was a move that in hindsight ... he should never have made.'[25] One of the most significant contributors of his generation to New Zealand's offshore yachting success, the unflappable Roy Dickson was well suited to liaise between sailors, builders, designers and management as sailing programme coordinator.[26] Yet despite their collective gifts, as it would prove for the sailors, so it would for the managers: at the height of their ascent, the learning curve was just too steep.

Following the delivery of *KZ-7* to Perth, Bruce and Russell spoke to one of the management team.

'Do you want more design support? Would you like us to come to Perth? What can we do?'

'Nothing, thanks. You've done a great job. Thank you very much. We'll take it from here.'

Of the many decisions made by Esquier, Dickson and Fay, that one would be the difference between *New Zealand* and *Stars & Stripes* — the reason, more likely than not, why their tiny country failed to achieve what might have been the sporting triumph of the century: the winning of the America's Cup at a first attempt.

> Bruce and Russell weren't really a part of that inner group there. And those who were, like Professor Jackson, weren't designers. That was a major problem with their programme: they didn't have a yacht designer who was full-time ... It was really a resources problem. A lot of wars have been lost that way; an attack pushes deep into enemy territory with no supply line ... [27]

On the other hand, the SAF design team kept pushing the frontiers of design and technology. They had used NASA-developed riblets to nurse their long heavy hull through the early light air rounds.[28] And they had a stadimeter, the only device of its type in the world which 'rendered range and bearing in digital form at the squeeze of a trigger'.[29] But now they made big changes to their boat to meet the boisterous weather predicted by their weather men. They changed their wings. They changed their ballast. They modified their rudder. And they built a livery of flat-cut jibs and small flat-cut mains.

On the other hand, New Zealand did nothing and went backwards.

By the end of 1986 Davidson and Holland had been dropped out of the loop and both Farr and Bowler were back in Annapolis. But sensing that the gap between *Stars & Stripes* and *KZ-7* was closing fast, Fay flew Farr to Perth. Bruce Farr:

> In ten knots of air we could easily beat *Stars & Stripes* [and held an advantage in the wind range above that]. Between eighteen and twenty knots there wasn't much in it. Above that Dennis may have had a slight edge. So I took the position that we should give up some of our light-air performance; there was no point in beating Dennis by large margins in the light if we weren't as fast as he was in the heavy. The cross-over for evenly matched yachts is around twelve knots; for *KZ-7* and *Stars & Stripes* that was between fifteen and twenty knots. Which meant we were way superior. So I strongly advised to put *KZ-7* into a heavier flotation with bow-down trim. But for some reason those in our camp making the decisions couldn't make that one — to give up their light-air advantage and concentrate on the heavy-air range where the bloke they had to

beat had pitched his boat. So that advice wasn't followed. Likewise with the wings. I asked them to test the long wings before I arrived back at the beginning of January 1987. Apart from the one light-air day when it was marginally slower than *KZ-5*, *KZ-7* proved more powerful and stable in heavy airs with these on.[30] But again, for no accountable reason these were taken off. In the end it was as if the team was too scared to change the magic that had been there from Day One. It was a huge pity because the Australians were dog slow and whoever won the challenger final was going to walk off with the America's Cup.[31]

Nor did it help that the sails the sailors selected for *New Zealand* were too full and powerful for the conditions. Or that those conditions were freakish. 'On only one other occasion [in the previous thirteen years] had the 3.00 pm wind speed off Fremantle exceeded 20 knots on three successive January afternoons'.[32] While all this was true, in the end *KZ-7* was beaten by a boat superbly sailed in the exact conditions it had been designed and optimised for. And given their fatal error in not changing *KZ-7's* configuration before the LVC finals, nothing but unseasonal conditions was ever going to alter the outcome.

But even more than failure to follow the designer's advice, even more than hitting the last mark rounding in the last race, perhaps there was another element of human malfunction that may have been the *coup de grâce* for the New Zealand challenge: a flawed moral imperative.

In 1986 the Bank of New Zealand, Fay Richwhite and Co. and Brierley Investments had each taken a 28 percent share in a quasi-banking operation based in Rarotonga: European Pacific Investments (EPI). Leaving a tantalising sixteen percent of stock available for public consumption, they launched their latest tax 'weapon' on the Auckland Stock Exchange just before the LVC final between *KZ-7* and *Stars & Stripes*. Such was the impeccability of their timing that EPI's euphoria-fuelled share price rocketed from two to forty dollars in a matter of days.

It would be another eight years before the New Zealand public learnt, through a Commission of Inquiry, the extent to which the BNZ Challenge's two main sponsors had been behind a widespread culture of deceit. Had the Winebox Commissioner not applied the wrong legal test to the winebox transactions (choosing to rule on form rather than going to substance), he might not have issued a preposterous finding — 'there is simply no evidence at all of ... fraud ... in any of the winebox transactions'[33] — that made a mockery of the overwhelming evidence of fraud that had emerged during the hearing. But that was hardly surprising. Had the switch not been flicked on the shadow side of Kiwi corporate culture by the 'commissioner's apparent errors of law',[34] and 'erroneous conclusions in his winebox report',[35] the public of New Zealand and those from overseas who had just spent billions buying state-owned assets off the New Zealand government would have been presented with a most fantastical scene: the government's own bank, along with much of corporate New Zealand, defrauding the government's own revenue on a massive scale.

If 'the detail of fraud is not so much interesting as incomprehensible to the non-criminal mind',[36] it was the farthest thing from the minds of those lining Auckland's Queen Street to welcome back their America's Cup heroes and the man who had done so much for New Zealand yachting and its international profile but whose companies had also helped members of corporate New Zealand abscond with vast sums of money destined for the public purse. But there'd be no *mea culpa* from Michael. Nor would there even be a sense of disquiet among some of the thinkers in the throng until the pieces of the puzzle fell into place. Then, ten years later, it would slowly dawn on some of them that just as yachting was beginning its rite of passage from amateurism to professionalism, their country was in the process of losing both its innocence and its moral heart.

Chapter 24
Hegemony

They are both very intelligent men, very good mathematically and in knowing what makes a boat go fast through the water ... One of the important things about them is that they are never closed off. They will talk and listen and then incorporate whatever's worthwhile into the new design ...

Einar Koefoed, Norwegian lawyer and yachtsman[1]

He makes his designs with his whole heart. His designs come from his character. He has very good character, and never gives up even when he has problems.

Harunobu Takeda, Japanese businessman and yachtsman[2]

By now Bruce and Russell were awash with work. It was as if they'd caught the quarter wave of their own designs and were on a perpetual plane. And having been champion yachtsmen, they knew instinctively how to drive their business and when to back off before they busted gear or did themselves some damage. Bruce Farr & Associates Inc was now not only a high-profile designer involved in high profile programmes like the Whitbread and America's Cup. It was also achieving successes in regattas all around the world. In the process it was building an increasingly multi-national client base. The move to Annapolis was finally paying off.

Harunobu Takeda had maintained contact with Bruce from the time they had worked the 1974 Tokyo Boat Show together. An importer of raw materials for the boat-building industry, Takeda was also a Kiwiphile and an enthusiastic promoter of Farr designs in Japan. He would eventually become chief director of the Japanese syndicate for the twenty-eighth America's Cup and order *Bengal V* from the Farr office for pre-Cup training on the Fifty-footer circuit.

Kaoru Ogimi first saw the Farr 727 design during the 1979 Quarter Ton World Championship sailed out of Sajima Bay, the venue for the 1964 Tokyo Olympics yachting events.[3] Two years earlier he had been successful in getting the ORC to reschedule the Far East's first level-rating Worlds from Singapore to Tokyo. It was the beginning of offshore yacht racing in Japan and, in hindsight, where the origins of the first Nippon challenge for the America's Cup lay. Sailing keel-boats like dinghies in ocean conditions was an eye-opener for Ogimi, using three-quarter rigs and crew as ballast revolutionary concepts for a Japanese sailor of heavy-displacement yachts.

> Bruce's designs captured the imagination of young sailors. They helped break the stuffy Anglo-American stranglehold on the sport and led to a break away from the influence of money and social status in offshore yachting. But while these ideas provided the initial breakthrough, they weren't his main contribution. Bruce was so obviously dedicated to questioning every norm, every standard, that he has gone on obtaining performance from everywhere it can be obtained.[4]

Kaoru Ogimi was so impressed with the mix of artistic and empirical in Farr designs that he ordered a 55-foot cruiser-racer from the Annapolis office in 1984, a commission handled through American Warwick Tompkins. By the time Marten Marine began building *Nakiri Daio* (*King of the Sea*) it had become the recipient of the technology and ideas that were flowing through BFA into the Whitbread and America's Cup programmes: composite technology, the ring-frame concept, weight-saving structures of Divinycell, Kevlar and carbon fibre.[5] The flush-decked *Nakiri Daio* may not have been designed as a race boat or to make gain within the IOR. But it performed to such high levels both on the wind and reaching that with 'Commodore' Tompkins and Kaoru Ogimi as co-skippers it won the inaugural 5500-mile Melbourne-to-Osaka Race in March 1987 by almost a day.[6] As proof of its pedigree and with New Zealander Ross Field in charge, it would win the race again in 1991, breaking its own record by over three days. And if it hadn't arrived 40 hours late for the start after rolling and losing its mast in Bass Strait, it would have won again in 1995.

***

It was through the tragic death of her flatmate that Mary Radcliffe met Geoff Stagg in 1984. It would be another four years before she would begin employment at Farr International — a year after brokerage and project-management had been added to its portfolio of activities — and another two before she would marry the confirmed bachelor and hard-driving yachtsman in June 1990. But whenever she could manage time away from her nursing duties, Mary would travel with Geoff as he sold custom Farr designs to grand prix clients or sailed them on the racetracks of the world. It was while they were in Brazil with a client, Erling Lorentzen, brother-in-law of H.R.H. Crown Prince Harald, that they learnt that bid papers were to be sent to leading yacht designers for the next *Fram*.[7] Those arranging the design selection and building of the yacht had been impressed with Lorentzen's Farr, *Saga IV*. They had been even more taken with the Spanish Navy's *Sirius II*, which Captain Juan Carlos Rodriguez-Toubes had steered to second in the 1986 One Ton Cup. The Norwegians had finished last in that event and were keen to procure a quick One Tonner for their gifted royal helmsman[8]. *Fram X* was to be a fiftieth birthday present for the Crown Prince as well as Norwegian industry's 'thank you' for his support. They knew how much he enjoyed sailing, how it gave him breathing space and a place where he could be treated as an equal and judged on his own ability. But a boat for H.R.H. was a delicate matter. Even more so given Norway's proud history of yacht design and that the last *Fram* had been designed and built in Norway. That, however, didn't stop a charging Geoffrey Stagg seizing the moment as soon as it presented itself.

> Fax to: Einar L. Koefoed
> From: Geoff Stagg Date: 24 October 1986
> Thank you for your interesting telex of 21/10/86 regarding Fram X for H.R.H. Crown Prince Harold. I do hope that you are not using us against Ivan Ambler, Judel/Vrolijk, and X Boats to beat their price down as we are desperately keen to have H.R.H. and your gang in one of our boats ...
> Good luck in your deliberations this weekend. Mary has her fingers crossed. Regardless of your choice, I would appreciate you advising us of your decision as soon as it is made.
> Best regards to your group,
> Geoffrey Stagg
> Vice President[9]

Having sent his opening price, two days later (a Sunday) he was on the phone longer than the Wichita lineman reassuring Oslo not only of his organisation's keen interest in the project but their impeccable design and sailing credentials and their connection with ('in my humble estimation') the best race boat-builders in the world. Einar Koefoed, lawyer and sailing manager for H.R.H. Harald, advised that they were only one of a number of groups who had received bid packages from the *Fram X* project board, and that Stagg should not delay with the balance of his reply.

Facsimiles followed from Farr International Inc on October 30 and 31, the latter transmission containing eight pages of construction details and fittings from Russ Bowler plus his explanation on the proposed Kevlar-Nomex and Divinycell construction.

Significantly, the quotes Geoff Stagg squeezed out of his design associates and TP Cookson, the Auckland boat-builders he'd selected for the contract, produced a total price not only among the lowest of the seven bidders but also a delivered product that was cheaper than an equivalent boat designed and built in Norway.

Sooner than expected — on November 4 — Stagg had his reply. He couldn't disguise his delight.

> Facsimile to: Einar Koefoed and Stein Foyen
> From: Geoffrey Stagg
> Date: November 4, 1986
> I have only just come down to earth. Please pass our sincere thanks to H.R.H. Crown Prince Harald for choosing the Farr/Ambler design. It will be a decision that he will not regret.
> Bruce is back from Perth, but has come down with a severe dose of the flu so we have been unable to schedule his trip to Oslo, although it will be some time around the 18th of November ...
> Cookson is delighted to be building the boat and assures me it will be the best he can do.
> I thank you and Stein personally for your help in assisting us win this very prestigious commission. Be assured of my best in making this a world beater. I would hate to face you guys in court.
> Mary has her bags packed.
> Regards,
> Geoff Stagg[10]

If Koefoed's reply was tongue-in-cheek, it was also prophetic.

> Fax to: Bruce Farr & Associates
> Attn: Mr Geoff Stagg
> Date: 5 November 1986
> Re: FRAM X
> I thank you for your telefax of 4 November and [am] pleased to note that I have received your unconditional guarantee that the boat to be built will be a world beater ...
> The contract was no doubt won by your good documentation work, and the fact that your total price package was very competitive. We trust that this will be the level of work and service you will provide also in the future ...

I have talked to H.R.H. today who stresses a meeting in Norway on 13 November because of other commitments. Could you not give Bruce a dose of anti-biotics?
Best regards,
Einar L. Koefoed[11]

The Farr International team was an unlikely combination — the studious Farr, the businessman Bowler, the gung-ho Stagg — but somehow it worked. And while there was a lot swinging on this pared-back contract, Bruce and Russ were as quietly confident as Geoffrey was effusive. Design 185 — the next *Fram* — would be the latest in a growing line of successful Farr 40s that traced their family history from Design 136 (*Pacific Sundance, Geronimo, Exador*) to Design 138 (*Sirius II* and *Swuzzlebubble*) to Design 158 (*Highland Fling* and *Bodacious*) and Design 168 (*Sirius V*). Add Bruce's refinements to its lines and rig, Russell's structural gains in weight and strength, and Cookson's growing expertise in building composite boats, and the Farr design and selling teams were more than confident that the boat scheduled to leave Auckland on 17 May 1987 — Norwegian Day — would be a winner. Mick Cookson:

> We were really the first combination of designer boat-builder programme from where you could order a custom race boat on the other side of the world, get it off the ship, put the rig in, tweak it and go sailing the next day at grand prix level. *Fram X* was the first New Zealand-built keel-boat exported directly into Europe.[12]

As it turned out, the *Benton's* departure was delayed, and its precious cargo arrived in Oslo a month later than expected. That left only six weeks to prepare for the One Ton Cup in Kiel. But with Mick Cookson and Geoff Stagg there to watch *Fram X* being unloaded, to tidy it up and motor it around to the Royal Yacht Club for the glittering welcome party, the new owners would soon learn how to drive it. It was an historic occasion. Not just for the Norwegians, but also for the New Zealanders.

Designing for the world's largest production yacht builder had been on Bruce Farr's mind ever since the ORC's political skulduggery had made it clear that self-interest was of greater interest to those who ran the sport than the sport itself. And although he had made contact with Chantiers Beneteau[13] as early as 1979, it wasn't until the French company began work on Design 198[14] for Bruno Troublé[15] that the first official introduction was made. By then — when Bruce and Russell sat down with Francois Chalain at the Annapolis Boat Show in October '87 — they were the rising stars of yachting's grand prix circuit with an engineering reputation to match. At last there was as much desirable about the Farr organisation for Beneteau as there was about Beneteau for Farr.

It was one of those projects that looked improbable at the start but might just produce a product that was even more than the sum of its high-profile parts. So removed from Kiwi culture that the *Rainbow Warrior* might have been a musical, Francois Chalain, the Beneteau Number Two, was nevertheless both passionate about sailing and a man of vision. For their first project together he suggested the inclusion of Pininfarina, the Italian design company noted for 60 years of head-turning cars like the Ferrari Testarossa. It was time to alter yacht design; time, he said, to place emphasis on style instead of being led by 'le nez': form following function. So that December, after signing a letter of agreement with Beneteau and drawing the lines of the first design, Bruce sat down with Pininfarina's Lorenzo Ramaciotta at the Paris Boat Show. After blending race-boat features such as mast-mounted readouts and deck control-line covers with a drop-down transom and modular interior, Russ took over the structural composition. Although Bruce became concerned that the styling might be too extreme for the nautical fraternity, he needn't have worried. The magical combination of the two design names and the sheer size of Beneteau's international dealer network meant things were fizzing before the stylish First 45F5 was unveiled at the Paris Boat Show in December 1989.

The risk and investment of time had been huge. The tooling alone took a year. But in the end it was worth it: forty First 45F5s were sold on the first day of the show, eighty by its end, and a total of one hundred were on the books by the end of the following month. It meant Beneteau had to dedicate most of one of its three European plants — with its five assembly lines and eco-friendly production systems — to producing the First 45F5. At eighteen per month, that meant there was one of these fast cruisers dropping off a production line nearly every two days. Bruce Farr:

> It was a huge job because the opinions of each of the three parties had to be carefully considered, making more work for everyone. That, however, ended up drawing us closer together as we three very different nationalities pulled together to produce a consensus on every aspect of the project.[16]

As spectacular as it was, the First 45F5 was just the beginning of European production yacht designing. Not only would Bruce Farr & Associates design all of Beneteau's First range, from 40 to 45 feet. They would also sign an agreement with Jeanneau, in July 1988, for Design 205: the spacious Sun Odyssey 51, with its five separate cabins.

Bruce and Russell were now in the process of realising the dream that had drawn them from New Zealand. Over the next six years, Beneteau would go on to produce 579 Farr-designed yachts including 236 First 45F5s,

191 Oceanis 440s, 33 Moorings 445s, 37 53F5s, 71 42F7s and 10 of their flag-ship: the Beneteau 64. To come were the First IMS 40.7, the Beneteau 50 (75 produced since 1996), as well as a number of Farr custom racing yachts and the one-design Mumm 36 for the European market. This combined with Jeanneau product meant that twenty million US dollars worth of ex-factory Farr-Bowler product was being sold per annum in France alone. Other production deals were looming. And, following on from the success of the pretty Farr 10[20], McDell Marine was now producing its new 40-footer, the Farr 12[20].

The irony, sweet as it was, was that by the time Bruce and Russell had the security of royalty income from what amounted to a significant share of the world's production yacht market, their race-boat income had become so important that they could have based their business around either. Wisely, they chose both. Each was of value to the other, the relationship circular, the lessons from the racetrack feeding into their fast cruiser designs even as the royalty income contributed to the ongoing cost of research.

***

> *By the late 1980s, [Bruce] Farr [& Associates] had beyond question become the dominant race designer ... [with] a lion's share of the market and ... boats ... winning more of the big prizes than those of other designers. Farr under-pinned his designs with a huge amount of research and sheer effort, but there is no doubt that he enjoyed the twin benefits of having Geoff Stagg ... feed[ing] back informa-tion from the race course, and of having so many boats sailing that each design change could be judged in what amounted to full-scale, real-life tank testing.*
>
> Timothy Jeffery, English yachting writer[17]

The idea of the Admiral's Cup was floated over Plymouth Gin. With the America's Cup not having been raced since 1937, and ocean rather than day racing more attractive to the RORC's Admiral, Sir Myles Wyatt, and its Commodore, Peter Green, a challenge by way of a letter was sent to American yachtsmen care of the CCA in 1956. 'It is the intention of certain British Owners to offer a Challenge Cup to be competed for between not less than three and not more than five privately owned yachts drawn from each side.'[18] It was proposed that the series, concluding with the Fastnet Race,[19] should be conducted under RORC rating and Royal Yachting Association (RYA) racing rules. An ornate 1830s lidded bowl was purchased, regilded and set on a sturdy wooden base. Donated through Wyatt's office of Club Admiral, it was to be known as the Admiral's Cup.

By 1975, the year of New Zealand's first challenge, the number of countries competing had increased every two-year regatta until there were nineteen national teams. Having finished fifth at their first attempt, fifth out of sixteen in 1981, sixth out of fifteen in 1983, and third out of eighteen in 1985, the Kiwis began a meticulous preparation for their 1987 campaign almost before they'd decamped at Cowes.

Because of the composite rating rule introduced after the 1985 regatta 'requiring the collective rating of each three-boat team to total 95 ft, it [was] expected that each country would enter one boat in the 34.4 ft range'[20] along with two One Tonners rating near the minimum 30 ft. It was the 'big boat' berth that Peter Walker was aiming for when he announced in August 1986 that he was heading a partnership to campaign a Farr 44-footer. A member of the four previous Admiral's Cup teams, and with no other 'big boat' in sight, Walker, now practising as an architect in Wellington, was odds-on favourite to make the New Zealand team. He would call the boat *Kiwi*.

When family friend Peter Tatham suggested he and Adrian Burr sponsor an Admiral's Cupper, dentist Bevan Woolley thought long and hard about who should design it.[21]

> It came down to Laurie or Bruce, and Bruce got the nod because he had a better track record over all. My friend Del Hogg and I ordered two identical boats from Bruce for a very modest fee at the time. But the choice wasn't easy. It never has been in this country because we've had a host of local designers who have dared to be different, who have dared to challenge. They've made our yachtsmen. They've designed boats that have given us not just the edge in boat speed but a psychological edge as well, something that's important when you come from a country as small as ours. And that's what happened with *Propaganda*.[22] Anyone could have sailed it. It was such an incredible boat once we got it going.[23]

However, during New Zealand's Admiral's Cup trials it looked as if Woolley, Burr and Tatham had made a substantial error of judgement. *Goldcorp* (ex *Mad Max*), the 1985 Davidson design now owned by Mal Canning, left the latest Farr One Tonners in its wake and with its handicap advantage kept the bigger *Kiwi*[24] honest.

Woolley, however, wasn't fazed. He'd cruised and raced Farrs since 1977 and regarded his Farr 38 as 'a boat fifteen years ahead of its time'.[25] The problem, he soon discovered, was the 40-footer's keel. Because the RNZYS had set a one-ton limit on their team's two smaller boats, they had had to remove 200 mm from *Propaganda*'s keel in order to make it rate. That had proved a performance disaster. But the replacement keel — built by crewman Jeremy Scantlebury — 'must have been the fairest keel in history because the boat was just superb after that'.[26]

But there was more than just attention to detail to this fifth New Zealand challenge. There had been, as Keith Chapman recalled, a critical shift in management's thinking: '[They] structured a team approach to the whole campaign, then carefully selected the right people and let us get on with our jobs'.[27] Everyone agreed: this time around the underlying culture should be 'the team'. The cult of 'top boat' would be left at the finish line of the last of the trials, the individual's wishes bent to the needs of the 36-man team. In addition, professionalism born of America's Cupping was finding its way back home from Perth. Not just by way of the Twelve-Metre sailors who, now recovered from their exhausting campaign, had joined the Admiral's Cup crews. But also by way of preparation. They practised on the Waitemata for twenty days before leaving for the UK 'and then had another twelve days sailing in British waters, with video analysis to bring crew technique and sail shapes to race-readiness. They even sailed an overnight practice race in the Channel, something few teams had contemplated before.'[28]

Forty-eight boats from sixteen teams arrived at Cowes for the 1987 five-race series. They came from Australia, Austria, Belgium, Britain, Denmark, France, Germany, Holland, Ireland, Italy, New Zealand, Spain, Sweden and the USA.

The first 24-mile race was sailed around marks in the Solent, the New Zealand boats drifting into sixth over all in the lifeless airs. But things were to change as the wind came in at fifteen knots for the start of the Channel Race. In the lively conditions around the 215-mile course, *Propaganda* finished first, *Goldcorp* sixth and *Kiwi* nineteenth. The team-first policy had paid out its first big dividend as the Kiwis leapt from sixth to second on the teams' points table. In the first inshore race, a 28-miler sailed on Christchurch Bay, *Propaganda* had its second win of the series. Not only was it 'starting to establish [itself] as something a little special',[29] with *Kiwi* third and *Goldcorp* tenth, New Zealand was now only one point behind the leading British team. After a lay day, the New Zealanders turned in another quality team performance in the second race on Christchurch Bay. The same couldn't be said of Britain. Sailing as individuals, with an almost insouciant air, they failed to cover the Kiwis. *Kiwi*'s fifth, *Propaganda*'s sixth and *Goldcorp*'s eighth scooped a massive 220 points to put the New Zealanders 111 points in the clear. For the first time in twelve years it seemed the Admiral's Cup was within the little country's grasp, provided, of course, they could keep Britain contained during the Fastnet finale. The Big One carried five times the points and could, as it had in the past, be a team's undoing. But not this time. Like Hawkins and Drake shadowing the Armada, the New Zealand boats clamped a cover on the British team right from the beginning. *Kiwi* clung to *Indulgence*, eventually beating it by 25 minutes. *Propaganda* duelled with *Jamarella* all the way home,[30] finishing just 65 seconds ahead of the British Farr after 605 miles of gripping ocean racing. And *Goldcorp*'s storming run back to Plymouth after rounding the bleak rock in twenty-seventh place meant it finished only 41 corrected minutes behind its opposite, *Juno*. And although the British turned in the best team performance of the Fastnet, with corrected time finishes giving *Propaganda* fourth, *Goldcorp* eleventh and *Kiwi* twentieth, New Zealand, 84 points clear of Britain, had at last achieved its goal: an Admiral's Cup win.[31]

If it was a big win for New Zealand, it was an even bigger one for the Annapolitan Kiwis. *Propaganda* and *Jamarella* (Design 184) had finished first and second in the 42-boat field, with *Kiwi* and the Australian Farr *Swan Premium II* also among the top ten finishers. It was heady stuff, but the season was just beginning.

Sailed out of Kiel three weeks later, the One Ton Cup attracted 34 entrants. Although not one of the pre-tournament favourites, *Fram X* had the pundits thinking after it finished second to the Spanish Navy's Farr One Tonner in Race One's fluky airs. In the next heat — the 308-mile ocean race sailed in a wide range of conditions — it said as much of the crew and royal helmsman as it did of the boat that the Norwegians finished 30 minutes ahead of *Sirius IV*, their more fancied Spanish rival. They finished second in the third, fifth in the fourth, and ended up on the wrong side of a major wind shift on the first beat of the fifth and final heat. Rounding that first weather mark in seventeenth place, it seemed a foregone conclusion that the Spanish would take the title, so far out in front was *Sirius IV*. The Norwegians, however, refused to follow the pack, setting out instead on what they thought would be a sounder, safer course. And it soon proved more than that. 'The only ... [yacht] in the fleet able to set a spinnaker',[32] *Fram X* sailed to leeward of the seventeen boats ahead of it, all pointing higher than the bottom mark, and stormed home to finish well clear of *Sirius IV*. If they were seen by some as the lottery winners in a foggy, flukey race, they had nevertheless been remarkably consistent throughout the fickle-aired series sailed on the Baltic. The win, of course, was a thrill for H.R.H., his crew and country, as it was for Bruce, Russ, Geoff and Mick and Terry Cookson.[33] And with the Spanish sailing into second, the Annapolis team saw their now famous Farr 40 once more finishing one-two in an international series, as well as having four of their designs among the top ten finishers.[34]

For one design office, these results seemed unbelievable. But as good as it was, BFA was only just hitting its straps.

The following year, *Propaganda* — now owned by Tim Bailey, Michael Fay and David Richwhite — became only the second boat in One Ton Cup history to win four races. That had not happened since Chris Bouzaid's *Wai-Aniwa* won the trophy in 1972. But *Propaganda*'s four-out-of-five race win in the 1988 event was arguably more comprehensive given the result of the long ocean race. Set up for winds that funnel down the bay past Alcatraz, the Farr 40 was struggling as the airs turned light out in the channel. Whether what followed was due to Rick Dodson's magic on the helm, their imported crew[35] or the unusually friendly breeze that blew through their new cobalt rigging would always be open to debate. But there was nothing moot about the finish. In fifth place and level-pegging with *Sagacious*, when the sun went down they not only lost sight of the Aussie Farr but the rest of the fleet as well. It was around midnight when *Propaganda* arrived back in San Francisco Bay, in a real peasouper. Taking a course close to the shore to keep out of the fast-ebbing tide, they weren't sure where they were until the Golden Gate Bridge suddenly appeared overhead. Nor did they know just how far in front of the fleet they were until they woke the next morning. Pulling back their motel curtains, they saw second-placed *Sagacious* finishing just over eight hours behind them.

> ... you're full of shit, now get lost. You're a loser, this is a winner's stage, get off the stage.[36]

Conner's sneering put-down of Farr during the final press conference of the twenty-seventh America's Cup said as much for the big man's mind as it did of his mouth. He couldn't have got it more wrong. Among the top ten of the 24 entries contesting the San Francisco One Ton Worlds that month were eight Farrs. One design group having its designs finish 1-2-3-4-5-6-8-10 in the world's premier level-rating event was an unparalleled achievement.[37] But if DC had lost the plot, DB had found a way to get more mileage for its-beer. 'BETTER BY FARR, AND THAT'S NOT JUST PROPAGANDA!' crowed Dominion Breweries' Wellington advertisement. They had it right. *Seahorse*, the RORC's official magazine, was about to vote the stunning Farr its yacht of the decade.

The pattern that had been established in 87 and 88 repeated itself in 89. For the third year in a row, a Farr-Bowler design won the One Ton Cup. Sailed out of Naples, the premier Ton trophy went to Italy's *Brava*, owned by Pasquale Landolfi and helmed by Bruno Finzi. The tried-and-true Design 185, also built by Cookson's, had done it yet again.

Thrilled with the success of his One Tonner *Jamarella*, Alan Gray was back for more, asking BFA to design a 'big boat' for the 1989 Admiral's Cup. Basing it on their successful new breed of Fifty-footers, Design 213, also called *Jamarella*, was the most consistent boat of the series. Its 1/3/2/3/2/4 result placed it first on the individual points table and helped the British team to its ninth Admiral's Cup win. Finishing in second place overall was the Cookson-built *Will* (Design 211), a 50-footer owned by Ryouji Oda of Japan. *Librah* (Design 216), a new Farr 44 owned by Michael Fay and David Richwhite, was placed third. While the New Zealand team hadn't fared so well — finishing third behind Britain and Denmark — the results for their compatriots in Annapolis were even better than before: twenty of the 42 1989 Admiral's Cup entrants had come from their office, with five of those finishing in the top ten and another six in the top twenty.

1989 was also the year the Farr designs began making waves in the big boat scene, formerly the preserve of Frers and Holland. The 50-footer circuit had recently grown out of a corner of the IOR when a group of wealthy owners formed an association to race, without handicaps, outside the ORC's control. The championship meetings held over a season — four days of intense racing held in exotic locations around the world — combined yacht racing with lifestyle. *Carat VII* (Design 203) — built by Cookson and owned by Wictor Forss of Sweden — won the first 50-Foot Association World Cup trophy in 1989 from *Windquest* (Design 206), owned by Michigan's Richard DeVoss.[38] And domination of the offshore race-boat circuit was made complete when Italian Gianni Varasi's *Longobarda* (Design 207) won the 1989 International Class A Yacht Association (ICAYA) Maxi World Championship. After only two weeks of sea trials, it went on to substantial wins in each of the three five-race regattas.

It now seemed an appropriate time for Bruce Farr & Associates, Farr International and their expanded team to move from their temporary premises in Fourth Street and Yacht Haven back under one roof — their own.[39] In September 1989 they moved into the three-storeyed 9000-square-foot building Bowler, Farr, Stagg and several other investors had had built on the corner of Third Street and Eastern Avenue.[40] Overlooking Back Creek was the perfect setting for the team of 'Kiwi Magnificos'[41] to prepare for the decade ahead. They had every reason to feel confident. Their designs had not only scooped the premier racetrack prizes. They were now so much accepted by the local community that as well as the half-models of their winners hanging on the entrance lobby walls, they now had a number of honours and awards.[42]

1. Dated 6th March 1987, a citation from the Governor of Maryland to Bruce Farr & Associates: '... in recognition of the skill, technical expertise and professionalism your company demonstrated in designing the hull of New Zealand's KIWI MAGIC, and as an expression of our appreciation for your significant contribution to Maryland's international reputation for excellence in sailing technology ...'

2. Dated 15th November 1988, and awarded from the State of Maryland to Bruce Farr & Associates Inc the title of ADMIRAL OF THE CHESAPEAKE BAY.

3. Dated 15th November 1988, the Distinguished Business Award from the City of Annapolis to Bruce Farr & Associates: '... in recognition of your many contributions to the sailing profession, the marine industry and the City of Annapolis ...'

4. Also in 1988, the first-ever Distinguished Business Award was awarded

from the Marine Trades Association of Maryland Inc to Bruce Farr & Associates Inc. Said its president: 'Mr Farr has managed to lift the Annapolis area back into the front ranks of boat design — not only through his work on ... New Zealand but also through the plethora of boat designs he has developed since locating here.'[43]

How things had changed in seven short years.

Interlogue 4

INT. YACHT CLUB COSTA SMERALDA. AFTERNOON. JUNE 22, 1987.

Malin Burnham is pleased with proceedings. Pleased to see the technology advantage moving away from New Zealand and back towards the USA. Where it belongs. Still, he's holding his cards closer to his chest than Kenny Rogers' gambler.

Star World Champion at seventeen and long time sailing buddy of Dennis Conner, it was Malin Burnham who dreamed up the Sail America Foundation for International Understanding (SAF/Sail Amercia) in 1985 and became its founding president. With the America's Cup now equal parts sailing and business, they had needed a vehicle from which to lasso corporates to fund their future campaigns. And Burnham knows big business. For over two decades he has been expanding the family real estate business founded by his grandfather. He knows the importance of muscle, the nuance of manipulation. He also knows there's no point in raising money if you can't control it. It was his idea that SAF become the financial steward of the America's Cup should his buddy bring it back from Perth. In return for the marketing rights for the next defence they would pay the bills and put up those troublesome guarantees for the San Diego Yacht Club (SDYC). It worked like a dream — until the silver ewer hit West Coast USA and arguments between SAF and the Cup's custodian erupted like mud pools round a crater. It wasn't only disagreements over the next regatta's venue or where the Cup was kept that sent the arrangement slipping into arbitration. It had slowly been dawning on the SDYC that Sail America's understanding of proprietary rights was keener than their own. What was the stewardship of the Cup compared with ownership of the right to stage its future events? Or control of the cash from sponsorship and television rights? That's when the lights came on for the SDYC members. Malin Burnham had just invented a minting machine for his Sail America Foundation. The only problem was he'd housed it behind their respectable yacht club doors.[1]

Doug Alford, SDYC's Vice Commodore, sits among the meeting's observers watching Burnham like a hawk. He's not sure that his attendance or his club's recent press release denouncing the commercialisation of the America's

Cup bothers Burnham all that much. Eventually the home-town boys will have to settle their differences and unite against a common foe. It's Farr and Bowler, he suspects, who bother Burnham more. Rumour has it that it was they who drew and engineered the fibreglass Twelve-Metre causing all the trouble. He knows Burnham doesn't much like their banker mate either, the one with the silver tongue and all that new money. Or his fresh-faced lobbying lawyer. Or the way the four of them talk and laugh together between motions put and carried.

The Chairman opens up the meeting for general discussion, suggesting that the next America's Cup be scheduled to avoid any conflict with the next Olympic Games.

Pajot: Monsieur Anderson, this is the question I must ask to Malin Burnham. When is it that the San Diego Yacht Club will hold the next America's Cup?

Burnham: When?

Pajot: Yes. This I like to know.

Fay/Bowler/Johns/Farr: Hear! Hear!

Pajot: This WE would like to know!

Burnham: Marc, it's not that easy. It's not a simple matter.

Pajot: Not simple? Not easy? It is many months since Dennis he win for you the Cup!

Howlett: Mr Chairman, if I may —

Anderson: Mr Howlett.

Howlett: Sir, I recall it always being a straightforward matter for the New York Yacht Club — the holder of the America's Cup for almost its entire history — to set the date and venue of the next match. In fact we'd be told the details before we left Newport. So it seems unusual that for no apparent reason this tradition has been dispensed with. One wonders if all is well between the current holder and the agent one hears it has contracted to run the next defence. Perhaps Mr Burnham, as president of that agency, would be good enough to advise us of the current position regarding this arrangement — yet another departure from Cup tradition, one would have to say.

Burnham: You can say it all you want.

Howlett: I think once is enough, Mr Chairman.

Laughter.

Burnham: Mr Chairman, I'd like to welcome Mr Howlett to the real world. Everyone knows those halcyon days in Newport have long since gone. Nor is it a simple matter today to work through and make the numerous decisions associated with a defence. For one thing, it is you designers,

Mr Howlett, who are saying that San Diego may not be the best venue for the heavy-displacement Twelves. So there's your first question: Do we modify the design or do we use a different venue? Is it Hawaii or a taller rig?

Johns: Stop kicking for touch, Malin! You know that's not the point.

Burnham: Matching boats to venue is very much the point.

Bowler: Ours seems to be doing okay in the light Sardinia airs!

Laughter.

Bowler: And we haven't re-configured it.

Anderson: Gentlemen!

Burnham: Don't misunderstand me. Our America's Cup Committee is in the process of being formed and we will be doing our best to finalise the venue and the dates just as soon as we can.

Fischer: It's been five friggin months since Freo, Malin. How many more do you want?

Burnham: At least another three, Syd. In the meantime I would ask you all to please be patient.

Howlett: Mr Chairman, I think the interested parties — and that's most of us here — have been exceedingly patient for quite some time.

Burnham: I assure the meeting that we are doing our damnedest to get the show on the road. But the bottom line is, gentleman, we hold the Cup and we will do what we think most fitting for this time-honoured contest. Why, given our light air conditions in San Diego we might even defend in a Maxi boat. We think they're great.

EXT. PORTO CERVO. DAY. JUNE 23, 1987.

The villa-dotted landscape, once the dust bowl of Sardinia, is now playground for the rich and famous. Centre of the Costa Smeralda resort area on the island's north-east coast, it is also backdrop to the Twelve-Metre Worlds. Many who competed in the last America's Cup are back for more. But here there's a difference. Here the sea drips an iridescent blue as the predominant light sea-breezes only whip it to an occasional Gage Roads frenzy when they turn offshore.

With different conditions and sponsors from Perth, there have been changes to liveries, wardrobes and wings for most of the eight teams competing in the Worlds: Dennis Conner's Stars & Stripes, Kookaburra II, *Syd Fischer's* Steak 'n' Kidney, *the renamed British boat* White Horse, *the former* Australia III *now* Bengal II, *Pelle Petterson's* Entertainer 12 *and the home-town favourite* Sfida Italiana. *But none of this has happened to KZ-7, still in its Perth trim and colours of white and black. However, with the airs consistently lighter than anticipated — eight knots or less — it has been something of a battle for skipper Barnes and his crew as they have fought their way through to the finals.*

But tough though it's been, it has already been repayment for their loss in Perth. Even after discarding 1400 kg (3000 lb) of lead and having smaller wings attached to Stars & Stripes, *and despite his rumoured million-dollar budget for the regatta, Conner not only suffered the ignominy of an eight-minute defeat by* Steak 'n' Kidney*', he also failed to win a race in his three semi-finals.*

EXT. COSTA SMERALDA SEASCAPE. DAY. JUNE 23, 1987.

The smart money on the revamped Bengal II *is looking even smarter as the easterly drops to just six knots at the start of the three-race final. But when Australian skipper Beashel misjudges his time-on-distance run to the start Barnes shuts him out. Forced to gybe away and head for the committee boat end of the line,* Bengal *is barely moving as the guns sounds. KZ-7 hits the line at speed, leaping to a twenty second lead before Beashel can get his lead-mine moving again. Game-set-match.*

EXT. COSTA SMERALDA SEASCAPE. DAY. JUNE 24, 1987.

There's a sudden burst of activity aboard KZ-7 as it heads towards the race track. Laurent Esquier has just radioed news of an approaching front from the Kiwi tender. A twenty knot westerly will be just perfect for their overweight boat. Pumped up, with sails changed, the Kiwis are now sure they can take the Worlds two-zip. But this time it is Barnes who makes the pre-start error. And although they sail a faultless race, KZ-7's two-minute win is all for nothing. The international jury confirms Barnes luffed past head-to-wind forcing Bengal II *to tack away.*

EXT. COSTA SMERALDA SEASCAPE. DAY. JUNE 25, 1987.

With the wind at seven knots and falling, no one but a small band of hopefuls thinks the Kiwi boat can win. And with Bengal II *almost a minute ahead by the halfway mark, the race seems all but over for KZ-7. But these are fluky airs and Barnes, tactician Davis and their crew work the four-knot shifts to perfection, putting the Magic back in Kiwi. Suddenly they are right on* Bengal II*'s tail as they charge towards the last turning mark. As both boats converge on port Barnes gybes to avoid a collision. Then the inexplicable: Beashel carries the Kiwis past the buoy. Way past. In fact they're already twenty-five seconds past when Davis unfurls his protest flag. And even though they fail to break the Japanese cover on that last windward leg, the New*

Zealanders are quietly confident that they have both the last race and the series in the bag.

EXT. PORTO CERVO. EVENING. JUNE 24, 1987.

The committee room lights burn long into the night. Five hours later the jury announces its decision. Bengal *infringed Rule 38.2 at the bottom mark.* KZ-7 *is the new world champion!*

EXT. PORTO CERVO PIAZZA. DAY. JUNE 25, 1987.

Amidst scenes of jubilation David Barnes steps forward to receive the winner's trophy from the Aga Khan.

Australian voice: Mate, how can you be happy with that? It's like winning a
    test with a penalty try!
New Zealand crew: Forget the protests. We'll take two-one on the water!

All that remains are the victory celebrations before the Kiwi crew head for home and the Johnses, Fays and Schnackenbergs fly by Lear to London. There there will be a night at the Phantom of the Opera *before the wives depart for Auckland and the husbands for Washington DC to meet with Farr and Bowler.*

Chapter 25
Immaculate Conception

It was the 1988 event that changed the America's Cup for ever ... The New Zealanders knocked out all that old technology, knocked out that strong base of experience the Americans had had for so long, and they did it in one fell swoop.

Clay Oliver, American naval architect[1]

You don't think of why you can't; you think of why you wouldn't.

Andrew Johns, Auckland barrister and solicitor[2]

The Problem
M. Burnham, D. Conner and The Sail America Foundation for International Understanding (SAF).

Ever since Conner had turned the America's Cup from the gentleman's 'summer in Newport' into a sponsor-fest with a two-year build-up, things hadn't been the same. That its intensity had increased with his loss in 1983 was to be expected. What wasn't expected were the off-the-water tactics of the organisation with the anserine name that Malin Burnham had formed and which, in September 1985, had cut that deal with the San Diego Yacht Club (SDYC). 'A bad case of seller's remorse,'[3] Burnham chortled when SDYC later took umbrage at his unilateral behaviour after his buddy returned Cup-in-hand from Perth. But while might seemed right to Burnham, he was playing both with history and the laws of trusts as he moved to turn the America's Cup from a challenger's into a defender's event.

The Solution
NEW ZEALAND — KZ-1
Designer: Bruce Farr & Associates

LOA: 132.8 feet (40.5 metres)
LWL: 90 feet (27.4 metres)
Beam: 26 feet (7.9 metres)
Waterline beam: 14 feet (4.3 metres)
Draught: 21 feet (6.4 metres)
Displacement: 83,000 pounds (37,648 kg)
Mast height: 153.5 feet (46.8 metres)
Sail area: 17,300 square feet (1607 square metres)

The concept
It was while Bowler, Fay, Farr, Johns and Schnackenberg were closeted in the Ritz-Carlton Hotel, Washington DC, that the following conversation took place. It was logically sound but failed to second-guess — at least by the end — the opposition's thinking. It began at 2130 hours on Sunday evening, July 12, 1987.

Fay: Under the Deed of Gift we have two choices in terms of maximum waterline length. Bruce, which would be the fastest: a ninety-foot sloop or the hundred-and-fifteen foot ketch?

Farr: Neither!

Fay: Neither?

Farr: The fastest wind-powered vessel would be a multihull. That would definitely give us our best chance of success.

Fay: Bruce, we couldn't do that! That's not the America's Cup!

Johns: And nor is it what the Cup's donors contemplated. It is clear from the Deed of Gift that the challenger is required to give the dimensions of their boat to the defender so that the defender can build a boat to match it. The competition would neither be friendly nor have any meaning if the defender used the challenger's dimensions to build a vastly superior vessel.

Fay: That wouldn't be a match. That would be a mismatch!

Johns: It's also legally safer for us to challenge in a monohull. I'm sure the San Diego Yacht Club, citing the Deed of gift, would reject a multihull challenge on legal grounds.

Fay: And if they try to defend in one, I'll take them on in the press!

Johns: Imagine if Michael challenged in a multihull — he'd be mutton back in New Zealand!

It was while this conversation bounced around the hotel room that Bruce sketched in a matter of minutes what would become *KZ-1/New Zealand*. There would be few changes from this rudimentary drawing to the real thing.

I had a notebook in which I was taking notes and I just kind of drew this thing out. And no surprise, it looked like a European Lake Boat because it had the same problem to solve: being the fastest single-hull vessel for a given size.

That meant quite long overhangs which would enable it to get longer when sailing. It also meant big downwind sails plus a bowsprit to make for ease of handling those sails. The difference, of course, was in the dimensions — with a maximum waterline length of ninety feet and a maximum draught of twenty-two feet this was enormous![4]

> Fay: Bruce, that's the biggest 18-footer in the world!
> Farr: Or the biggest Lake Boat!

The next day Bruce, Russ and Tom returned to Annapolis and for the next 48 hours devoted themselves to fine-tuning the Big Boat's dimensions on their in-house VPP.

By the end of the third day, July 15, they had established the dimensions required for a challenge under the Deed of Gift (rig; length on load waterline; beam at load waterline; extreme beam; draught) and despatched them by facsimile to the Beverly Wilshire Hotel, LA. Andrew Johns completed the certificate — adding these dimensions to the name and rig — which Michael Fay signed. The same day, Wednesday, Peter Debreceny arrived in LA with the signatures of the Mercury Bay Boating Club's Commodore and committee member on the challenge documents Johns had faxed to Auckland before leaving Washington. The following day they phoned Fred Frye, Commodore of the San Diego Yacht Club (SDYC), and arranged to meet him and Vice Commodore Doug Alford for lunch on Friday, July 17.

'When Michael Fay awoke that [Friday] morning ... he was surprised to see a darkened sky, with driving rain and low clouds racing across the horizon ... "Hardly a day for flying," he muttered to himself, as he crossed the suite to the room of Andrew Johns ... to ask him to cancel their flight to San Diego and instead to arrange a car and driver. Promptly at 8.30 am, with the challenge in hand, they departed for the San Diego Yacht Club'.[5] It was while they were waiting for their hosts that they noticed the silver tray awarded to *KZ-7* for the challenger's series in Perth. Delighted as they were to find the missing trophy in the lavish yacht club cabinet, the discovery was to prove a portent of things to come.

It was another month before Fay and Johns were satisfied that the SDYC had accepted no prior challenge and that Bruce Farr & Associates could be instructed to proceed with the design.

The team
Concept: Russell Bowler, Bruce Farr, Tom Schnackenberg
Design: Russell Bowler, Bruce Farr
Design coordinator: Russell Bowler
Design team: Russell Bowler, Mike Drummond, Bruce Farr, Richard Honey, Richard Karn

Electronics: Richard Morris
Builder: Steve Marten
Facilitators: John Wade, Peter Walker
Sail design: John Clinton, Tom Schnackenberg
Strategy: Russell Bowler, Laurent Esquier, Bruce Farr, Michael Fay, Andrew
    Johns, Tom Schnackenberg
Skipper: David Barnes
Tactician: Peter Lester
Sailing master: Rod Davis

Bruce and Russell immediately began their consultative process by arranging for a meeting in Annapolis with the design and sailing teams. The 'cast of thousands' spent time refining the boat's concepts, including the deck layout, interior and rig. That left Bruce two weeks to absorb the many suggestions and complete the hull design process.

> We had only been using computer modelling for little over a year when we began the Big Boat project. In many ways the same skills are used — developing shapes, plotting them out — as when you are draw a standard set of lines by hand. But with the computer you avoid the tedious process of erasing and redrawing by simply changing the 3-D shape on the screen. It also enabled me to move the design into manufacturing much faster because the intricate modelling within the actual lines plans — the big radiuses around the sheer and the clipping of the hull and deck surfaces together — could be done much faster.[6]

Once again Russ Bowler had a lot of ground to break as the project's engineer: no one before him had engineered a hull this big or a mast this tall (and thin: 3 mm in places) solely from exotics, or attached a 21-foot keel to a boat with a fourteen-foot beam. Like *KZ-7*, it was another world first. It was also a tribute to his gifts that the experimental structure stayed together — apart from the thrice-breaking ('It's only a piece of engineering!') bowsprit.

The time-frame
Time — or lack of it — was the programme's biggest constraint. From the date of the Notice of Challenge there were only ten months in which to design, build, test and ship the world's largest exotic sloop. And by the time the programme proper got under way that had shrunk to nine, making even more surreal this design and engineering feat. Nor did Bruce and Russell have Gustave Eiffel's luxury of time to plod through the complexities of the concept or its structure. Pure logic and their accumulated knowledge was all they had to go on. There was no time to invent new technology, only to use

and push to the limit that which already existed. Bruce did not have time for tank-testing, Russ insufficient hours to complete the engineering before construction began at Marten Marine after the full-sized lines drawings arrived there on 21 September 1987.

The structure
Like the law, engineering relies on precedent. But with nothing like this ever built before there would be no comfort zone for Russell to crawl back into. The closest were the J-Boats, with their similar dimensions. But their masses and materials were completely different — their wood, galvanised steel, Egyptian cotton, Italian hemp a far cry from *New Zealand's* hull of Nomex honeycomb coated with a wafer-thin skin of T800 carbon fibre. Just 35 mm thick, it would be unable to carry the massive loads without another Bowler alloy frame fitted inside the shell. Designed to weigh just a quarter of *Ranger*, the last great J-Boat built 50 years before, *KZ-1/New Zealand* would be light years ahead of its predecessor.[7] It would even make the three fibreglass Twelve-Metres, built in the same yard just two years before, look like yachting fossils. Steve Marten:

> Depending on how you compare it — twice the length, three times the volume, one hundred and fifty percent of the surface area — *KZ-1* was in excess of twice the size of *KZ-7* but almost identical in weight. And that included the mast, hull and keel![8]

Construction
'Steve?'

'Russ?'

'Morning.'

'Evening.'

'We're about to sign a contract with Dalts for a Whitbread Maxi. We'd like you to build *Fisher & Paykel*. Plus we've got this long skinny one.'

'Tell me more.'

'Can't say much right now. But keep some space free. It could take up quite a bit!'

When Marten received the dimensions, they seemed almost beyond belief. So did the project.

> The whole programme was one of wonder and amazement. We worked on it for six months — with security guards at the door as we signed ourselves in and out — not being able to talk to anyone about this boat of extraordinary proportions we were creating. It was just a wonderful time. Every day we'd walk in and look at this thing knowing that no one in the world knew what we were doing.[9]

As much as the support that kept arriving down the phone and fax lines from Annapolis, as much as the high-tech systems and controls they had refined since building the fibreglass Twelves, it was the people employed by Marten Marine who enabled this technological marvel to be built in just 24 weeks. Not surprisingly, it was for his staff and those in management who supported them that Steve Marten reserved his warmest praise.

> We ended up working shifts 24 hours a day. The speed at which we built it was phenomenal. There was a huge group of people — between 40 and 60 throughout the period and all of them working very long hours. Some were averaging 80 hours a week, which meant the only time they had left was for sleeping. And we worked them so hard anyway they had little energy left for their social life. Many of them were yachtsmen and it was a real sacrifice for them to give up their activities on the water. But to their credit, to a man, they did it.[10]

Centre of Gravity (cg)

With a waterline beam little more than ten percent of its length and nine percent of its height, *New Zealand* looked about as stable as a minority government dropping in the polls. Scaled down to twelve feet it would have been difficult to sit on. Its cg would therefore have to be much lower than a standard racing yacht if it was to be stable enough to carry its huge sail area and 40-person crew.

That meant a keel, replete with trim tab, which prevented *New Zealand* from sailing in two-thirds of the Waitemata. Solid aluminium at the top, hollow aluminium for three-quarters of its length, the keel's tip was a nineteen-ton lead casting. Not surprisingly, it had a lower-than-normal grounding speed, with a crew member assigned full-time to checking the water depth below it.

> It was an on-the-limit structure which required some degree of maintenance to keep it on! One of the interesting things was that you could feel the reaction of the keel quite independently of the boat. With its long flat ends the hull would bang and bounce as it leapt off a wave. A second later you'd feel the keel catching up and it would actually kick the boat around quite significantly.[11]

Its marginal waterline beam also meant a lightweight rig. At fifteen storeys high it would have to be if *New Zealand*'s low cg were to be maintained. It was here that the Farr-Bowler small-boat experience — as well as Chris Mitchell's computer program — proved invaluable, as they worked on scaling laws to determine the loads. They also had to assume that they would be up against the best the US could produce and therefore couldn't over-design it with conservative engineering.

They had three choices:
(a) a conventional aluminium spar with soft fabric sails;
(b) a carbon fibre spar with/without fairings onto the sails;
(c) a rotating wing mast.

Leaving Chris Mitchell, Richard Karn and Peter Jackson from Auckland University's Yacht Research Institute to investigate option (c), Bruce and Russell spent time on BFA's computer assessing more conventional mast options. With early indications showing that (b) would be half the weight of (a), (a) was dropped. The aerodynamic advantages of (c) made it the preferred option. But the time needed to design and engineer such a structure — as well as the negative contribution of its weight to the boat's cg — were against it. With (b) now preferred, it was hoped that fairings from the mast onto the sail would go some way towards the aerodynamic properties of a wing mast. They did, but gaining only a few seconds per mile the sailors would eventually discard them. To compensate, extra time was spent on the giant spreaders,[12] with each one given 'a different angle and ... shape along its length to present the lowest profile to the wind.'[13] With Bowler and Honey responsible for engineering the laminates and Mitchell the engineering detail, two conventional masts were built, the second an improvement on the first and some kilos lighter. With the tube weight of around 600 kg and a total weight with rigging and spreader of approximately 1300 kg, the 153-foot mast weighed substantially less than a Maxi's, although it was half as tall again.[14] Such was the experimental nature of the boat that Bruce and Russ had 48 gauges[15] fitted to the rig and hull to measure strain. The safety of the crew was paramount and this information, fed back to the on-board computers, would help provide it.

To help lower the cg further, all the winches were placed below deck. With eight crew to a winch, the cross-connected coffee grinders packed enormous power, hoisting the sails at astonishing speed and trimming the massive gennakers[16] with the same speed and efficiency as those on a much smaller boat.

From the time *New Zealand* was barged to Auckland's naval dockyard in the face of cyclone Bola, and launched two weeks later, on 27 March, before 130,00 people, it was a boat destined for headlines. Like *America* before it, in the year the world's great nations were displaying their wealth and power at the Great Exhibition, *New Zealand* was also symbolic of a small nation's entrance on the international stage. For New Zealanders — even Americans — to be embarrassed, shocked or angered by the audacity of its challenge was to miss its technological significance and its pivotal place in history. For far from turning the world's oldest sporting trophy into a 'tarnished cup',[17] *KZ-1/New Zealand* was about to save it from those who would suck every last silver dollar from its beaten silver lips.

The Response

It was only with hindsight that Andrew Johns could say he and Michael Fay had 'had a wild time trying to keep San Diego on the straight and level'.[18] Despite his brilliant and typically Kiwi answer to San Diego's intransigence, what he, and Fay, had failed to comprehend was that their Deed Boat challenge would be up against a powerful group of amoralists engaged in lie-based propaganda. Even in the America's Cup you play by the rules, and the rules were in the Deed. So thought Johns and Fay. Not so thought Malin Burnham; he made his own rules. Besides, for him there were only three important people in his charter of aggrandisement: I, me, myself. Six, if he included the Flesh Made Word. 'Sportsmanship is non-existent, because professional sports — money and sports — and sportsmanship do not go hand in hand. An Australian friend of mine who plays the horses has a saying: "Bet on self-interest — it's always running." I'll tell you one thing, there has never been any sportsmanship in the America's Cup. Anyone who thinks so is kidding himself. Check the history books.'[19]

In the heat of the moment it was easy to be seduced by Conner's non seq. logic, to accept his boorishness as normative. While the America's Cup had always involved large amounts of money, until Conner came along it had never been about financial gain at the expense of the principles on which its trust had been founded: friendly competition. And if you read the history Conner now espoused it taught that the Cup could never be controlled, although it might control you.

But for the moment SAF so controlled the event it now 'owned' that it left the SDYC anchored on the fiscal equivalent of a lee shore. And just how fast Conner's self-interest was running in the new race for the cup can be seen in the contract purportedly entered into in 1988 between SAF and C&M Partners (a Conner-Marshall partnership). As US attorney and former NYCC point man understood it: '[under this arrangement the partners were to receive] a monthly stipend of $75,000, payable for a period of four years, for their services in organizing and carrying out the catamaran defense. It was well into 1992,' Michael suggests, 'when that four years ended and Dennis Conner paid John Marshall his $37,500 share of the final monthly payment. By then, they had collectively received $3,600,000, and the San Diego Yacht Club was looking at creditors with unpaid bills for the America's Cup totalling $3,500,000'.[20]

But even before the depth of greed was known, it was clear that no amount of persuasion was going to turn this tide. The Deed Boat challenge had already raised a bruise on the fragile psyche of the world's most powerful nation. And it was into no-man's-land — that place between rational legal argument and nationalism — that the SAF kept planting its public relations landmines. New Zealanders challenging for the America's Cup in a Farr-Bowler boat were akin to Japanese in Zeros heading for you-know-where. Suddenly the Mayors of New York City and San Diego and the New

York Attorney-General were endorsing Malin Burnham's plans for controlling the Cup while simultaneously linking New Zealand's anti-nuclear policy with 'sea wolves in lambs' clothing'.[21] SAF's executive Vice President, Tom Ehman, didn't mince words: 'We're going to jimmy the rules to win this thing.'[22] Conner joined the patriotic frothing at the mouth with talk of 'sneak attacks' and 'legalised ambush'.[23] With a total abrogation of their duties as trustees of the America's Cup they parleyed their greed into a cause célèbre by linking national honour and security with the defence of a trophy that fortuitously bore the same name as their country. However the inexplicable is explained, SAF's public projection of self on 'the other' plagued the US psyche like a double dose of dengue until it ran amok in the New York judiciary.

The Law

The court finds that the intent of the donor, as expressed in the Deed of Gift, was to exclude a defense of the America's Cup in a multihulled vessel by a defender faced with a monohull challenge. The challenge provision would be rendered meaningless if the defender was provided with the specifications of the challenging vessel and then afforded ten months to produce a vessel with an insurmountable competitive advantage. To sail a multihulled vessel against a monohulled yacht ... is ... to create a gross mismatch and, therefore, is violative of the donor's primary purpose of fostering friendly competition ...

Therefore, whether the court limits its inquiry to the trust instrument or accepts extrinsic evidence, it is clear that a catamaran may not defend in America's Cup competition against a monohull. Accordingly, San Diego shall be disqualified in the September 1988 competition ...

San Diego was well aware of the risk it ran when it chose to follow the unprecedented course of defending in a catamaran. Barely paying lip service to the significance of the competition, its clear goal was to retain the Cup at all costs so that it could host a competition on its own terms. San Diego thus violated the spirit of the Deed.[24]

Not only had Carmen Beauchamp Ciparick delivered her decision with a clarity of logic. She had pricked the pride of all those involved in SAF. Ten days after delivering her judgement on 28 March 1989, the bespectacled New York judge entered an order pursuant to her decision awarding the America's Cup to the Mercury Bay Boating Club (MBBC), ordering that the trophy be handed over to the challenger within 30 days.

Neither Ciparick's decision nor the fact that New Zealand had won the America's Cup at only its second attempt surprised the yachting world. Despite San Diego having access to its country's aerospace technology, as well as its best naval architects and boat-builders, Johns, Farr, Bowler and

Marten had so out-thought, out-designed and out-built them that the Cup-holders knew they couldn't match the Kiwis boat for boat. That, they thought, left them with only one option if they were to mount a successful defence: to cheat, and hope that as their policy grew legs their country would run along with them. Clay Oliver:

> I got a phone call during Thanksgiving Dinner 1987 from John Marshall. We all had to meet at La Guardia, at the airport, in several days' time. There we sat and decided the quickest way to knock this thing off was a multihull ... The catamaran was initially chosen because it would clearly beat any monohull and would be significantly cheaper. And ... at the time of our first meeting we didn't consider we had time to properly design and build a monohull.[25]

That SAF had resorted to low-life tactics didn't shock the international sailing community; Conner had made an art form of it.[26] What did was the ruling of the Appellate Division of the New York Supreme Court handed down on 19 September 1989. Responding to San Diego's appeal of Judge Ciparick's decision, the Appellate's majority opinion was not only 'written with a grim determination that no foreign yacht club was going to take the America's Cup away from an American Yacht Club, come what may.'[27] It had 'lifted all but a minor portion of its passages substantially verbatim from the briefs of the prevailing parties'.[28]

How could this be? How could Ciparick's logic have been rendered null and void by a 4-1 decision from the Appellate Division?

As defendant-appellant San Diego took the position that if the Deed of Gift didn't say you couldn't, you could. It was thinking so simplistic it would have failed the Playdough course at preschool. In the process San Diego eschewed the Deed's intent, its wording and its history — except where they could twist it to suit their own ends. To wit:

(a) Just because the Deed of Gift stated that the America's Cup shall be 'perpetually a Challenge Cup for friendly competition between foreign countries'[29] did not mean that the defender had to be either friendly or accept that the America's Cup was a challenger's cup. In fact, unbeknown to Andrew Johns or George Tompkins,[30] and prior to MBBC instituting legal proceedings requiring San Diego to accept its July 15 challenge, 'SDYC and SAF had been preparing ... [a] petition to amend the Deed of Gift',[31] 'a petition [which] also asked for a declaration that ... [the SDYC] was not required to accept the New Zealand challenge'.[32] If it had not been for MBBC's happenstance filing four days earlier,[33] no one, including the New York Yacht Club (NYYC), would have had an opportunity to challenge San Diego's amendments and 'the America's Cup would no longer ... [have been] a challenge cup, but rather one dominated and controlled by the defending club'.[34]

(b) If the Deed of Gift does not expressly exclude multihulls, a catamaran can be used in an America's Cup defence, despite one never having been used in its 137-year history and in spite of George Schuyler's strenuous efforts when writing the Third Deed of Gift to ensure that the defender would meet the specifications of the challenger's monohull.[35] Thus, for the SDYC's counsel, an express urging of the Deed of Gift was precatory not obligatory, its major tenet an irrelevancy, its words old ink on dusty pages.

The plaintiff-respondent (MBBC) used as the basis of its case the Deed of Gift in its legal, historical and intentional settings. Prepared by Andrew Johns and George Tompkins and argued by Tompkins, it stated that SDYC's use of a catamaran did not produce 'a match' as intended by the Cup's donors and was a decision 'designed not to compete but to guarantee a win at all costs against a challenger that had no potential to win'.[36] Not only was there no historical precedent for using a multihull in an America's Cup defence; its use was a blatant violation of the donors' wishes.

Such was the potential embarrassment of this decision to the state's judicial system and its danger to future America's Cup regattas that San Francisco attorney James Michael decided that the concerns he and many of yachting's illuminati had needed to be made known to the court before MBBC's appeal was heard.[37] The way to do that was through an amici curiae ('friends of the court') brief. Michael's list was impressive.[38] It included every American skipper (except Briggs Cunningham and Dennis Conner) and every living foreign skipper (except Australian Iain Murray) who had competed in the America's Cup between 1958 and 1987. Together they stated their profound disagreement with the Appellate Division's decision, endorsing both MBBC's position and Judge Ciparick's decision.

But their opinions were to have little effect on the five judges of the New York Court of Appeals who heard MBBC's appeal of this decision. Russ Bowler saw it coming.

> I remember going on the day the parties were making their presentations. And the Chief Judge, Sol Wachtler, asked MBBC's counsel: Where in the Deed of Gift does it say you cannot race a catamaran? You could tell from that it was all over.[39]

And so it was, on April 26 1990, that the Farr-Bowler boat was voted out of history by the New York Court of Appeal's 5-2 decision. By failing to rule on the fiduciary duties of the trustee of a valid trust within its jurisdiction, the highest court in New York State had not only found in favour of the SDYC, it had completely missed the point. 'It was San Diego's conduct as a trustee which was the conduct in issue [not Mercury Bay's as the challenger]. Examining that conduct it is inconceivable that San Diego should ... [have been] exonerated from violating its trust in meeting a monohull with a catamaran for the express purpose of turning the event into a mismatch and aborting a challenge judicially held to be valid.'[40] So flawed was the majority's decision that it deliberately distorted MBBC's description of a mismatch 'to mean the exact opposite of what was actually said'.[41] And when it came to the friendly competition between foreign countries [phrase] 'the majority, like SDYC, ... [sought] to put Mercury Bay in a bad light for doing what the Deed contemplated it could do'.[42]

As for the amici's opinion, it whizzed right over the head of the appeal court's presiding judge. After stating that the 'case has little or no significance for the law',[43] Chief Judge Wachtler concluded that 'it must be the contestants, not the courts, who define the traditions and ideals of the sport.'[44] If Sol seemed oddly out of touch, America would soon know why. In November 1992 he 'was arrested by the FBI and charged with blackmail and extortion against his ex-mistress ... A few days after his arrest ... [he] resigned from the Court of Appeals [and] later ... pleaded not guilty to the criminal charges by reason of insanity'.[45] Ironically, it would be the lucid Carmen Ciparick who would soon ascend to the bench of the New York Court of Appeals.

Resolution

It was ironic that for all the invective it produced, by soundly beating the USA in the design and technology race this extraordinary boat had shown the way ahead. A quantum leap from *KZ-7*, it screamed the obvious: Twelve-Metre yachts had had their day. What was needed was not a modified version of the outmoded heavy-displacement design but a millennium machine like *New Zealand*, a new breed of yacht that embraced the latest technology.

Farr and Bowler had not only produced the fastest displacement monohull ever, a hugely experimental design with very few failures. They were also making the play onshore. It was during the week of the racing that they organised and paid for a conference which they asked Bill Ficker to chair. Their aim was to get designers thinking about a new America's Cup class, talking to each other instead of the press, reaching a politics-free consensus, one that was best for the sport. In the belief that rating rules detracted from the art and science of design by producing loopholes, Bruce and Russell advocated a class with simple rules. They recommended large dimensions to retain the allure of the America's Cup. But they also suggested a limited trade-off between sail area and length to allow for minor adaptations to light and heavy air after the boats had been built. That, they believed, would make them much closer in design and produce a more level playing field, one in which the sailors would exert the greater influence on the outcome of the competition. And while the San Diego meeting did not stay completely free of politics, and the new International America's Cup Class (IACC) rule[46] was ultimately produced by a committee of five elected from syndicates meeting in Southampton, it was still the critical turning point for the grand event.[47]

Despite the bitter controversy surrounding its campaign, *KZ-1* had shown Farr and Bowler at their ground-breaking best. The giant sloop had side-stepped Stephens, out-winged Lexcen and even out-glassed Bowler, providing, in the process, an impetus for a brand new type of boat to sail the America's Cup into the twenty-first century. Yet despite its significance and the energy it absorbed, the Big Boat challenge was still not as important to offshore yacht design and the Farr-Bowler business as the research and development programme it had just interrupted.

PART III

Chapter 26
Clean Sweep

We had decoy sail plans. We had sail plans on the wall.
We were getting quotes from sailmakers for sloops of this
size. We were doing everything possible to spread misin-
formation, and not really telling lies, exactly, but pretty
close to it sometimes. And so what if it helps us win the
race.

Peter Blake, skipper, *Steinlager 2*[1]

The difference in the two yachts is clearly illustrated by
the two captains. Peter Blake MBE, 41, has raced in every
Whitbread so far and won none. Grant Dalton, 32 ...
won as a crew member on Flyer *in 1982. Blake ... would*
get a gold medal at the PR Olympics. Dalton, who some
regard as the better sailor, wouldn't.

Steven O'Meagher, New Zealand journalist, reporting on
the 1989–90 Whitbread during the Auckland stopover[2]

It's the victors who write the histories, the spinners with the gold who make
the myths to sell their products who have the final say.

'New Zealanders are never going to forget this. It's like Hillary climbing
Everest. To win the Whitbread and win all six legs is one of New Zealand's
greatest sporting achievements. Blake and his crew will go down in
history.'[3]

Given that the country's largest brewer 'claimed to have made an estimated $60 million in free advertising ... for an outlay of only 20 percent of
that amount',[4] it was little wonder that Lion Nathan's chief executive had
dollar signs for eyeballs. Not only was his a crass over-simplification — four
other skippers had won the race before Peter Blake, and Edmund Hillary
had neither the massive backing of a brewer nor the comfort of a hi-tech
structure from which to escape the elements. It also conveniently failed to

comment on the even greater Kiwi achievement in this the fifth Whitbread: Farr-Bowler designs had finished one-two-three. If Butterworth's crash-gybe had not saved *Steinlager 2*'s rig on the sixth and final leg another Farr-Bowler boat would have won. And if *Fisher & Paykel*'s mizzen mast had fallen over the side a second time, another Farr-Bowler Maxi was close behind waiting to claim first prize. Peter Blake's win after seventeen years and 400,000 ocean miles of trying was indeed meritorious. But if it was like Hillary climbing Everest, then BFA's results were the equivalent of putting the entire *Steinlager* crew on the moon before Neil Armstrong. As they had for the last two decades, Bruce Farr and Russell Bowler had provided New Zealand sailors with the fastest, safest yachts with which to achieve dominance on the oceans of the world. But like Ernest Rutherford, their achievements went largely unheralded in the country their designs had done so much to promote. The pair may have been possessed of a formidable intelligence quotient. They may now have assumed the mantle of the world's best race-boat designers. They may have both been vastly superior helmsmen to both Kiwi ketch skippers. But none of that seemed to matter. At a time when free-market mania was manifest as mana, when New Zealand was looking for an icon in a landscape other than the mountains, only one person had the genes to fit the ad-man's bill.

Although Blake had tried and failed four times to win the Whitbread,[5] suddenly this high-profile yachting failure had become 'a man destined to go down in history as the 20th Century's supreme sporting seafarer.'[6] 'He doesn't have fantastic sailing skills, or fantastic technical skills', said his navigator.[7] But placed behind a wheel in a mountainous sea, he was an archetypal Viking the corporates' quick-thinking ad-men could sail across the nation's TV screens for many years to come. And the corporates' full-blown hero was right there with them, blowing more spin on a self-serving video in what should have been a straightforward telling of a great Whitbread race.

> What sailmakers had designed sails for the back part of this boat? Nobody. In fact, some of the sails that the sailmakers wanted to give us, that Bruce Farr had designed, we didn't even ... allow them to make because I knew from my experience it was a waste of time. The sails we got them to make were the sails that we wanted, that they said won't work, were the reason we won the race.[8]

It is hard to comprehend why anyone would want to denigrate the designer who had given them a winning boat. It is even harder to understand why they should do so by way of sophistry purveyed to an unsuspecting public. For not only is Blake's statement a simplistic nonsense. Bruce Farr had not even designed a single sail for the corporeal boat. The closest he'd got was to assume some sail measurements for his VPP studies, without which he would have been unable to 'sail' Design 190 around the world in his computer. The Farr design team had also detailed maximum dimensions on *Steinlager 2*'s sail plans in order to establish the maximum rated area for the mizzen staysails, 'free' sail area Bruce had discovered for his clients while studying ketch rigs within the IOR.

> There was a lot of discussion on sizes and dimensions of various staysails, on where to tack and sheet them, and, it's fair to say, we were a part of Peter's 'team' that developed those ideas. We were also involved in changes to the mizzen rig, including the decision made by 'the team' in March 1989, to make it taller. At their request I sailed on *Steinlager 2* in Auckland during the development period. I made various observations and suggestions, some of which, like the sheeting of the headsails around the wide chain-plates, were enthusiastically accepted. Working with them as a team, we were also involved in other aspects of their sails and I naturally made comment when they asked for my advice. But to say they didn't allow the sailmakers to make the sails I had designed is simply untrue. I/we did not design their sails; we gave them starting dimensions which they went away and developed. Where they did do well was in the development of their big mizzen staysails; they were spinnaker-like in shape and differed from the norm. But not even they were the reason they won the race. Their win, a fantastic one, was far more complex than that.[9]

***

> *When you boil it down, if you don't go to Bruce and Russell because you'll have the same as everyone else and go elsewhere, then there's a very high chance you'll end up slower than everyone else.[10]*

For Grant Dalton it was like choosing between freeze-dried or fresh, the Concorde or a 747. Others, like Davidson, Frers, Nelson/Marek, were good, damned good. But Farr and Bowler were the best in the world, their business the complete one-stop shop. They had cutting-edge flair, their own VPP, a tank-test programme, and they combined a proven record with outstanding engineering. The first business call Dalton made after his white-goods sponsor confirmed its backing was to Annapolis. And 1988 proved just how correct his 1987 decision had been. Farr designs had won the America's Cup, the One Ton Cup, the Kenwood Cup, the Japan Cup and the North American One Ton Championship.

It wasn't even hard for Peter Blake. Four failed attempts and a blood nose from *Lion* meant he'd have to do a complete one-eighty if he were to have any chance of success in what looked like his last Whitbread race.

Gone for the moment were his weasel words about others 'getting the advantage of our thinking'.[11] Fortunately for him, all this had flown right by Bruce and Russell. They welcomed him into their spacious Eastport offices as they had Skip Novak, Grant Dalton and Pierre Fehlmann. It would become an irony of yachting history that Farr and Bowler's professional integrity, maligned by Blake and Sefton in *Lion: Around the World Race in Lion New Zealand*, would be so complete that not even a whisper of the strategy they devised to give Blake a winning boat would reach the other syndicates.

Before BFA had been approached by any syndicates, Bruce had already planned tank-testing of his latest Maxi shapes at the Wolfson Unit in Southampton. Spreading the cost of the research between these parties, by agreement, was one thing. Customising it was another. To do that each syndicate was asked for the route they planned to sail around the world and the weather patterns they expected to encounter on the way. By sailing these prototypes in their VPP along these theoretical routes, in these theoretical conditions, Bruce and Russell, together with their clients, could arrive at design solutions to best match those results.

But the character of the fifth Whitbread had been dramatically redrawn by the race committee. Because of the political unrest in South Africa, Punta del Este, not Cape Town, had been chosen as the first stop after Southampton. Gone was the 3000-mile beat from the Doldrums into the Cape of Good Hope. Added were a further 6000 miles, a much longer Southern Ocean leg, and thousands more miles of reaching. Now 'the burning question faced by skippers and designers was ... how much the boats should change'.[12]

With less upwind and a lot more downwind sailing in the race, Bruce and Russell considered that a masthead sloop option, with its greater downwind sail area relative to rating, would be the way to gain most benefit from the altered course. After three months of weather research, Pierre Fehlmann agreed. But quite independently of each other, American Skip Novak,[13] Grant Dalton and Peter Blake suggested that Bruce and Russell put ketch rigs on their prototypes and test them in their VPP. The results surprised the designers: around the new Whitbread course a ketch looked as though it would be faster than a sloop. Bruce Farr:

> At that point we felt our hands were tied. Our job was just to do the research, to present them with the answer and let them make their decisions. Which is why we couldn't make the suggestion to anyone else — even though without that input it might have eventually crossed our minds. And because he hadn't brought the idea to the table we were effectively prevented from offering the ketch option to Pierre.[14]

Because the results were unexpectedly good, Bruce questioned the mathematics of this part of 'Fast Yacht'. How correctly did the programme treat wind angles for a ketch? How well did it handle the blanketing of one sail over another? But the more he talked to George Hazen, the more he realised George's life-long interest in ketch rigs had resulted in a clearly thought-out aerodynamic model.[15] Novak, Blake and Dalton had been right to push them towards a ketch-rig option.

It was at this point that Bruce's gift for interpreting the rating rule under which the Whitbread would be raced would become of great benefit to the ketch-rig syndicates. Studying how the IOR rated what it viewed as an old-fashioned rig, he discovered square metres of free (unrated) sail area. Within the IOR the mizzen sail was rated slightly less than the mainsail, primarily because the original framers of the IOR considered it to be less efficient due to the turbulent air flowing onto it from the forward rig. But the mizzen staysail, the sail between the two masts, was rated at only one-third of its actual area. And because the IOR combined the two as one by taking the larger of the mizzen and mizzen staysail as the area for calculation, if both sails were the same size and both sails were up, a ketch would be only 'charged' for one. And because of a small correction in the IOR, the more the rigs were moved apart the cheaper the mizzen staysail became. Suddenly the Maxi ketches were looking even faster than 'Fast Yacht' had predicted.

Bruce then turned his attention to the Mark IIIa version of the IOR, the rule amendment that had heavily penalised his light-displacement yachts a decade before. Because he knew that ketches were seen by the rulemakers as an older type of yacht, he guessed there might be further rating 'help' under Mark IIIa he could extract for his clients. What he discovered was that the IOR rated mizzen sails even more cheaply under the anachronistic Mark IIIa than it did under Mark III. Being able to rate the boats by whichever was the most favourable of the two versions of the rule meant the amount of penalty avoided (or rating saved) under Mark IIIa could be transferred into hull length, giving the ketch-rigged maxis more waterline and therefore more speed than their sloop-rigged counterparts. Unwittingly, Novak, Blake and Dalton had sent Farr on a theoretical journey of discovery that had turned up a rating treasure trove. In return he gave them designs which were longer and carried considerably more sail area relative to their handicaps than the Maxi sloops.

It was late in the customised design programme when Peter Blake dropped a casual comment into one of the conversations he was having with the designers. Knowing he needed an edge over Dalton, he suggested one of Bruce's most significant innovations.

'What about a fractional rig?'

'Not a good idea, Blakey.'

'Why?'

'Because we don't think a ketch rig will be able to reach very deeply

before the mizzen blankets the mainsail and you're left with only the relatively small spinnaker you get with a fractional rig set-up.'

'Perhaps, Bruce, but let's try it.'

'You really want to?'

'Yes.'

'Okay.'

It wasn't just the theory that caused Bruce to hesitate. Building deadlines
were looming and time was running out. As simple as the idea sounded, it
required a lot of work to prove it. It wasn't just a matter of keying in a ketch
rig and having the sail force model drive it round the world. A new rig had
to be designed, the structures to support it estimated, and the load and
weight calculations redone in order to establish the new vertical centre of
gravity. But the VPP results seemed to support the suggestion. Marginally
better than for a masthead ketch, they sent Bruce diving back into Mark IIIa
searching for still more rating benefit he might be able to extract, this time
for a fractionally rigged ketch.

> What we discovered was that the fractional rig resulted in the boat getting an
> even lower rated sail area. Because now the mizzen was comparatively bigger
> than the fore-triangle on the forward rig which, by some stupid twist of the
> mathematics, actually gave it more benefit from the Mark IIIA. But it only really
> worked mathematically if you made the boat bigger and heavier. So we con-
> verted to fractional and ended up with several more feet of rating. That's how
> *Steinlager 2* was born — as quite heavy-displacement and very long.[16]

The results also presented the designers with their biggest ethical dilemma
of the programme. They could not offer the fractional ketch rig arrangement
to any of their clients unless they were specifically requested to conduct
such a test. So when Dalton later asked what they thought about a fractional
set-up, Bruce could only give the original answer he had first given Blake.
Dalton's reply was swift; he'd had the same reservations: 'So let's not even
run a VPP on that option.'

The irony wasn't lost on Bruce and Russell. One of Grant Dalton's
strengths was his quick decision-making. But unable to breach their own
code of ethics, they wondered if this time it might not prove to be his
downfall. Conversely, Peter Blake's throwaway line had added a metre to
his boat's mizzen mast, four feet to its length and 450 pounds to its keel. It
was now a stiffer, longer boat with more sail area for its wetted surface than
Pierre Fehlmann's *Merit*, developed from the same design package. At 84
feet, it was also two feet longer than *F&P*.

Russell's approach to structures and ballast was no less ingenious than
Bruce's reading of the rule. By adding more of the brittle but lighter carbon
fibre[17] to the large Kevlar panels he'd developed for *UBS*, he'd made the Maxi

hulls so much lighter than the ageing IOR had ever envisaged that he had a massive amount of 'excess' weight to put somewhere in order to meet the rule's displacement requirements. First tried in their recent Fifty-footer designs, like most Bowler innovations the practical answer to a complex problem seemed almost simple at first glance. Knowing that speed in waves is lost if a yacht's ballast is not placed closest to its centre of pitching — and with Russell horrified by visions of lead ingots coming loose in the Southern Ocean — Bruce drew a large hole in the bottom of the hull. Into that they'd place the excess ballast: a 2 by 1.5-metre slab of lead. It would double as the bottom of that part of the hull and be the structure onto which the keel was bolted. An idea first used by Britton Chance in *Resolute Salmon* but not tried in boats of this size before, it would prove so efficient and successful that it would become *de rigueur* in all large, light-displacement yachts rated under the IOR.

No one knew just how fast the Kiwi ketches were, or which was the faster. The first time they met — in the 100-mile Noel Angus Memorial Race — the white masthead ketch proved a narrow winner in torrential rain and flat seas but was beaten by just three minutes in the 1989 Fastnet. Between the only two races in which the Kiwi ketches met before the Whitbread, Dalton and his crew looked formidable,[18] claiming a succession of records: the Auckland-to-Russell monohull record in 11 hours 58 minutes, the Auckland-Sydney monohull record in five days three hours, and a new transatlantic record by winning the Newport-to-Cork Race in July 1989. Covering the 2600 miles in eleven days and fourteen hours, *F&P* finished 400 miles ahead of the latest Ron Holland Maxi design, *NCB Ireland.*

Suddenly Farr and Bowler were under attack again, this time from an unexpected quarter. Ron Holland, the charming and normally affable designer who had grown up in Auckland's Torbay, had popped his cork. And he had a supporter in England's Rob Humphreys. 'The dispute flared up again on January 28 when [he] circulated a letter to other competitors claiming the four new Bruce Farr designs in the fleet were seeking to exploit a loophole in the 1988 version of the International Offshore Rule to gain "a significant and unfair advantage".'[19]

More than a storm in a teacup, it showed again that designing the fastest race boats wasn't the only thing you had to do to stay at the top of the monohull design tree. If you weren't coping with owners of slower boats putting pressure on the ITC to recommend changes to the rating rule, or having former clients taking pot-shots at your integrity, you had fellow designers crying foul. BFA not only demanded a retraction of the Holland letter. They insisted that the Whitbread organisers keep to the original conditions under which they'd said they would be running the race.

The problem had arisen because the ORC, which controls the IOR — the rating rule the Whitbread race committee used for measuring its contestants — had decided to use the hull-measuring machine[20] being used by the up-and-coming IMS.[21] That, thought the ORC, would mean all new yachts racing under its auspices would be rated for both rating rules by the same method at the same time. Unfortunately, not only did the IMS measurement machines in Europe and the USA produce slightly different results. They calculated the boats to be heavier than they were, producing a maximum 0.4 percent rating advantage over the hull-points hand-measurement method used by the IOR. Under pressure from the ITC to rectify the anomaly, the ORC voted in November 1988 to immediately remove machine measuring as a method of rating. The news shocked the Whitbread organisers. In order to facilitate boats being designed and built in time for the race, they had announced in January 1988 that the 1989–90 Whitbread would be raced under the 1988 IOR. Only after the ORC assured them that there would be no significant changes to the IOR had the Whitbread race committee written their Notice of Race notifying designers and contestants that the Whitbread would now be raced under the 1989 IOR (including the no significant change proviso).

Having designed their new maxis to the original conditions, Bruce and Russell were more than a little bemused by Butch Dalrymple-Smith's barely disguised broadside. 'Why should people with the intelligence to see the changes coming be penalised?'

Truth to tell, the Holland office had made another miscalculation. As the maxis were racing each other in this Whitbread without handicap, it was imperative that each reached the maximum IOR rating of 70 feet. But even under the 1989 IOR, *NCB Ireland* was nearly a foot below the rating maximum. And the 400 miles by which *F&P* had beaten it across the Atlantic was rather more than the 0.3 ft of rating the Farr maxis had gained under the machine-measuring system. Whichever deck you stood on, the difference was massive.

In the face of Holland's letter and his partner's plea for enlightenment, Rear Admiral Williams stood firm. At a press conference in Auckland in February 1989 he announced that his race committee's decision was based 'on the following facts ... not opinions. A: All yachts designed and built for the 1989 Whitbread race were either afloat or very near completion when the ORC decided to make the relevant change to the IOR. B: The change to the IOR is a departure from previously stated ORC policy. C: The International Technical Committee of the ORC recommended in November that the implementation of the rule change should be effective in one year's time. A one-year delay would have avoided conflict with preparations for the 1989 Whitbread race. D: Notwithstanding the recommendation of the ITC, the change adopted by the ORC was implemented without notice. E: The change, if allowed to stand, would adversely affect the rating of those yachts already designed and built or nearly completed. In fact, it would have put them over the limit.'[22]

Bruce and Russell breathed sighs of relief. Not only had their company survived another attack on its reputation. In the process, they had helped save the Dalton, Blake and Nilson syndicates expensive alterations to rig and flotation and the loss of performance for their carefully optimised charges.

Steinlager's break on the rest of the fleet as it headed for the Doldrums was the most decisive of the race. It became even more so when *Fisher & Paykel* lost its mizzen because of a faulty mast fitting, and a lot of time as well. Mike Quilter's intensive study of the first leg weather had definitely paid off. *Big Red* finished with over eleven hours on *Merit* and 31 on *F&P*. Sadly, though, the drama and excitement of Leg One soon gave way to tragedy. Alexei Grischenko, co-skipper of the Russian entry *Fazisi*, hanged himself, and *The Card*'s extrovert Janne Gustavsson died in a motorbike crash. There were, however, lighter moments, supplied mostly by the *F&P* crew. Arguably the highlight was Keith Chapman's strip show, the table-top performance ending with his wet weather gear floating in the harbour. But this was just the beginning of a wild ride around the world. On the gale-swept second leg the race was to become the most dramatic of the Whitbreads sailed so far. Six men and one woman were swept overboard, with all but one surviving.[23] At various times over the rest of the course whales were hit, bones broke, rigs collapsed and the Frers-designed *Martella OF* inverted in fourteen seconds after its radical keel fell off.[24]

But it was also the Whitbread with the most spectacular racing, most noticeably between the two Kiwi ketches. Having led for 7000 of the 7700-mile second leg, *F&P* slipped back to fourth as the big red ketch, the Humphreys-designed *Rothmans* and *Merit* slipped past it into Freo. Thereafter, there were only two boats in it. Blake managed to hold off Dalton by just six minutes into Auckland, after 3400 miles of duelling across the Tasman and down the Northland coast. Leg Four was more of the same, the 84-foot ketch just 22 minutes ahead as it strode into Punta 6300 miles later. On Leg Five Blake again had the better of Dalton, fetching into Fort Lauderdale 34 minutes in front after 5500 miles. *Steinlager 2* began Leg Six with an overall lead of 35 hours. But on the fourth day out of Florida, it could all have come to naught had it not been for a pair of very fast hands. Butterworth knew by the sound of the crack that they were in serious trouble and ordered the aft sails dropped. But no sooner had they hit the deck than the port mizzen chain-plate exploded. In that split-second he crash-gybed the boat, transferring the loads to the opposite shrouds. Had 'Billy' not done so, with the main mast's runners also attached to the fitting, both fore and aft rigs might have gone over the side, taking *Big Red*'s record and Peter Blake's career with it into the choppy Atlantic. But with some Dean Phipps' bravery, roping the mast to the deck, they made it in the end, gliding down Hurst Narrows and into Southampton Water 36 minutes in front of Dalton's *Fisher & Paykel*.[25]

'One of the most breath-taking yachting triumphs in the annals of ocean racing'[26] trumpeted the brewer as it videoed up a storm to celebrate the victory. Lawyer Andrew Johns saw it without the brand name: 'Blake won because he had the best boat';[27] with the fastest in the fleet it would have been hard to lose. But as the accolades deservedly resounded for Peter Blake and his crew,[28] the first to win all legs on elapsed and corrected time, there was an even more astonishing statistic being written in the record books: of the top ten Whitbread finishers six were Farrs.[29]

Bruce would be honoured for that and his long list of yachting achievements by being awarded an OBE award in the Queen's Birthday Honours list, announced on 15 June 1990.

> We all hired a bus, put a cooler of beer in the back, and headed up the highway to Washington DC. It was a wonderful occasion at the embassy, and very important to Bruce. Tim Woods, the Ambassador, was very generous in his praise. But more than anything else I think it was the realisation that this had come from New Zealand that touched Bruce. He was emotional and close to tears, and it made everyone feel the same way. He had this wonderful small-town quality about him — chuffed with the recognition, but sharing it with all his mates.[30]

That high summer month was a particularly happy one for the expatriate designers and sailors living in Annapolis, the award coming just six days after Geoff Stagg's marriage to Mary Radcliffe. It also made an auspicious beginning to the century's last decade.

July would produce yet another Ton Cup winner as *Lone*, with a 1/1/1/1/2 result, defeated a fleet of 26 boats to capture the World Three-quarter Ton championship. It not only went on to win the Three-quarter Ton Class but also fleet honours in the King's Cup regatta at Copa del Rey. Meanwhile, on the Mediterranean, the all-Farr Italian team was winning the Sardinia Cup by 45 points from nineteen yachts representing seven teams and five nations. Of that three-boat team, *Mandrake* won both the 50-foot Class and was first over all, *Larouge* was the top Two Tonner, and *Brava* second in the One Ton Class. In 1991 *Larouge* would win the World Two Ton Cup and David Clarke's *Vibes* the World One Ton Cup. Those, together with *Merit*'s win in the Offshore Maxi Yachting Association (OMYA) Worlds, had taken the number of world championship wins in Farr-Bowler designs to a total of 21 over the last 21 years.

The New Zealand public may have been only vaguely aware of the international dominance achieved by Farr and Bowler from 40 to 80 feet. But there was no doubting the impact Farr boats were having on the country's yachting culture when close to a thousand men and women lined up on 125 Farrs in a regatta sailed in Bruce's honour. Crewing on the aptly named *Hard Labour*, the designer raced against his old Moth Class mate Rob Blackburn, himself the owner of *XTC*, a Farr 10^{20} named in honour of the little boat that had started the light-displacement revolution. Watching from a VIP boat that wet April day in 1991 was the retired Gil Galley and Bruce's mum and dad. 'I am very proud of Bruce,' said Galley, speaking for all those sailing, and for Jim and Ileene. Proud, too, was regatta co-organiser Kim McDell, as he watched many of the boats his companies had built, including eleven of the sleek new match-racing MRXs.[31, 32]

Steinlager 2 had won the Whitbread. But it was still only one of thousands of Farrs giving pleasure to yachties around the world, and only one of hundreds winning.

Chapter 27
Sacrificial Lambs

I'm not interested in being here. I'm not interested in play-
ing the America's Cup game ... I cannot believe people
want to spend so much money and waste their time sail-
ing the America's Cup. It's just a bullshit competition.
Peter Blake, New Zealand Challenge Limited General
Manager, 1991–92[1]

Some stones are so heavy only silence helps you carry them.
Anne Michaels, Canadian poet and novelist[2]

The loss of *Smackwater Jack* in 1980 and *NZL20*'s loss to *Il Moro di Venezia*
in 1992 are two of the most emotionally charged yet least discussed topics
in New Zealand's modern yachting history. In different ways each is an
event of great sadness. But for the lingering sense of loss the former has
been laid to rest, those tragically lost doing what they loved in the weeks
before their yacht went down.[3] And although his was a life cut short, Paul
Whiting had already made a major contribution to Australasian yachting, not
the least of which had been in providing Bruce Farr with some of the
strongest competition of his early career.

The San Diego loss is different. No lives were lost. Only friendships and
reputations, as those who distanced themselves from failure traded soul-
searching for slander, critique for controversy, dignity for diatribe. Like most
of the challenge team, Bruce and Russell had given their all to winning the
Cup for New Zealand. The two-year campaign had been the most exhausting of their lives, that final month, April 92, the darkest of their careers.
'Probably the most radical boat that had ever been ... in an America's Cup',[4]
NZL20 had been the fastest throughout the challenger's series and had most
wins. But like the Plastic Fantastic before it, it failed to make it in the end.
To share responsibility for that loss is one thing. To be blamed for the
programme's failure by its chief malcontents is another.

Sometimes it is better that the stones are dropped, the silence broken, the rock reduced to fragments.

***

The boat

New Zealand ... failed to lift the Auld Mug when they had the fastest boat there.[5]

That was the view of English yachting writer Bob Fisher. Others like Russell Coutts took a different tack and were just as emphatic: 'Farr suggested we could take *NZL20* back to the 1995 Cup and, with a few modifications, we ought to do pretty well! That seemed to me to be the highest form of fantasy'.[6] Not only is this quote a fantasy in itself, used out of context and time. It is not what Bruce Farr said — nor what any sane designer would have said at such an early stage in the development of the International America's Cup Class.[7] Besides, is it even valid to imply comparison with *NZL32*, a boat designed for a completely different course? And can *NZL20* be assessed in isolation from the sailors who sailed it and the managers who managed it?

Although the 1992 and 1995 America's Cup regattas were sailed off Point Loma, there was one big difference: the former was sailed around a course which had a 'Z' leg of two 135-degree reaches and a 110-degree reach, while the latter was a windward-leeward course.[8] All their experience, which their research confirmed, pointed Bruce and Russell towards a medium displacement boat of moderate beam as the fastest for this course. While they knew the lightness of their hull would give them good speed reaching and make them quick downwind, they also knew that lightness and beam had drawbacks. Under the 1988 IACC rule,[9] if a yacht was not designed to maximum-displacement it could not carry the maximum sail area permitted under that rule. This meant theirs might struggle on the wind in tactical battles against larger sail carriers, although its downwind speed would not be affected as full rig height could be maintained. Which was why Russ had gone to work on weight, removing as much as possible from the hull, including most of the side decks, and transferring the saved weight to the bottom of the keel where it would lower the boat's cg and increase its stability. Quite unintentionally, the bowsprit and the deckless look gave the little red boat a radical appearance. It was a look that not only underscored its blistering early speed. It also caused opposition syndicates to question the design direction they themselves had taken.

But there was yet another advantage — this one unseen — fitted to the bottom of this Farr-Bowler hull: a twin foil tandem. Painstakingly researched, meticulously designed and brilliantly engineered, it seemed not only to have untapped potential. It gave the helmsman the extraordinary ability to dial leeway in and out, as well as producing attractive wave-making and end-effect benefits. By moving each foil towards opposite ends of the hull, Bruce and Russell had surmised that the wave-making drag created by a fin keel attached to the most voluminous part of the hull would be reduced by their unusual configuration, and be more than enough to offset the negative effects of the boat's extra beam.

Although it was the smallest, lightest boat in the fleet, BFA seemed to have it right. Racing the month it was launched, *NZL20* was the fastest boat out of the box — so fast, in fact, in these opening rounds, that Marc Pajot was heard to say that he considered racing against it a hopeless task, that it was just too fast, especially reaching, and simply could not be beaten.

Race One: 1992 Louis Vuitton Cup Finals

The Kiwis seemed as much together as a team as the Italians were out of sorts during the pre-start manoeuvres. And after crossing the start line eighteen seconds ahead, they continued to build on that lead until, by the time they crossed the finish line, they were a minute thirty-eight in front.

NZL: 1

ITA: 0

Shape

We've never been scared to try things. So it's not surprising that in America's Cups our designs have seemed revolutionary, radical or just plain different. That's a product of Russell and I not feeling bound by convention, of always being willing to look beyond accepted boundaries for solutions.[10]

During 1990 Bruce and Russell had taken the unusual step of conducting their own tank tests without an America's Cup commission. Keen to explore combinations of shape, size and displacement within the new rating rule, they then sold their study as starter kits to New Zealand, Spain and Japan. It was not until the end of that year that they were commissioned as the New Zealand Challenge designers.

Built by Marten Marine, *NZL10*[11] and *NZL12*, the first boats built, were identical in every way, including their metre bows.[12] They were used for testing and tuning, for building a database of results and making comparative adjustments. Although there were no sizing or weight adjustments made to the third boat, *NZL14*, there was one significant change: the Cookson boat was given a destroyer bow.[13] Bruce had drawn a near-upright stem in order to lengthen the waterline, to save weight and reduce windage. Shipped immediately to San Diego, the bow tested as well in real life as it had in the tank, producing less water and noise than its predecessors.

Race Two: 1992 Louis Vuitton Cup Finals

Described by commentators as one of the greatest races in America's Cup history, it certainly was the closest; there were never more than four boat lengths between *Il Moro* and *New Zealand*. Although making gains on the downwind legs with their French-designed sails, the Italians were consistently out-sailed by the Kiwis on the first two windward legs. But the tables were turned on the third when skipper Rod Davis missed a slam-dunk.[14] Suddenly *Il Moro* was nineteen seconds clear. *NZL20* closed to within inches on that last downwind run. But as the two boats surged for the line, *Il Moro* gained clearer air to ride just one second clear.

NZL: 1
ITA: 1

Displacement

We put additional ballast in one of our two identical boats. It was quicker around the course in all conditions above five knots of air, without the extra sail that went with increased displacement. So we were stunned when Bruce came back to us with *NZL20* at the same displacement. All our on-the-water testing pointed towards a bigger heavier boat with more sail area.[15]

That, though, was at odds with the results Clay Oliver's VPP was making of Bruce's extensive tank-testing. As was Clay's analysis of beam. Bruce Farr:

We didn't find benefit in narrow boats. And it was much the same with displacement. We had two serious shots at sizing and came up with the same answer both times: a 21,000-kilogram boat with a moderate beam. In retrospect that might have been a mistake. But I've never reviewed that decision from the point of view of the course. It wasn't just a windward-leeward course. It went windward-leeward-windward, reach-reach-reach, windward-leeward. So it was a course that had a lot of reaching and that always favours lighter beamier boats.[16]

If *NZL20*'s size-displacement ratio was a weakness, German Frers didn't think so; he doubted his design could win the LVC Finals. But what wasn't moot was the tactical disadvantage of the smaller sail area. It was harder for *NZL20* to slam-dunk a larger yacht which could sail through the tack with its bigger sail plan. And punching through a void left by a large sail area was also problematic for the smallest boat.

Race Three: 1992 Louis Vuitton Cup Finals

It was not until the second windward leg, during which Cayard failed to cover, that *NZL20* made up its nearly two-minute deficit and charged out to a half-minute lead. From that mark until the end Davis sailed the perfect tactical race, sticking like a limpet to *Il Moro di Venezia* to win by 34 seconds.

NZL: 2
ITA: 1

The keel

> The tandem keel had huge benefits. It was an amazing engineering feat. If it had been up to me, it would have been the starting point for the next campaign.[17]

The tandem keel was revolutionary, as much in its conceptual design by Bruce, Russell and Steven Morris and its engineering detailing by Neil Wilkinson and Russell, as in its untapped potential. It also made the red boat rudderless. To helm it required a different range of skills from those needed for a conventional keel-rudder. It also demanded a different mindset to extract its unique advantages.

The structure consisted of two narrow, near-vertical fins attached to an elongated sixteen-ton lead bulb. Steered from two separate wheels mounted one behind the other, both foils turned through 25–30 degrees and worked at high angles without stalling. Testing of the only full-sized tandem keel in the full-size testing programme against their other keel types took place in August–September 1991. Chosen by the sailors, the tandem keel was fitted between Round Robins One and Two, replacing the L keel and rudder. It took just four days (and nights) for the selfless shore crew to rebuild the support structures for this very different keel and the mast's new position, moved forward by nearly a metre to accommodate the altered centre of lift.

There were a number of reasons why Bruce and Russell pursued the tandem keel options when the other keels worked well. The reduced volume the twin fins required to perform the turning and side force functions was much less than a conventional arrangement in which the keel mostly carries the side force and the rudder is used for steering. Less wetted surface meant more boat speed, and the smaller foil ends meant the tip vortex problem was more easily tidied up. Tank tests showed that the tandem suppressed the bow wave and midships trough, drag-inducing features which a a fin keel augments because of the added volume it brings to the fattest part of the hull. It was an important discovery not made by those who tested their tandems in wind tunnels only. And to be able to 'dial in' leeway, a concept arguably even more revolutionary than Lexcen's winged keel, gave a huge tactical advantage. It meant that when sailing on the wind with both foils set a few degrees up from the centre line, the yacht could climb to weather without a significant drop in speed compared with an opposition yacht with a standard configuration. Likewise, the tandem also offered greater gains from lifting puffs.

> Rather than sheeting everything on and unloading the rudder to steer the boat up, you'd crank both foils over and the boat would leap sideways. But unless you used both foils together, its benefits were lost. If you steered it like a normal boat using the aft foil as a rudder, it was just a normal boat.[18]

But the tandem did have weaknesses. Because the twin foils were heavily loaded up from carrying the side force as well as creating lift, there was a lack of feel for the helmsman. The tandem was also more difficult to steer. With less space between the foils than between a conventional keel and rudder, there was less leverage to create turning moment. And in a seaway or downwind the boat would sometimes 'wander'. But the positives seemed to outweigh the negatives. At least that's what the sailing team decided when they opted for Keel Nine and decided against reverting to the L keel and rudder — perhaps the faster configuration for reaching — later in the regatta.

Other syndicates had designed and fitted tandem keels but with much less success. The *Stars & Stripes* version produced such unmanageable weather helm that it was discarded by Dennis Conner after just three races. And while *Spirit of Australia*'s lasted rather longer, Peter Gilmour still likened it to 'steering a fourteen wheeler'.[19] That the Farr-Bowler tandem worked with great efficiency was a tribute not just to its design and engineering but to the thoroughness with which the designers had conducted their research.

First we asked Ian Court to build two three-foot radio-controlled models. We put a conventional keel on one, a tandem on the other and, in the middle of winter, found a pond near Southampton in which to test sail them. When we tried the boat with the conventional keel we found we could steer, tack and gybe it. But we couldn't steer the other; it kept tacking itself. So we changed the areas of the rudders to the point where we could almost sail it. Then we went back to the Wolfson Unit with an eighteen-foot model. I jumped in and steered it with its little tiller down the big 300-metre-long tank. That's when I discovered it was like a car with oversteer, like driving a rear-engine Porsche on a wet road when you start to turn.[20]

Bruce also conducted on-the-water tests with the sailing team, often steering the yacht himself. He was keen to assess the tandem keel's capabilities near its handling limits (slow and downwind) and to find ways of increasing its efficiency. He believed its theoretical advantages outweighed both the 'T' and 'L' keels. And he also felt the tandem's inherent difficulties could be overcome with time and practice. But would there be enough of both?

Race Four: 1992 Louis Vuitton Cup Finals

Forced to the left-hand side of the course after the start, Davis and tactician Barnes were not unhappy with where they found themselves. There was more pressure out left, and as the breeze began to build so did their lead. And their crew never once faltered, putting in a perfect team performance to have *Il Moro* nearly two and a half minutes astern when they cruised across the finish line.

NZL: 3

ITA: 1

The bowsprit

I cannot understand for the life of me why there should have been an argument against [the] bowsprit on the [New Zealand] America's Cup boat. A bowsprit is a time-honoured feature of sailing vessels everywhere.[21]

There was no good reason. Even more so given that bowsprits were permitted under the IACC rule,[22] and had been declared legal by the Chief Measurer.

Initially Bruce had drawn *NZL20* without a bowsprit. But things had been happening out on the water during the first IACC World Championship held in May 1991 and the preliminary fleet racing organised by the Challenger of Record Committee in December 91–January 92. What the Kiwi team had noticed was that the other yachts, with their long metre bows, were not using their spinnaker poles, although the IACC rule required that they did so. Instead, the foreguy was short-circuiting the spinnaker pole by going to the tack of the sail, with the afterguy slack but attached through the pole to the tack.[23, 24] This exploited a loophole in the racing rules which said the spinnaker pole only had to be in close proximity to the tack of the sail, not attached to it. Following the decision to do a short bow, Russell thought that this emerging trend could be improved on in the Kiwi boats by adding a bowsprit to get the foreguy out directly under the end of the spinnaker pole. That would make the foreguy a vertical rather than an angled line going to the spinnaker pole, thus lessening the load on the spinnaker pole and foreguy. But the sailors went one step further. Before Round Robin One they decided to take the foreguy directly to the sail as other crews were doing, to make gybing quicker and easier. It wasn't hard to figure why. Or to understand why they'd stick like glue to their decision after Cayard and Gardini had used the elegant apparatus to create the ugly affair called Spritgate: gybing on a second foreguy from the bow (not the foreguy from the bowsprit) was not only legal, it was the exact same method used by all the other competitors, including the Italians, for gybing their spinnakers from the end of their long bow overhangs.

But with *NZL20* showing blistering early speed, things were bound to change. So what if the Louis Vuitton Cup Jury had ruled as legal New Zealand's use of the bowsprit after hearing French and Italian protests in Round Robin One and that Andrew Johns had worked with the jury to help keep that option secure? This had long been a contest where money and morals don't mix, and egos, like Gardini's, match the money they spend (rumoured in Gardini's case to be in excess of US$150 million for the 92 event). A small part of that cost was Laurent Esquier. Now working for the Italians, *Le General* knew only too well from Glassgate what rattles the Kiwis' cage: accuse them of cheating. As well, New Zealand's management failed to appropriate the lessons of the past, offering only minimal opposition in the court of public opinion to counter the illegal-use-of-bowsprit claims. Needless to say, the pressure intensified after *NZL20* won all its races in Round Robin Two. But even then the Kiwis were incredulous that the America's Cup (or Match) Jury, which adjudicates the America's Cup but has no jurisdiction over the challengers' event, could make a public announcement just two days before Round Robin Three — '[New Zealand is] using the bow-sprit improperly'[25] — that contradicted the finding of the Louis Vuitton Cup Jury.

It was, of course, a nonsense. In order to agree with the Italian protest, the Match Jury had had to conclude that a bowsprit wasn't a bowsprit but an outrigger, as bowsprits were permitted within the IACC rule and New Zealand's use of theirs 'was ... in a proper fashion'[26] according to the Louis Vuitton Cup Jury. In other words, the Match Jury members had had to conclude that a bowsprit wasn't a spar protruding from a yacht's centre line but a device running parallel to a hull. The announcement not only defied historical precedent. It was based on a semantic impossibility.

Nonetheless, the announcement sent lawyer Johns scrambling for clarification. It also saw Blake, Johns and Farr in informal meetings with the sailors trying to find a way to use the bowsprit that would accommodate the threatened change. They suggested, for instance, that the sailors take the foreguy through the spinnaker pole, in the traditional manner. Farr even suggested they forget about the bowsprit; the boat worked well without it and its minimal advantages weren't worth the risk or this much distraction. But the sailors were convinced they could comply with a rule change by using a second foreguy or provided the afterguy wasn't released while the foreguy was carrying load if the bowsprit was in use. It was, though, a dangerous path to be taking in this climate of bowsprit hysteria. Even more so when the possibility of the Match Jury's announcement impacting on members of the Jury turned from rumour into decision. But after much toing and froing and shifting of its position, Andrew Johns finally received a definitive answer, an edict given to all the challenging syndicates 'concerning one of the rules [condition 8.9] governing the races of the Louis Vuitton Cup ... [which] states, 'When a spinnaker boom is not in use (including the period during which the spinnaker is hoisted, gybed or lowered), a line which is attached to the spinnaker and is controlling the setting of the spinnaker and is led over or through a bowsprit, infringes ... International Yacht Racing Rule 64.4. ... part of [which] is: Use of Outriggers a) No sail shall be sheeted over or through an outrigger. An outrigger is any fitting or other device so placed that it could exert outward pressure on a sheet or sail at a point from which, with the yacht upright, a vertical line would fall outside the hull or deck planking'.'[27]

Now, not only had 'the bowsprit ... been ... defined as an outrigger but its use ... allowed when the line from the end of it exerts downward on the spinnaker boom only'.[28] Central to this ruling was another leap of logic: a guy was now a sheet. That is, the foreguy — a line which acts as a down-haul and controls the luff of the spinnaker by attachment to the spinnaker pole or the sail's tack (the lower forward corner of the sail where the luff meets the foot) — had, by this ruling, turned into a sheet — a line which controls a sail by attachment to its clew (the lower aftermost corner of the sail where the leech meets the foot) (or a boom when a fore-and-aft sail is attached). This meant that between the bowsprit and the tack the foreguy had morphed into a sheet going to a clew although it was the same length of rope (a guy, not a sheet) attached at the same place on the sail (the tack, not the clew) as it was on all the other yachts in the competition. It also meant that the use in the LVC of this 'time-honoured feature of sailing vessels everywhere' was now based on two transmogrifications or two semantic impossibilities.

It was a decision that mocked the lot. It mocked a tradition that has long separated sheets-to-clews from guys-to-poles (at the forward end of the sail). It mocked the *Oxford Companion to Ships and the Sea*, which defines a clew ('where fore-and-aft sails are not ... laced to a boom') as 'the corner of the sail to which the sheet, by which the sail is trimmed, is secured'. And it mocked IACC rule 33.6 which says (and said in 1992): 'A headsail (genoa, jib, staysail, spinnaker) shall not have ... more than one sheet or any other contrivance for extending the sail to other than triangular shape.' In other words, although the IACC rule said that a sail forward of the mast could only have one sheet, the LVC Jury now decreed that *NZL20* had two, even though by legal definition and according to tradition it only had one, the sheet, controlling the sail at the clew and the other, the guy, still attached to the tack like every other yacht. Bruce Farr:

> In the end it wasn't a bowsprit issue but a use-of-bowsprit issue. And in Race Five at Mark Five the trimmer had a momentary lapse of concentration and released the afterguy a few seconds early. Which meant for that brief period of time the spinnaker pole had no effect on the setting of the gennaker. And there were the Italian photographers all going click-click-click.[29]

In truth, it wasn't even a use-of-bowsprit issue. It was a brilliantly orchestrated campaign of intimidation and deception that should have seen the Italians ousted from the competition.

Not only were the IACC class measurers convinced that the bowsprit's use was legal, even with a slack afterguy. IYRU jurors not on that particular jury would later comment on the decision of their peers as being one that should never have been made. Indeed, the very decision was revisited at the November 1992 meeting of the IYRU and the ruling overturned.

Race Five: 1992 Louis Vuitton Cup Finals

At the end of the first beat *New Zealand* was 30 seconds behind *Il Moro*. The Italians showed better downwind speed, and although the Kiwis gained on the reaching leg, they were still almost a minute behind when the second windward leg began. *NZL20* closed to within a boat's length when *Il Moro*'s gennaker trapped itself below the hounds. But when the wind dropped to four knots the bowsprit boat was in trouble again. Four minutes twenty seconds behind at the final turning mark, it seemed the race was over for the Kiwis. But Cayard failed to cover. And when the two boats reconverged the New Zealanders had turned their massive deficit into a two and a half minute lead.

NZL: 4
ITA: 1

New Zealand was now just one win away from being the next America's Cup challenger. Or so its crew thought until they saw a protest flag flying from Cayard's backstay. He'd timed his hoist and his heist to perfection.

It took a third jury, the International Jury — called in to give a ruling on the IYRU's rule 64.4 — nearly six hours to deliver their decision on Cayard's elaborately prepared protest. In this long line of nonsense there was still more to come.

Although the Jury agreed with Cayard's claim that *New Zealand* had infringed LVC condition 8.9 by gybing its gennaker without using its afterguy, because it was for only eight seconds of their 2:38 win it "had no significant effect on the outcome of the match" ... The previous time ... [the Jury] had added that to any of their decisions ... they had allowed the result to stand. This time, with their usual inconsistency, they "deemed the most equitable arrangement is that NZL20's win ... shall be annulled".[30] It was a decision unprecedented in America's Cup history.

NZL: 3
ITA: 1

Management 1

'Where [the Kiwis] made their biggest mistake was in their handling of personnel.'[31] Key players became displaced from their areas of expertise, their effectiveness diluted by degrees of separation from what they did best.

To run the 1992 America's Cup campaign, New Zealand Challenge Limited was formed. Its directors were Jim Hoare, a Fay Richwhite functionary, and Richard Green and Andrew Johns, partners in law firm Russell McVeagh McKenzie Bartleet & Co. Its chief source of funding was a $25 million sponsorship deal with Television New Zealand (TVNZ).[32]

Bruce and Russell, along with their fifteen-strong research team, were employed by New Zealand Challenge Limited as its yacht designers. They were responsible for the concept, design and delivery of the four challenge yachts. They neither ran nor controlled the campaign. That was the role of Michael Fay who, from August 1990 to August 1991, delegated the responsibility co-equally to yachtsman David Barnes[33] and sail designer John Clinton. Barnes was also in charge of the sailing programme, boat-testing and feedback to the designers, and Clinton sail design and development. The Barnes-Clinton management style may not have sparkled with Fay charisma. But it got the job done with an honest efficiency. That in spite of the early problems of having the country's top sailors racing in the Whitbread or preparing Olympic campaigns.

It was no easier for Bruce and Russell. Old hands were hard to get: Richard Karn was chasing time, Chris Mitchell pasta, and Tom Schnackenberg his hat as it flew from Bengal Bay to Alan Bond then Iain Murray's camp when the other two fell over.

But that in a sense was minutiae; the problem lay at the top. Sir Humphrey Michael Gerard Fay was a Clayton's 'chairman of the board'. In San Diego for only the last six months of the contest, his mind had been more on banking than Cupping. Not only was his group balanced precariously between its 30 percent stake in the Bank of New Zealand (BNZ) and the struggling state bank's startling over-exposure to its new minority owner.[34] In the Auckland spring of '91 secret BNZ papers were doing the rounds. The Serious Fraud Office (SFO) was first made aware of these documents on 17 September 1991, its director cognizant of their potential impact on New Zealand's America's Cup challenge.

> Mr Lunn told me that the documents in question contained evidence of fraud committed by Sir Michael Fay and Mr David Richwhite against Americans ... [who] were waiting for Michael Fay to win the Louis Vuitton Cup ... The timing for the proposed release of the documents was to enable the biggest possible impact when the information was made public.' [35]

True or not, and whether or not Fay had any inkling of that, it didn't much matter. Walking the financial high wire with a billion-dollar balancing bar and these damning documents somewhere beneath him was not just a tricky act. With his business life this revved up, the America's Cup must have seemed little more than a sideshow for New Zealand's newest knight.[36] In little over a year, in October '92, parliament would be in an uproar over a winebox of documents and the country later stunned by what they revealed.[37]

Fay wouldn't make it to San Diego on a full-time basis until the end of '91. But whenever he was in town he gathered an informal group around him, not to make decisions, but to chat about events of the day and discuss options and strategies. Typically it would include some or all of the following: managers Barnes and Clinton, lawyer Andrew Johns, designers Farr and Bowler when either one was in town, coach Eddie Warden-Owen, consultant Clay Oliver, and sometimes even journalist Alan Sefton. But still at sea in yachting circles, and under pressure back home, Fay continued to feel the need for someone to stand in his stead, someone more high-profile than the gifted helmsman Barnes and the sail-designing Clinton. Casting around for options, he finally took the advice of the septuagenarian Tom Clark, a fellow knight still living out his blue-water dreams through his 'great hero', Peter Blake.[38]

> Mike had so many irons in the fire and the whole thing was so involved politically within his own organisation [that] I pushed like hell to get Peter up there ... [And] when it was all over up there, he and I sat down and had a little talk: unfinished business.[39]

Management wasn't keen because Blake had nothing to offer in terms of America's Cup experience. He had also failed four times in Whitbread campaigns, twice as skipper and syndicate head. But if a figurehead was needed and Blake was Fay's choice, then he might have a role to play if only with the sponsors. So despite Blake's minimal qualifications for the job, syndicate chairman Fay made the expensive appointment, the Kiwis issuing their press release on 18 May 1991: '42-year-old Blake has joined the New Zealand Challenge management committee and will have the responsibility for the day-to-day management of the New Zealand effort to win the Cup.'[40] It would prove a wholly uninspired decision, the task of managing a large campaign seemingly beyond Blake's level of skill and management expertise. Soon his lack of leadership became a double negative, apparently inducing a sense of entropy, and his self-admitted displeasure at being there a dissolving of team spirit. True, he'd inherited an existing structure when he arrived in San Diego in August 1991. But nothing could disguise his lack of heart commitment. 'I could be earning four times as much money back in England.'[41] If that was pure farce — the now-dubbed Million-Dollar Man[42] had been previously unemployed — his remarks about a gifted Kiwi yachtie sailing for Japan were proof of his lack of wisdom, his complete absence of tact, and a reminder of his own serious lack of talent as a competitive yachtsman. Calling 'Chris Dickson a mercenary and suggest[ing] it might be better if he didn't come back to New Zealand'[43] was like calling the kettle black. Indeed it was both arrogant and stupid given that Blake had long lived in England and used Kiwi dollars to fund his personal yachting exploits which he tagged with the words 'New Zealand'.

Race Six: 1992 Louis Vuitton Cup Final

After splitting tacks at the start, it wasn't until near the first windward mark that the two boats reconverged, with *NZL20* a minute behind *Il Moro*. The New Zealanders fought back throughout the race to be just nineteen seconds behind at the last turning mark. But with their illegal downwind sails,[44] the Italians again had the better of the last-leg run and had more than doubled that lead by the time they crossed the finish line.

    NZL: 3

    ITA: 2

Management 2

> Peter treated the shore crew like the scum of the earth and told them that. 'The sailors are the rock stars. You guys are just here to do a job and I don't give a shit about you.' But they were the people who were doing all the work. Without them you didn't have a campaign ... Peter's so-called people skills were hopeless.[45]

Blake expressed it prosaically. 'There was rarely a time ... when the NZ team didn't have huge personality problems of one sort or another, be they wives, girlfriends, sailing crew fighting for places on the race yacht, or a well meaning but often grumpy and cynical shore team'.[46] And when members of his Whitbread crew began receiving special treatment, suddenly there were two sets of rules: one for the 'Steinlager' clan and one for everyone else. The Steinie boys received new vehicles and apartments. There were jobs for their wives and girlfriends, while others in the challenge had to pay for their partners to be there as well as sharing accommodation. All of which was made worse when the pay disk was left unattended and pay differentials discovered. Not only was one 'Steinlager' couple found to be receiving a six-figure per annum sum. One sailor joining the team late in the day had been given false information about wage levels and, as a consequence, negotiated a lower salary level than those in like positions. Under Blake, it seemed, divide-and-rule took on a whole new meaning.

    Blake's presence also ensured that the behaviour of Coutts and Butterworth went largely unchecked. From the moment he turned down the role as Rod Davis's tactician and left camp for a few days, Russell Coutts's behaviour was undermining the programme. If he couldn't have the A-team helm he'd sooner play golf with 'Billy'. Or take a club, a golf mat and a few balls out on the boat and, with his mate, hit them off the aft deck. Or fake sleep on the boat's aft deck. Or, on maintenance days, work on his Soling in the busy Kiwi compound. If 'no one in the syndicate tried to sort out the problem'[47] that was hardly surprising. 'Billy' was worth his weight in gold: a year before he'd saved *Steinlager 2*'s rig and probably his skipper's career. It was a happy conjunction of place and supply for the plenipotent Peter Blake.

Is this where, one has to wonder, the following lies were born and began their race for print and tape? Not only is this scapegoating of others consistent with Peter Blake's history of behaviour. As the failed manager of the 1992 campaign, he had every motivation to lay the blame at someone else's feet, especially if he were to secure the New Zealand Challenge sponsors for the 'unfinished business' he was discussing with Tom Clark in San Diego. Said Coutts to Becht in *Champions Under Sail*: 'I think [Sir Michael Fay] made a decision to run with the Farr office, and if you look at the whole [1992] programme it was run by them'.[48] Mumbled Andy 'Raw Meat' Taylor to the *Sunday Star-Times*: 'Designer Bruce Farr drove that challenge. All decisions basically went through his office. It was very difficult for the crew. Farr would talk to Fay and both talked to the crew through the managers'.[49] 'Meaty' gave vent to more of his thoughts on *Born to Win: The Inside Story of Team New Zealand*: 'Well he [Farr] was given the reins right from the start. So, you know, he was told to do the job, and that was that.'[50] Even the talented Laurie Davidson was banging away on *Born to Win*: 'The whole programme was once again ... too much designer-driven rather than driven by the crew ... It was Farr's programme and well, frankly, he just didn't do it right.'[51] So was Greg Clarke in this free-for-all: 'The designer was Kiwi Bruce Farr. He'd been with New Zealand in the '87 and '88 challenges. Rightly or wrongly, 1992 was his baby.'[52] Parroted Richard Becht: '[Blake and Coutts had] seen the 1992 programme run by the Farr and Associates office; they believed there was another way'.[53] Claimed Coutts in *Course To Victory*: 'Although Fay, Barnes and Clinton were nominally in charge of the campaign, when Blake was hired there was little doubt among most of us that Bruce Farr was really controlling the programme'.[54] Clay Oliver:[55]

> That's completely incorrect. Absolutely wrong. It's totally unfair. It's a lie. Blakey ... was the figurehead. He was responsible. He was accountable. And he had the authority. He had everything you're supposed to have except he wasn't, in my opinion, committed to the programme. So how does he defend against [the] lack of leadership that existed there? There's only one way to go and that's to say the thing was run by someone else ... that he wasn't really in charge. I mean, why did the Kiwis fall apart in 1992? The boat was doing well. They were four-zip. They were getting ready to go into the America's Cup. What happened? That wasn't a boat problem. That was a management problem.[56]

To suggest that either Bruce or Russell could ride roughshod over one of New Zealand's wealthiest men (Fay), a five-times Whitbread veteran (Blake), a four-times world sailing champion (Barnes) and his co-managing sail designer (Clinton) is as silly as a soup-only diet for 'Raw Meat'. The Team

New Zealand rhetoric becomes even more inane given that the designers were in San Diego for less than 30 percent of the time, and for ten of those days Bruce lay seriously ill in hospital after a fall sustained on the boat. During his six weeks' recovery from a lacerated kidney and cracked ribs, he had little enough energy for work let alone conducting a management coup.

Race Seven: 1992 Louis Vuitton Cup Finals

Left behind at the gun, *NZL20* was soon three lengths behind *Il Moro*. That turned into a minute fourteen deficit by the first windward mark and two minutes by the second. A wind shift on that beat went *New Zealand*'s way. But although the Kiwis clawed back *Il Moro*'s lead, they were unable to get closer than 51 seconds by the end.

NZL: 3

ITA: 3

The skipper

> The final move was a total disaster. Swapping the skipper and tactician for the last two races ... made cheap scapegoats of Rod Davis and David Barnes.[57]

Grown up in Coronado, married to a Kiwi, an Olympic Soling gold medallist, 1992's number three on the world match-racing circuit, Rod Davis appeared to be the perfect choice to helm *NZL20*. He'd steered it with distinction in the first two round robins, not dropping a race in the second, and had won the LVC semis with a 7-2 score. But as convincing as these were, they had been accompanied by an attitudinal shift as big as the change made to the keel. If *NZL20* was now 'a weird boat to steer'[58] according to Davis, management failed to read the text beneath his thinking. It wasn't a question of whether he had the talent but that the talent he had was born of West Coast fundamentalism. If Davis remained the skipper, then the sailor's choice of keel should have been overridden; if the tandem stayed, the skipper and tactician should have reversed their roles. With that arrangement, Barnes and Davis had won the Twelve-Metre Worlds on *KZ-7*; there was no good reason why it couldn't have worked again. The tandem was different, very different, and needed the lateral-thinking and more scientific Barnes in charge. Steering from the aft foil with the front foil fixed meant Davis failed to exploit the tandem's tactical advantages. Barnes, on the other hand, had an innate understanding of its engineering and relished steering the twin foils from their twin carbon fibre wheels.

But there was more than just panic in the Kiwi camp as the scores drew level. There was the thud of factional self-interest colliding with the science of pre-start tactics as Blake and Fay and those they called on for advice tried to come to grips with the rapidly disintegrating position. Although Davis had been sailing well around the course, there was now no doubt that Cayard had his number on the start line. For some time there had been talk of the

Californian-Kiwi being replaced for the starts. But it was not until an exhausted Davis went off with Johns and Farr to hear Cayard's protest that Blake made his move. Bruce was still in the protest room when he got the GM's call.

'Well, what do you think?'

Bruce repeated his already known opinion, one that had not always gone down well with members of Fay's informal circle.

'Put Coutts on the boat in place of another crew member and have him do the starts. He then changes places with Rod, who sails the boat for the rest of the race.'

He certainly would not have been in favour of exchanging Barnes for Butterworth had he known what Blake was thinking. He had never held to the notion that the skipper and tactician were an inseparable pair.

Russ Bowler was in Annapolis when he received a call from a female member of staff. Suddenly he was on a conference call with San Diego, Fay asking the question.

'What do you think of replacing Rod with Russell?'

Russ would go along with some sort of change for starting the next race if that was the group's thinking. But again there was no mention of Butterworth joining Coutts. Perhaps with good reason; when Blake had suggested earlier in the campaign that 'Billy' should be on the boat, Bowler had made the observation that Butterworth had been lazy and uncommitted to the programme.

It has never been recorded — or at least not for public consumption — exactly when Blake and Fay made their startling decision, or why. But what is known is that sometime between the time Davis, Johns and Farr left for the protest room and midnight, it was. And when it was delivered, the timing was nothing short of appalling. Emerging at midnight, and utterly spent, Davis was handed a note: 'Phone Peter Blake.'

[The following morning] I had to settle the boys down on the boat because it was a huge shock for them. They had got up at 8 and ... [gone] down to the boat, rigged it up thinking, 'let's go 4-3 up ... and get this thing under control' only to find out they ... [had] a whole new back end. It was a huge meltdown for them.'[59]

Blake had delivered his decision by phone but left it to the deposed Davis to break the news to the sailors: he and Barnes were off, Coutts and Butterworth on. Two of the campaign's chief malcontents now controlled the boat. And when one of the shell-shocked sailors arraigned Peter Blake, he too was put off the boat.

Race Eight: 1992 Louis Vuitton Cup Finals

Coutts, aggressive on the start line, had *NZL20* in the lead until a hundred metres from the first mark. Then his lack of practice told: he failed to execute a slam-dunk properly and Cayard sailed through to leeward. Within seconds he'd made another error and *Il Moro* was away, increasing its lead to 34 seconds at the leeward mark. While *NZL20* took thirteen seconds out of the Italian lead on the second beat, Coutts made a third error, this time in sail selection, and *Il Moro* doubled its lead on the second reach. The New Zealanders fought back bravely but were still twenty seconds behind when they crossed the finish line.

NZL: 3
ITA: 4

The sails

Their downwind sails, which at one time were the envy of all, were, by the semi-finals, generally considered to be well behind those of the other top three challengers ... That may well have accounted for New Zealand's demise from the competition.[60]

Bruce despaired at the lack of sail development, even getting John Clinton to study *Longobarda*'s medium and heavy-air genoa designs. But it was in their downwind sail programme where the weaknesses were most glaring. These were never corrected and the once-good sails never improved upon. After the semi-finals Bruce even suggested to Peter Blake that the sail designers talk to Chris Dickson about *Nippon*'s downwind sails. But given Blake's public demeaning of Dickson that was a lost cause.[61] Instead, it was the Italians who gained access to both the French and Nippon downwind sail knowledge. And preferring to see an America's Cup sailed out of Venice than Auckland, the French offered their sail designer to Cayard and Gardini. From being twenty seconds slower on a run, *Il Moro* became fifteen seconds faster than *New Zealand*. It was a blatant breach of the rules and the Americans later made public their intent to protest the Italians if they used their French-made sails in the America's Cup contest. They wouldn't, and their 4-1 loss to Bill Koch's Cubens really surprised no one.

Race Nine: 1992 Louis Vuitton Cup Finals

Il Moro di Venezia hit the line at speed, five seconds ahead of *NZL20*. Coutts was controlled by Cayard all the way to the first leeward mark and found himself a minute thirteen behind. That was the race and the series. The Kiwis couldn't break Cayard's cover, and although they closed the gap they were still twenty seconds behind at the end.

NZL: 3
ITA: 5

Once again in an America's Cup, New Zealand had 'snatched defeat from the jaws of victory'.[62] True, they had faced a fierce psychological war waged by a businessman who, fifteen months later, would blow his brains out. But in the end there were no excuses: the Kiwis defeated themselves. Fay, as a leader, had been *in absentia*. They had made poor management choices and managed their people poorly, spreading not focusing skills. They had failed to harness the technological advances Farr and Bowler delivered to keep improving boat speed. They had failed to improve their sails. In all, too many poor decisions were made. And having been called for cheating throughout the competition, management even failed to protest Italy out of the Challenger Finals for breaching the good sportsmanship rule. David Barnes:

> One of the jury told us afterwards: 'We were just waiting for you to put in the Rule 75 because we were prepared to chuck the Italians because of their behaviour.'[63]

In fact Fay, at Farr's behest, had lodged a protest under Rule 75, the good sportsmanship rule, before the last LVC race but withdrew it when Gardini publicly apologised. Farr was upset — and not a little bemused given the constant string of Italian abuse the Kiwis and their boat had been subjected to from Round Robin One.

Whether this protest was dropped because Fay was a 'gentleman challenger' who wanted to win on the water, was still suffering from the Big Boat legal blues, or had become more concerned about the need to protect himself from the risk of exposure presented by a winebox of documents than about winning the Louis Vuitton Cup is open to conjecture. Whatever the reason, there seems to have been an uncharacteristic reluctance on his and management's part towards the end of the series to seize every opportunity to get New Zealand into the America's Cup. Even protesting Italy's illegal downwind sails might well have proved successful.[64] For some, a question mark will always punctuate the displacement choice text. But given that *NZL20* showed blistering speed at the start and had most wins at the end,[65] it is doubtful that that is where the real problem lay. Indeed, the LVC Final's record shows otherwise: *Il Moro di Venezia* had margins of 0:01, 0:43, 0.53, 0:20 and 1:33 in its five wins over *NZL20*, a total of 3 minutes 30 seconds; *NZL20* had margins of 1:32, 0:34 and 2:26 in its three wins over *Il Moro*, a total of 4 minutes 32 seconds (7:10 counting the race that was annulled). No, the problem lay not so much with the boat or with its core crew, but with a flawed and fractured leadership and a once successful team divided deep within.

But just how deeply divided was the New Zealand challenge?

That remains an unfathomable question, although certain things are known. Fay, for one, had a lot more on his mind than winning an America's

Cup — like a raft of $100 companies hiding from the Reserve Bank his group's massive overexposure to the state bank he part-owned. It is also a matter of public record that in spite of injunctions lighting up the corporate sky like skyrockets on Guy Fawkes night, the contents of the winebox eventually blew open its lid, revealing weighty secrets. And it is now clear that Blake's lack of interest and managerial incompetence had a debilitating effect on the entire campaign. Telling two of the sailors soon after he arrived in San Diego, 'I'm not at all interested in America's Cup yachting, I'm here purely as a favour to my sponsor Steinlager,'[66] was deeply destructive of morale. As was the meeting he stormed out of when members of the shore crew challenged his decision preventing them from wearing their team blazers to the Louis Vuitton Ball, the discovery from the wages disk that his young nanny was being paid more than sailors on the boat,[67] and his nepotic goings-on with the Steinlager clan.

Given this surfeit of subtext, Fay's early absence, Blake's lack of commitment and their change-of-afterguard decision ('a recipe for disaster'[68]), what happened to the little red boat wasn't at all surprising — that after showing vastly superior speed to all in the challenger series, it suddenly and not so strangely, after being 4–1 (then 3–1) up, lost the next four races in a row and, along with them, another chance for New Zealand to compete in an America's Cup.

It was a dispirited group of Kiwis who departed San Diego in May 1992, their bags packed with broken dreams. A few, sadly, packed theirs with soiled sheets they'd soon be airing in public. Those who had been the campaign's cancer were about to begin a campaign of their own — against *New Zealand*'s designers — as they plotted to secure a multi-million dollar purse in a state broadcaster's pocket. They might not have thought it then or even acknowledge it now. But they were about to embark on one of the most reprehensible episodes in New Zealand sporting history.

Chapter 28
Full Circle

*There is a certain irony that Farr's two decades at the top
of yacht design have seen the rules almost do a complete
circle to produce the boats he would have liked all along.*
Shane Kelly, New Zealand yachting writer[1]

*If you look at the IMS boats today and put them alongside
some of the Young, Townson, Spencer and Farr designs
of the early seventies, there's not a lot of difference. In
our IMS boats we still use a style of shape similar to* Titus
Canby *and* Moonshine — *a long fine bow, the knuckle
clear of the water at the bow, a broadish stern, moderate
beam, a low centre of gravity keel. So the concept that
developed in New Zealand as a general keel-boat shape
and style ... is still completely valid today.*
Bruce Farr[2]

Socially elitist, historically odd and sitting outside the mainstream, the
America's Cup still commands a following disproportionate to its contribution to the sport of keel-boat racing. While the money it attracts and
research it engenders make it impossible to ignore, there is still an ocean
more to yachting than the oldest sporting trophy in modern sporting history.

Unlike some designers who are free to attach to a syndicate and devote
themselves completely to its cause, an America's Cup programme rarely
absorbs more than 40 percent of Bruce and Russell's time. So while *NZL20*'s
loss in San Diego was a huge personal blow, it was still business as usual
for the expatriate pair and their busy Third Street team.

In 1990, Jennifer Smith, ex Beneteau USA, had joined Bobbi Hobson to
help with the burgeoning administration. Dave Ramos and Ed Frank, who
left that year, had been replaced by Mick Price, a designer-builder and avid
sailor, and Patrick Shaughnessy, a nineteen-year-old who had dropped out
of college. Hired part-time after showing Bruce samples of his work, he was

soon full-time draughting, his heart set on designer status as he was eased into keel and rudder design and construction detailing. Perhaps more than any other member of staff, he reaped the benefits of the Kiwi culture still fertile in the Farr-Bowler business after ten years in the USA. There was more than just an echo of Jim Young's spirit in Patrick's contract of employment.

> I'd answered an ad for a draughtsman and had no idea of the stature of their office when I applied for the job. Nor was I in a good space in my life — without a college degree it's almost impossible to get a design job in the USA let alone one at the top place in the world. So it blew me away that neither Bruce nor Russell minded that I had no formal education and were prepared to give me a chance, that they were so open-minded. It's magic working here, and that's nine years on. I get up each morning and can't wait to get into work. Even after one of those days when things go completely wrong. The honest-to-God truth is that if my wife and I didn't need the money I would do this work for free; this place and the people we work with are just the best in the world. I owe Bruce and Russ. Not just because they went out of their way to help me make something of myself, but because they made me feel valued, made me feel a part of the team. I just hope the growing pride I have in myself reflects a little in my work.[3]

The new beginning for Shaughnessy — as well as New Zealander 'Tink' Chambers,[4] who joined in 1991, and New Yorker Harry Dunning, hired in 1992[5] — also marked a new beginning for Bruce Farr & Associates: a new decade with new rules. But BFA had more than just survived the turbulent eighties. Throughout its peaks and troughs Bruce and Russell had not so much sought breakthroughs as refining their ground-breaking trends. It was a policy that worked. Even as businesses around the world slid into oblivion theirs continued to grow. As national economies faltered, as tycoons came under scrutiny or found themselves behind bars, they were riding high on the successes of their record-breaking maxis and the enduring Farr 40 — shapes about as good as you could get under IOR.

But the ageing rule had just about run its course. For twenty years it had produced exciting racing around the world. That was its strength. Its greater weakness was its linchpin: the concept of a 'base boat' by which it rated those it measured.[6] Anything that went faster than this theoretical cruiser-racer was brought back into line with rating penalties — or, if slower, encouraged back towards it with rating assistance. Which was why IOR had become a type-forming rule, producing hulls with more than a hint of a beer-drinker's belly about their mid-point girths. But not only had IOR pushed shape and displacement in the fat and heavy direction. As the nineties dawned, an IOR racer had become so costly that international

offshore yachting was fast out-sailing all but the very wealthy. That, however, was about to change.

In 1975, under the auspices of the United States Yacht Racing Union
(USYRU), later US Sailing, work had begun on the Pratt Project at the
Massachusetts Institute of Technology (MIT). Its primary aim was to produce
a comprehensive computer-based handicapping system for a wide variety of
yachts racing in a wide range of conditions. With IOR the de facto grand
prix rating rule, the members of the Pratt Project set out to write a handicapping system for cruiser-racers. In the process they hoped to foster fairer
competition between non-IOR boats, to reduce obsolescence and above all
to avoid producing type-forming boats that raced to its rules. In 1976 the
USYRU approved its use as a handicapping method to be used alongside IOR. In November 1985 the Offshore Racing Council (ORC) adopted
the Measurement Handicapping System, renamed it the International
Measurement System (IMS) and, as they had with IOR, brought it under the
control of the International Technical Committee (ITC).

It would have been impossible to develop IMS before the invention of
the microchip and high-speed computers. And its implementation would
have been equally impossible without the Velocity Prediction Program
(VPP) and Hull Measuring Instrument (HMI) produced by the Pratt Project.
But once these tools were developed, the new handicapping system offered
enormous advantages. Error-free, the hull-measuring machine could do in
hours what took IOR measurers weeks. Gone was the hand-measuring of
predetermined points along a hull. Instead, reading the hull's shape, measuring the dimensions of its rig and sails, and performing the inclining test
to measure its righting moment, the HMI converted the information into a
form the IMS VPP could use to produce a hull elevation, calculate its hydrostatic properties and take it for a sail.

Calculating the forces involved in 'sailing' the computer model — balancing the aerodynamic forces on the sails against the hull's resistance and
the yacht's stability — the IMS VPP produced speed predictions of the yacht
expressed in seconds per mile and as functions of true wind speed and
angle. Like handicapping a golfer against the golf course (expressed in
strokes per hole), these were converted into time allowances (expressed in
seconds per mile), over one of six course types[7] and one of seven true wind
velocities[8] chosen by the race committee from predicted or recorded conditions on the day of racing. Racing against itself in a race against other
yachts, the subject's performance could now be measured against the time
its measurements had told the computer it should have taken to sail the
course. While the method itself is complex, the results are simply expressed.
To determine a race winner 'the program compares each yacht's elapsed
time with its speed predictions for the prescribed course and derives an
implied wind strength. First place goes to the yacht whose implied wind is

greatest ... the one which made the best use of the conditions on the course and sailed to the greatest percentage of its potential.'[9] Here at last was a handicapping rule that didn't demand shape distortion to reduce rating penalties if a boat didn't meet the girth station measurements IOR thought best for offshore racing yachts.

Bruce would later say: 'One of the wonderful things about designing to IMS is the relative freedom that a designer and client have in choosing their type of boat.'[10] But because it began life as a rule for cruiser-racers and had still not developed into a grand prix racing rule by the end of the eighties, it was not for BFA. At least not until 1991, when Bruce and Russell could no longer ignore the number of their competitors involved in IMS. Seeing their designs compete against the best in the world was what made them tick, performance measurement the incentive that kept them pushing the frontiers of design and construction. They also liked the empirical base of IMS. That, they thought, should free them from the prejudices of shape and stability inherent in IOR.

It was Geoff Stagg who convinced them to fund their first IMS design: 266. He'd sell it, he reckoned, no sweat. Which is what he did — to Christian Schmiegelow, entrepreneurial owner-president of Astillero del Estuario, a boat-building facility in Buenos Aires.[11] Its construction would be overseen by Farr International's (FI) Bryan Fishback, with FI managing its first campaign.

Straight out of the box *Gaucho* was hot. The Farr 44-footer won its inaugural race — a 160-mile feeder race from Fort Lauderdale to the one-time hang-out of literary greats Ernest Hemingway and Tennessee Williams and singer (and ageing pirate) Jimmy Buffett. But that was nothing compared with the way *Gaucho* bolaed Key West Race Week in January 1992. With American Jim Brady steering, Geoff Stagg calling tactics, FI's Australian Andie Ogilvie, Kiwis Joe Allen and Steve Wilson, Christian Schmiegelow and his factory manager as crew, the new Farr demolished the 119-boat fleet from twenty-two states and six countries. *Gaucho*'s five wins from five starts won it the performer of the week award. That it won with margins no closer than three minutes thirty-two against boats feet longer astonished all who saw it. Things may have gone bang among competitors as they suffered knockdowns, shredded sails and broken gear in gusts of 30 knots. But the biggest bang of all was 'the bang that *Gaucho* made ... breaking through the International Measurement System's equivalent to the sound barrier'.[12]

They'd done it again. With no history in IMS, BFA had created another breakthrough boat. Unlike IOR, there were no trade-offs in IMS. Bruce had drawn the hull he'd wanted and the weight Russell saved in construction went into a keel 8.75 feet (2.67 metres) deep. That not only greatly enhanced *Gaucho*'s stability. With a displacement-to-length ratio bordering on the ultra-light and a high sail-area-to-displacement ratio (the equivalent of motor racing's power-to-weight ratio), it was devastating upwind. Beautifully balanced and small for its rated size, like key Farr boats before it *Gaucho* pointed the way ahead.

Successful *Gaucho* may have been, but it was also business as usual in the secateurs department. Complaints that it hadn't been designed to the spirit of the rule formed the early Key West scuttlebutt. Unable to be sustained after a public showing revealed *Gaucho*'s interior exceeded the minimum interior requirements of IMS, that didn't stop the US IMS Owners' Committee swinging into action. It was a case of history repeating itself as they moved swiftly to protect the year-old yachts that a single Farr had made obsolete overnight. And they went for the jugular. Without any scientific justification they placed penalties for deep keels and arbitrary pitching moments into the IMS VPP. Bruce Farr:

> The logic went like this: the IMS must be encouraging deep boats because deep boats are winning. That means there must be something wrong with the IMS because it's a rule for cruiser-racers [and cruiser-racers have never been deep in this part of the world]. That led to an arbitrary penalty for draught which was carefully engineered not to go right across the board; it didn't take effect until it was relatively deep and then became very steep. That is, the new formulation only applied to the top five percent of the fleet but did so so effectively that it killed its competitiveness.[13]

At the heart of the politics were the 'heresies' for which Bruce had been keel-hauled by the ITC at the beginning of his career: beam and stability. They came from the same people, the same place: the American northeast. Even after two decades of being beaten by Farr-Bowler boats, this powerful cruising contingent still hadn't seen the light — that while deep keels don't stop you getting your Boston boat shoes wet, they do give you more time to enjoy your gins and tonic once you get to where you're going. This time, however, Bruce didn't pack his tent. Not even when the US connection sold their complaint to the ORC as a necessary correction to the IMS formula. Instead, he and Russell employed the services of VPP consultant Clay Oliver and wrote a heel-drag formula which proved that the US Owners' Committee correction was flawed. It helped that when the ITC vote was taken there were more technicians than politicians on the committee who understood the Farr-Oliver formulation. It also helped that Bruce was now a member of ITC, establishment almost, no longer an unknown young New Zealander timorously seeking to be heard in snow-covered Newport.

Further key decisions were taken at the ORC's council meeting in November 1992. The first — with vocal encouragement from Bruce and

several other observers — was to separate IMS into two divisions: one for
flat-out racers with minimal interiors, the other for less spartan cruiser-racers. The second was to improve the IMS VPP drag formulations with input
from Holland's Delft University and America's Cup research programmes.
And two years later, after an immense amount of work from Bruce and others on the ITC, a set of formulations with which IMS could move with
greater certainty towards the new millennium had been completed.

> To get it right in IMS you have to prove it scientifically; only then will you
> have a rule that will allow yachting to develop in a right direction. If a handi-
> capping rule is artificially tipped to favour a particular boat type you will get
> biased evolution, a branch of design that will eventually fail because it doesn't
> do the job and because most people don't want it. You can't have a rule that
> says a boat ought to be slowed because it is winning races.[14]

This was basic Farr, the same message in 1995 that he'd delivered in 1973:[15]
his politics-free commitment to making yachts fun and fast to sail.

***

> *The Mumm 36 ... designed by Bruce Farr [& Associates] ...*
> *is now built in five different countries by five different*
> *companies, and is the first design of the modern era to*
> *have cut across national preferences and prejudices.*
>   Stuart Alexander, UK Sailing Correspondent of
>   *The Independent*[16]

With the economic downturn still biting and IOR gasping its last, Jim
Allsop's 50-footer customers were wanting to downsize but could find nothing smaller that enjoyed the same high level of competition. The Annapolis
North Sails' manager mentioned the problem to Stagg. Soon FI's phones
were running hot, Geoff Stagg and 'Tink' Chambers totting up the inquiries
they had for a mid-sized IMS Farr for club and international competition. It
didn't take much to convince Bruce and Russell to fund the design. Or for FI
to put together a package with Carroll Marine[17] and three investor-buyers to
fund the tooling.

Design 299 would be 35.86 feet (10.93 metres) with a draught of 7.38
feet (2.25 metres). It would of course be classic Farr: fractional rig, destroyer
bow, fine entry, broad flat after-sections, a wide transom, a wedge-shaped
deckhouse dropping to a long clean cockpit. It would also incorporate those
ideas that had recently sent the Farr name racing to the top of IMS: light-
displacement, high stability, a long waterline-to-overall length and powerful
sail-area-to-displacement and displacement-to-length ratios. Conceptually it

would be a development of *Cookson's High 5*, a Farr IMS 40. To keep both weight and cost down it would be production-built in epoxy resins and E-glass over a PVC foam core, with balsa core in the high-load areas, the hull vacuum-bagged and post-cured in an oven.

Stevenson Kaminar III had a good idea of what he'd be getting when he ordered hull Number One and named it *Predator*. So did New Jersey's David Clark when he bought one of the first.[18] He'd already won the One Ton Cup with *Vibes* in 1991 and was hoping for more of the same in IMS with *Pigs in Space*.

> I told Bruce and Russell that I didn't want a breakthrough boat, that you don't have to re-invent the wheel to have a fast car. I'd originally chosen them for two main reasons: their boats were winning and they came highly recommended by *Brava*'s Pasquale Landolfi. What I discovered was that they are very good when given specific mandates and are both engineering oriented. They pay immense attention to detail and deliver a finished product, fully tuned. That was perfect for someone like me because it meant I didn't have to spend hundreds of hours finishing and preparing it for competition.[19]

If the Farr IMS 36 wasn't a breakthrough boat, it was the ultimate refinement of the Farr-Bowler thinking that had changed the shape of speed. Gleaned from their database of success, it would soon be among the winning stats surfing in faster than ever from all around the globe.

By the end of 1992 boats by BFA had won five of the year's international regattas. Italian Pasquale Landolfi's *Brava Q8* had won the World One Ton Championship, the sixth Farr to do so. New Zealand's Neville Crichton had finally turned to Farr and been rewarded with *Shockwave* and the World Two Ton Cup. Mick Cookson's *Cookson's High 5* had taken first overall in the Honolulu-hosted Kenwood Cup, winning every race in the IMS division and finishing 60 points clear of the second place-getter. In that same regatta the Farr Two Tonner *Larouge*, owned by Italians Dr Giuseppe DeGennaro and his son Davide, was first in its class and overall winner in IOR. *Califa 3*, a Farr IMS 44 owned by Arturo Arrebillaga of Argentina, was the best of 36 yachts in the inaugural Commodore's Cup sailed at Cowes. *Merit Cup*, skippered by Pierre Fehlmann, won the OMYA World Cup for the third year in a row against eighteen other Maxis. And the 1989 Three-quarter Ton design, *Xacobeo 93*, the renamed *Lone* owned by Jan Bonde Nielsen of London and helmed by Spain's Pedro Campos, had achieved what no other yacht in ORC-sanctioned level-rating racing had before it: winning a World Ton Cup three years in a row. There were other regatta victories for Farr designs that year. *Mandrake*, an IOR 50-footer owned by Giorgio Carriero and helmed by Francesco de Angelis, won the Mediterranean Cup. *Lanjaron* (ex-*Juno*), another Farr 50, won the King's

Cup in Spain. *Cha Cha II* (ex-*Vibes*) won the IOR B class at the Japan Cup and its class in the Corum Cup Kansai Championship Series. *Flash Gordon*, the new Farr IMS 39 built by Cookson Boats and owned by Helmut Jahn of Chicago, won every race boat-for-boat in the 40-foot Class Championship at Harbor Springs, Michigan. Touichi Okada and a team of Nippon Telephone and Telegraph employees sailed *Dreampic*, a Farr IMS 44 from Ian Franklin's New Zealand yard, to a win in the IMS division of the Japan Cup.[20] And sailing out of Portofino, Italy, *Morgana*, a Farr 72-footer fast-cruiser owned by Dr Francesco DeSantis, was first in the IMS OMYA fleet. *Morgana* was also judged the most elegant yacht of the series, belying the spurious notion that Farr-Bowler boats were only racing machines. Chartered by David Clarke after the Kenwood Cup, *Cookson's High 5* won the Atlantic IMS Grand Prix class in the San Francisco Big Boat series with a 1/1/1/2/1 result. And for the third year in a row the Royal Ocean Racing Club (RORC) awarded their prestigious Yacht of the Year title to *Bounder*, another Farr-designed Beneteau-built First 45F5.

To top the year off, Geoff Stagg was invited to join a small steering committee established by the RORC. The worldwide recession, the growing cost of racing yachts and the collapsing IOR meant the RORC had to take drastic action to ensure its Champagne Mumm Admiral's Cup (CMAC) remained the world's premier offshore fleet regatta. The first and obvious decision was to begin phasing out IOR and phasing in IMS. The more difficult task for the six-man committee was to decide on the three classes to sail the 1995 and 1997 CMACs. The ILC 40 was the easiest one to pick; it was the up-and-coming class.[21] The big-boat decision turned into a hedged bet: IOR handicapping for 44- to 50-footers in 1995 to be replaced by ILC 46s in 1997. It was in fact the small-boat decision that was the biggest surprise of all and showed just how committed to change the RORC really was. 'In an attempt to keep costs down, make it easier for countries to send teams and create an "entry level" boat, the steering committee decided to recommend the small boat in 1995 and 1997 be a one-design, a most radical decision.'[22]

Geoff Stagg's appointment had been a singular honour. He sat there not just because the trustees wanted to hear from yachties. Or because of the contribution Bruce and Russell had made to offshore yachting. He sat there out of respect for his commitment to the sport and his standing on the international racing circuit. He sat there with Luc Gellusseau, head of Corum Sailing. With Baring's Bank's John Dare. With Alan Green who, for four decades, had been involved with the Admiral's Cup. With RORC rear-commodore Stuart Quarrie and the ORC chairman, John Bourke. The outspoken Wellington yachtie, who had never really been accepted in Auckland circles, sat there at what could not have been a more perfect time for the Farr design and sales group. Not long before *Pigs in Space* had shown up as a potential one-design IMS racer. At its first outing, the Chicago-Mackinac Race, it was first in class and second in fleet and soon after won both fleet and class in the Round Long Island Race.

> The committee said: What about a 36-foot international one-design boat? I said: You're crazy! But after a lot of discussion and John Dare's ruling that the decision would have to be unanimous, I said: Fine, I've got the perfect boat for you ... and pulled it out of my briefcase. Soon after I withdrew; there was now a clear conflict of interest. Besides, there was a hell of a lot of work to do in preparing our IMS 36 tender document.[23]

Six proposals were received by the RORC's International Mumm 36 Selection Committee.[24, 25] All were carefully examined. Included in the committee's due process were questions asked of the designers and visits made to the boat-building facilities allied with each proposer. With the investigative work done, the committee took all day on 26 April 1993 to make its decision. It was just before midnight when John Dare emerged from the meeting to read the RORC's statement of choice: 'Selection has been extremely difficult due to the quality of the submissions, but in the final analysis the Farr proposal most closely matched our criteria.'[26]

What the RORC had chosen was, in fact, a series-built one-off, the shape of things to come. It was a decision that not only fuelled a rush of interest in the next CMAC. It was to send FI's sales into orbit. As at April 1993 they had sold eight Farr IMS 36s. By the end of 1994, 86 Mumm 36s had been sold to buyers in Australia, England, France, Germany, Italy, Japan, New Zealand and the USA. These were being built by Farr-licensed builders in Argentina (Astillero del Estuario), France (Chantiers Beneteau), New Zealand (Cookson Boats), South Africa (Robertson and Caine), and the USA (Carroll Marine). Chosen as Boat of the Year by *Sailing World* in March 1994, the numbers surprised no one. 'Over the last 10 months [the Mumm 36] has won nearly every major open IMS contest in the world — Key West Race, the Southern Cross Cup, and the St Francis Yacht Club Big Boat Series among them; it's the case of a production boat upstaging one-offs built at twice the expense.'[27] Built to a set of strict one-design regulations, with Class rules written by Russ Bowler and FI's Bryan Fishback, the Mumm 36 had become the fastest-selling one-design keel-boat of its size in the world. Admittedly it helped boost the percentage of Farr-Bowler boats in the 1995 CMAC to 66.6 percent of fleet. But sixteen out of twenty-four[28] was still an astonishing statistic compared with the 42 percent of fleet Sparkman & Stephens had achieved at its peak in 1973 and 1975.

Things had come full circle. For over two decades Bruce and Russell had disproved the theory 'that a displacement sailboat in flat water cannot move a fraction faster than the product of 1.3 times the square root of the boat's waterline length'[29] — and now an extraordinary number of sailors believed them. 'Not to take away anything from the designers', said Bill Schanen, editor of *Sailing*, 'but the underwater shape of *Pigs* ... seems obvious. This must be the way the god of sailboat design wanted hulls to look like. We just had to endure centuries of drawing complex curves resembling the shape of a claret glass to get there ... But it would be a mistake for adherents of traditional design to assume that light-displacement boats give anything away in seakeeping ability. Crews of the Whitbread 60s [W60s], which are essentially larger versions of the Farr 36, were ecstatic at the ease with which they were able to manage the boats in extreme conditions in last summer's transatlantic race. It figures: boats that don't have to move tons of water to achieve forward motion are not only faster; they're easier to sail.'[30]

Schanen was right. Just as he was about the Whitbread Offshore Rule (WOR) that had of late also been demanding Bruce and Russell's attention. The W60s, products of this new rule, were about to test their theories to the max in what would amount to a last-ditch battle for yachting's soul. No one would have seen it quite like that — sponsors as government bankers, organisers as strategists, designers as military observers, boats as weapons, sailors as soldiers, the fundamentalists and evolutionists the real gods of war in yachting's Armageddon. But as the Maxis lined up against the W60s, that is precisely what the 1993–94 Whitbread Round the World Race would be about.

Chapter 29
Evolution

Far from being a debate about whether an 85-foot yacht should be faster than a 64-footer, the performance of a Whitbread 60 against the Maxis is an indictment of the International Offshore Rule and the slow, cumbersome leviathans that it has encouraged designers to produce.
Richard Gladwell, New Zealand yachting writer[1]

The idea of a smaller Whitbread class wasn't new. But it took a cocktail party during the Uruguay stopover of the 1989–90 Whitbread Round the World Race before the idea was floated by writer Bob Fisher to those who mattered and a general meeting called for the second-last stop, Fort Lauderdale. Attended by sailors, designers and race officials, the Fort Lauderdale gathering chose a date (June 1990) and venue (Goodwood, England), and a group of designers it charged with producing a design that would be 'faster, safer, more exciting to sail and less expensive than their IOR counterparts'.[2] The new rule was not to rate existing boats or calculate their handicaps. Instead, like the recent IACC rule, it was to produce a restricted class of yacht with minimal variation between the primary performance parameters. A provisional version of the Whitbread Offshore Rule (WOR) was circulated late in 1990 before the final version was released in January 1992. It had settled on an overall length of 59–65 feet (18–20 metres), a displacement of 29,700–33,100 pounds (13,500–15,000 kg) and a maximum draught of 12 feet (3.75 metres). It set a maximum working sail area of 2200 square feet (200 square metres) with the mast and fore-triangle set at predetermined heights. Significantly, it also permitted a maximum of 2500 litres of salt water (2562 kg/5649 lb) to be carried in one of the two side tanks at any time. This meant an extra 5500 pounds (2500 kg) of ballast or the equivalent of 30 crew sitting on the windward rail.

The beauty of the WOR was that it encouraged stability and placed virtually no limits on speed-producing factors, the principles a young Bruce Farr had first used in *Titus Canby*. But the Whitbread race officials were not yet ready to let go of the rule that had run offshore yachting for the last twenty years — and in the form of its progenitors for fifty years before that.[3] It was a decision based less on sentiment than on cold, hard cash. The IOR had, after all, been responsible for producing the glamour boats of their fleets: the Maxis. They, in turn, had attracted the sponsorship which had transformed this once-obscure adventure sport into an international event with more worldwide interest than the America's Cup.[4] Using the two rules would undoubtedly create a win-and-place comfort zone for the Whitbread race organisers. Yet by having a race within a race and a prize for first around the world, they were also turning the 1993–94 event into a show-down between the two schools of thought that controlled the shape of keel-boats: those who projected their vision of what an offshore yacht should be into designers' calculations via their mathematical formulae and those who encouraged straight boat speed. Even if unwittingly conceived, having this battle in the famous race was not a bad thing. But which school would win?

It was a simple question that had no simple answer simply because superstition still held hands with science in yacht design. While the shape and personality of Farr-Bowler boats had made them winners all around the world, there was still little understanding that the Farr-Bowler principles of design and construction were as fundamental to freeing yachting from heavy-displacement thinking as Copernican theory had been in freeing science from Ptolemaic fallacy. But here for all to see, in the forthcoming Whitbread, would be the two chief ways yachting measured its universe. And testing the two hypotheses would be the sport's master designers, with samples of their work sailed by both schools of thought. In that sense it could not have been a fairer fight — this battle that had grown out of the complex task of measuring different sizes and shapes of yachts so that they could race against each other on equal terms; a battle that had raged for years and was fiercest when rulemakers grappled with stability and its effect on monohull speed.

And it was precisely on this issue that the two measurement systems to be used in the next Whitbread were poles apart. One encouraged motor racing's equivalent of horsepower, the other penalised it. By refusing to measure stability — indeed, by encouraging it through the use of water bal-last — the WOR ignored the convenient but unscientific notion of a 'base boat' as the basis for both measurement and handicapping. Instead it embraced a foundational principle used by its more advanced cousins, the aerospace and motor-racing industries: generating as much power as possi-ble within the lightest possible body. IOR, on the other hand, penalised stability. It did this by giving the more stable (stiffer) boats a higher rating (penalty) than the less stable (tender) boats because the greater stability of the stiffer boats made them faster than the fictional boat by which it measured performance. It further disguised its lack of logic by giving the less-stable boats a rating bonus to bring them up to speed. The problem was compounded for Maxi yachts because their formula in this rule already made them relatively heavy.[5] But as a result of IOR's penalties for stability — as it sought to equalise the boats it measured by trading off (rather than encouraging) the primary performance factors which make monohulls go fast: length, displacement, beam, draught and sail area — they ended up with a centre of gravity one metre higher than the W60s. That is, because the speed-producing potential of weight held low (not weight per se) was restricted by IOR's view of stability, the maxis were a lot less powerful than their smaller opponents.

Bruce saw it mathematically. 'The Whitbread 60s have 60 percent of the sail area and 85–90 percent of the stability of the Maxis. In sailing trim, with full ballast, the W60 is just over half the weight of a Maxi. The W60, in full ballast, will generate about 80–90 percent of the sail forces of the Maxi but its hull has only half the drag, or resistance — which equates to significantly higher speed relative to its length. The effect of the filling of the water ballast tanks is for the heel of the yacht to be reduced by 10 degrees, or for the stability to be increased by one third, in round numbers. The combination of reduced hull drag, high stability and lighter weight goes a long way to overcoming any length deficiency between the two classes.'[6]

As they had for the 1992 America's Cup campaign, Bruce and Russell embarked on extensive research of the new rule before they had client contracts — in fact, before the WOR had been completed. They twice tank-tested eight-foot models in a variety of flotations and trims to obtain data to produce their own W60 VPP. Then, after receiving contracts and their clients' decisions on weather and waterplane width, they set about refining their prototype. They massaged hull shape, especially around the water ballast tanks. They minimised hull and deck weight within conservative practice. They refined the 100-foot rig. They expended time and money in keel and rudder reduction, the slender new shapes making significant gains in drag reduction though not so much as to lose the gains to inefficiency in countering leeway. So much effort went into designing these early models that a third generation had been produced before *Yamaha*, their first W60, was built for New Zealander Ross Field.[7] 'All the time the helm was very easy. It was just about one-hand-on-the-wheel stuff. The faster it went, the easier it became,' said the ecstatic skipper, describing the handling characteristics of his waterborne Porsche during an early test.[8] And as the numbers Field amassed in his on-board computer were fed back to Annapolis, it soon became apparent that the W60s, with their high stability and dynamic bursts of speed, might even beat the maxis boat-for-boat.

Their speed was officially confirmed when Dennis Conner's 65-foot *Winston* broke the transatlantic record that had been held since 1905 by the 185-foot schooner *Atlantic*. Averaging 10.5 knots, *Winston* reduced the monohull time for the 3000 miles (New York Harbour to the Lizard) by ten hours. 'On three days we did more than 335 miles; any yacht that can do one thousand miles in three days is impressive.'[9] Conner might have added that he no longer thought Farr 'a loser'. Being on the same side as those who had been his greatest threat in 1987 and 1988 had, it seemed, turned the angry talent into a model and grateful client.

Already this Whitbread was shaping up to be an extraordinary race. Not just because of the electrifying speed of the W60s or the designers' efforts to make the giant maxis even faster than before. But because the race was turning BFA into what looked like the Annapolitan branch of the United Nations. Field and Conner weren't the only ones with boats from the former Auckland boys. The premier of Galicia had announced their Javier de la Gandara-skippered entry of a Farr W60 to promote the culture, cuisine and industry of the north-western Spanish province. Roger Nilson, backed by the pan-European debt collector Intrum Justitia, had also given notice that he would be representing the European Community in one. New Zealand's prodigious talent, Chris Dickson, in his first Whitbread, would, like Field, be in with Japanese backing and one Farr design in his two-boat campaign.[10] After surviving the last Whitbread on *Fazisi*, Eugene Platon was back, this time promoting the Ukraine in *Hetman Sahaidachny*, another Farr W60. And the first of the two *Yamahas*, it was later announced, would be raced as the *US Women's Challenge* headed by Nance Frank.[11] There was a similar international feeling among the three new maxis for the 93–94 Whitbread, all Farr-Bowler ketches. Grant Dalton was determined to take line honours in his fourth Whitbread with *New Zealand Endeavour*. Pierre Fehlmann would be flying the Swiss flag from his new *Merit Cup*. And Daniel Malle was back on the job with his fellow French posties on *La Poste*.

It seemed fitting for a race that would be its own voyage of discovery that Bruce, like his great-great-grandfather, would also draw a vessel named *Endeavour*. Eighty-six years earlier the retired Captain Clayton had used brushes, oils and canvas. For Design 274 Bruce used a screen, a mouse and a VPP. The differences were just as great in the materials used to build the two boats. Cook's *Endeavour* used wood, metal, cotton and hemp; Dalton's Nomex, carbon fibre, Spectra and Vectran. The cat-built bark, although only 21 feet (6.4 metres) longer, carried five times the crew and displaced twelve times as much as the upmarket Maxi. 'It's an eggshell on steroids,' said Steve Marten of the exotic hull and deck which, at 2000 kg (2 tons), weighed little more than a Toyota Camry sedan.[12]

But there was more than just weight saved in these latest Maxi ketches. They had a narrower hull planform. They had larger mizzen masts, only two feet shorter than the main mast. There was also a much greater separation between the two fractional rigs. This meant little overlap and a much improved airflow which enabled the efficient mizzen staysails to be carried at closer angles to the wind. As with the W60s, a huge effort was put into fine-tuning appendage shape, with the hull weight saved being placed in the narrow foil and its streamlined bulb. All three had extreme clipper bows.[13] This narrow platform at the bow, another Farr-Bowler innovation, enabled crews to get the downhaul directly under the end of the spinnaker pole to help get more luff tension on the large gennakers the Whitbread race organisers had permitted the maxis to use, in the hope that this would give them a Maxi line-honours win..

New Zealand Endeavour quickly proved its lineage by being first across the line in the 1992–93 Sydney–Hobart.[14] And after winning the Maxi division from *Merit Cup* and *La Poste* in the 1993 Round Europe race, Dalton took on the Fastnet with Farr and Bowler on board. It was the first time Bruce had raced the Fastnet since crewing on *Gerontius* in 1975. Russell never had. *Endeavour* was his sentimental favourite for the Whitbread although he, like Bruce, knew the maxis' days were numbered. It was a message written in lights when *Endeavour* finished fourth in the Fastnet behind three Farr W60s.[15] Perhaps that's why, back in Annapolis, Russ put pen to paper.

The crew of the *Endeavour*,
That mighty ketch-rigged boat,
Asked me to go racing
And supplied a hat and coat.

The start was quite a spectacle,
A Solent full of sail.
We set out for the rock
With *Fortuna* on our tail.

She did not stay there long,
She vanished in the fog.
Perhaps she needed tuning,
Or perhaps she's just a dog.

My watch began at midnight,
I went looking for my gear.
In the dark I could not find the stuff,
I was in complete despair.

Some prick had stole my coat and hat,
Essential when you're old.
Some prick was keeping nice and warm
While I was freezing cold.

In great distress I stole a hat,
I took it off the rack,
But the owner recognised it
And took the darn thing back.

It may amaze and shock you,
It caused me to be sick.
The exposure in windy weather
By some thoughtless prick.

I think the ship *Endeavour*
Will win the Whitbread race,
But watch out for the hat thief
Lurking around the place.

I won't say who I think it is,
It's not my place to squeal.
I hope he's read this plaintive tale
And feels a proper heel ...

I won't record his name in print
But I feel compelled to Grant
A warning to my crewmates,
His first name rhymes with aunt.

His second rhymes with Bolt On,
Which is what your hat should be.
Use the wire to secure them
Before you go to sea.

Good fortune in the race,
Wear the bastards down.
I think you guys can do it,
Be first in every town.[16]

While not first in every town, *New Zealand Endeavour* was first in three of six. But while Grant Dalton and his crew achieved their goal of a line-honours win, they did so by a mere nine hours from Ross Field's *Yamaha*. And even that might not have been had *Tokio* not lost its mast on Leg Five. Or if the W60s had been permitted to use the masthead gennakers on more than Legs One and Five.

It was another superb double for New Zealand and the Farr-Bowler team. But the reality was that the maxis had been outperformed by their

20-foot shorter rivals. In the fierce winds and massive following seas of the Southern Ocean, the WOR had shown just how important stability really was. It was on the second leg, from Punta del Este to Fremantle, that *Intrum Justitia* set a world monohull record, covering 428.7 miles in 24 hours.[17] Averaging 12.4 knots, it took 39 hours off *Steinlager 2*'s time for the 7558-mile run. And after leading around Cape Reinga, *Tokio* was beaten into Auckland by *Endeavour* by just two minutes after 3272 miles from Freo. The fiercely focused Dickson may have suffered an error of judgement when, in front of 30,000 New Zealanders, he said: 'If they [the *Endeavour* crew] had been racing in the W60 class boat for boat they would have been half a day behind *Tokio*.'[18] But too easily forgotten was his public apology, and too easily missed was what his statement revealed: that a new era had dawned; only halfway around the world and the race was over for IOR. Besides, he also sensed what lay ahead as the fleet headed for Southampton via Cape Horn. In one six-hour run *Tokio* would cover 126 miles at an average speed of 21 knots and come within 1.1 miles of breaking *Intrum*'s 24-hour mono-hull record. And surfing down the Solent at 25 knots, *Tokio* would sink *Steinlager 2*'s record for the 3818-mile final leg by four hours four minutes. By contrast, the best 24-hour run achieved by *New Zealand Endeavour* was 405 miles at an average speed of 16.9 knots. It was a highly commendable effort from a highly disciplined crew. But despite Dalton demolishing Blake's circumnavigation record by eight days and four hours, the obvious was clear to all: the aged IOR had just been buried at sea by the racy WOR. Evolution had won.

And so had Bruce and Russell. With the first eight finishers, with nine in the top ten, and ten of the total fourteen boats to their design, and with the first seven of those breaking *Steinlager 2*'s record, it could reasonably be said that they had made the greatest race of all their own.[19]

But that's not how it was with the America's Cup. And nor would it be if 'Blakey from Bayswater'[20] had anything to do with it.

Chapter 30
Gang New Zealand

*Evil ... [is] defined most simply as the use of political
power to destroy others for the purpose of defending or
preserving the integrity of one's sick self.*
M Scott Peck in *People of the Lie*[1]

Give a lie a twenty-four-hour start and you will never overtake it. But give it
five years in the desert of deceit and you just might run it down. And run it
down we must if we are to uncover just how much Peter Blake and his
gang-mates have deceived the public by disseminating their calumny
through the media. But unravelling prevarication can be an exacting business. So before examining the epistolary evidence we should first examine
the multi facets of the fraud.

That Bruce Farr could not or would not work within a team environment
The Strip, **February 1994:**[2]
'[Mike] Spanhake admits the team was a little disappointed when Farr signed
an exclusive contract with Dickson's challenge. "In a perfect world we
would have Bruce Farr involved too, but it didn't work out. We began talking, but drifted apart and didn't get around to pursuing it further. This time
round we had decided the design was to be a team effort and that's possibly
a concept which didn't suit Bruce".'

John Matheson in the *Sunday News*, February, 1995:[3]
'Farr ... turned down the chance to work with Team NZ ...'

Andrew Sanders in the *Sunday Star-Times*, October 1996:[4]
'Farr has long been tagged as New Zealand's yachting design guru, and has
a string of international successes to his credit. But Coutts had major doubts
... He met with Farr early in the piece and explained how the campaign

would be managed.[5] As a result of the meeting, Coutts says Farr sent the team a letter which he said was in the form of a report card.[6] It was a critical analysis of the management aspects of the New Zealand campaign and gave grades. The main grade was a D ... What the Farr letter did do was show that he couldn't work in the team-driven, question-everything environment planned by Blake's syndicate.'

Russell Coutts in *Course to Victory*, 1996:[7]
'We also believed unanimously that Bruce Farr would have difficulties working within our team environment.'

This from the person who had been so destructive of team spirit in San Diego. This about the first person to create a team environment for yacht design in New Zealand — the person who had been a part of the four-man design team for New Zealand's first America's Cup, had formed design teams for New Zealand's next two and had worked with every member of Team New Zealand's (TNZ's) design team except David Egan, as well as giving most of them their first America's Cup opportunities.

That Bruce Farr failed to make a commitment to Team New Zealand by the nationality deadline
Suzanne McFadden in *The New Zealand Herald*, circa February 1994:[8]
'The Farr office in San Diego [sic] was approached by Peter Blake's Team New Zealand campaign to be part of their design team for the 1995 challenge, but did not give them a decision before the nationality deadline last May.'

Jan Corbett in *Metro,* January 1995:[9]
'Bruce Farr, the doyen of yacht designers, was initially approached to head the design team, but Farr, now an American resident, was slow to commit and the May 6 1993 deadline for naming a designer who wasn't resident in your country — a requirement of America's Cup rules — was edging precariously close. "That put us in an invidious position," says Sefton. "If we'd waited for Farr we wouldn't have been able to recruit anyone else [to the design team] other than New Zealanders".[10]

Russell Coutts in *Course to Victory*, 1996:[11]
'Farr also said his firm ... would not commit to any syndicate until just before the nationality date.'[12]

Andrew Sanders in the *Sunday Star-Times*, October 1996:[13]
'The other worry for Coutts was that Farr wasn't prepared to commit to any syndicate until just before the nationality date.'

This about a person (and a business) who was given nothing by Blake, Coutts or TNZ with which to make a commitment before the nationality deadline date. This about a person who is regarded by clients and critics alike as being professional and punctilious in all his business dealings.

That Bruce Farr would only sell his services to the highest bidder

Andrew Sanders in the *Sunday Star-Times*, October 1996:[14]
'There was a feeling in the New Zealand camp that Farr would go with the most lucrative deal and that wouldn't be coming from the Kiwi coffers.'

Russell Coutts in *Course to Victory*, 1996:[15]
'Within our small band of Cup hopefuls, we agreed that if Bruce was seeking the most lucrative deal, then the money he was seeking wouldn't be found in our coffers.'

This about the person who, with Russell Bowler, had won many contracts for their company through a policy of mid-range fees. This when TNZ failed to make any financial offer to BFA, and paid Doug Peterson a design fee so large it might even have covered the cost of hiring the entire BFA staff!

That Bruce Farr had written a letter to Peter Blake critical of his management approach

Soundtrack from *Black Magic The Team New Zealand Story*.[16]
Peter Montgomery: 'At some stage in the three previous New Zealand challenges the Kiwis had been drawn into controversy. Designer Bruce Farr had been a key player in all of New Zealand's previous challenges, but for 1995 he was not involved in Team New Zealand.'

Peter Blake: 'We just didn't really agree on how the project was going to go together.[17] And I stuck to my guns.'[18]

Anonymous interviewer: 'So did he, though.'

Peter Blake: 'Fine. But he hadn't won the America's Cup. Doug Peterson had. And we wanted to have this very open approach that I think was important. And I don't particularly like being told that my, um, my management approach — i.e., that the set-up of the campaign — was going to get a D, oh no, maybe an E for concept. I didn't think we were going to get on too well at that point.'

This after Peter Blake had told Bruce and Russell that they would be 'the only ... [designers] who would be acceptable to sponsors, public, and yachtsmen alike'.[19] This about the management advice both Russell and Bruce had given Peter Blake at his request, advice he had agreed with, and advice that soon became the basis of TNZ's much-vaunted team approach.

Russell Coutts in *Course to Victory*, 1996:[20]

'This was brought home in spades when Peter received a letter from the Farr office after I'd gone to Annapolis and met with the designer. I had outlined our thinking and our objectives and I'd explained how we saw the management and administration of the campaign.[21] The letter was in the form of a 'report card' which gave letter grades giving a critical analysis of the management aspects of our programme. The letter gave us a D, and then suggested that, considering recent communications, the grade might be downgraded further.'[22]

But Coutts had not just taken off with the fraud. He had flown it to such heights that he used it to distort historical time in *Course* — by falsely claiming at this meeting that he 'outlined our thinking and our objectives and ... explained how ... [they] saw the management and administration of the campaign'.[23] In fact, Coutts did none of the above because, when this meeting took place,[24] he was just setting out on his 'post-graduate education in America's Cup history, yacht design, velocity prediction programmes, computational fluid dynamics, wind tunnels and tank testing and all sorts of arcane subjects.'[25] But not content with playing with time, Coutts dupes his readers further by suggesting that the 'report card' letter was the chief reason for not using Farr when, on his own admission earlier in the chapter, he explains why and when the prior decision not to use BFA was made: the appointment of Tom Schnackenberg (at least a month before Russell Bowler's fax was sent).[26]

If this was self-absorption preening self-belief, the conceit became full-blown when Slick Willy-like Coutts added: 'We were astonished. I could understand why he might have a bone or two to pick with me because I'd questioned some of the design aspects and explained what the consensus view was.[27] But what I couldn't understand is how or why he would upset Peter. Peter was the one who enlisted Farr to design *Ceramco*, the first competitive Farr Whitbread boat. Peter also went to Farr for the *Steinlager 2* challenge. There was no doubt that Farr had provided great boats on both occasions, but there is also no doubt that Peter provided great, well-sailed record-breaking programmes which one would have thought were of great benefit to Bruce Farr and Associates, particularly in their infancy.'[28]

Having committed to the fraud, Coutts asks his readers to do the same by swallowing his nonsensical myth of 'infancy'. The reality was, Blake had gone to BFA in order to win the Whitbread. And neither the first nor the last time Blake commissioned Bruce Farr & Associates was it anywhere near its 'infancy'. By the end of 1981, the year *Ceramco* raced the Whitbread, designs by Farr and Bowler had won nine world titles. By the end of 1990, *Steinlager*'s 'year', that number had grown to seventeen. And by December 1992, the month these meetings took place, it had increased to 25 — more world titles than any other design office in yachting history. By way of the comparison Coutts makes, Blake had sailed off a high (8 percent) handicap as a P Class skipper, had won no provincial, national or interdominion titles in any class of yacht, held no world titles, and had only ever won four real races as skipper — a paltry record of wins for over half a million miles and a lifetime of sailing.[29] In fact, Blake's CV is so empty of wins he could have made admiral of the Armada.

So what lies behind this noisome fraud? Russell Coutts explains: 'Our choice of design co-ordinator meant cutting ties with Bruce Farr and Associates ...'[30] 'From that day [when Tom Schnackenberg[31] called to say 'he was in'] until the end of the last race more than 30 months later, Tom and I communicated with each other ... almost daily.'[32] In other words, the defining appointment of Schnackenberg — which 'meant cutting ties with' BFA — had been made at least one month prior to Blake's first official fax of 9 December 1992 asking Farr and Bowler to be Team New Zealand's designers.

So why would Blake not say: 'Thanks but no thanks. We've had a change of plan; we want to use other designers this time.' And why would he and Coutts and other TNZ members continue the charade for another seven months and the full-blown fraud after they had won the Cup in May 1995?[33]

There were, most likely, a number of reasons, not the least of which was the need to gain access to the technical information BFA had compiled while producing the four NZC designs for AC92. And given that Laurie Davidson, according to Peter Blake, 'doesn't even use a computer',[34] that need might well have reached crisis proportions until they secured Doug Peterson and his *America³* data with which they could begin originating their designs. But there was yet another, greater problem. Aligned with any other syndicate, Bruce Farr & Associates represented TNZ's greatest threat. So having decided not to use them, there were three important things TNZ had to do. First, they had to keep BFA on side in the hope of gaining technical information should they not secure a technically proficient IACC designer. Second, they had to explain to their sponsors (and anyone else who mattered) why the world's greatest racing yacht designers, New Zealanders, would not be part of their New Zealand team. Third, to secure the tacit backing of the state and the multi-million dollar backing of their sponsors they had to convince all and sundry that they would be New Zealand's only entry in the America's Cup, at least in the beginning. That meant keeping Farr and Bowler off the market for as long as they possibly could. And what better way to do that than by inviting them to the party without sending them invitations?

But that's not how this fraud began. 'Capable of deviousness in order to outmanoeuvre the competition'[35] Blake might be. But like most sophist-opportunists, in the beginning — just one month after the 1992 America's Cup had concluded in San Diego — he was coveting support for a new campaign and, as ever, searching for ideas to call his own. In fact, not only was he getting freebies from Bruce Farr. He was also angling for a job.

June 15, 1992

To: Bruce Farr and Associates

Attention: Bruce Farr and Russell Bowler

Subject: Jules Verne and America's Cup

Dear Bruce and Russell,

Many thanks for the drawings and data re the 80 day yacht ... The concept looks pretty much as I imagined it would be ...

Now, to a different matter.

The 1995 America's Cup.

I have had Sir Michael and Sir Tom talking to me about heading an America's Cup Challenge for NZ.

My reply so far has been guarded and that I want to give it some thinking time and do further research.

I have talked to many people that I consider to be vital ingredients to such an effort, and have asked them various questions, such as:

1) Does the NZ public want to have another go? ...

2) Does NZ stand a chance of winning against the technology and money potentially available to the US and European syndicates? ...

There is no place in a winning team for the in-house personal problems and egos that distorted many of the excellent aspects of the 1991/'92 campaign. There was rarely a time during the last AC when the NZ team didn't have huge personality problems of one sort or another, be they wives, girlfriends, sailing crew fighting for places on the race yacht, or a well meaning but often grumpy and cynical shore team.

3) Would NZ (and other) sponsors foot the bill?

I have asked a number of key sponsors from the last effort for their views. Most were very pleased with the exposure they gained, although Steinlager has it's doubts about the way the PR was handled and about the way they were treated. But these doubts can be overcome, I am sure ...

4) Who should design the next series of yachts? ...

Bruce, you and Russell are the only ones who would be acceptable to sponsors, public, and yachtsmen alike. They all have the highest regard for your expertise and lateral thinking ...

And now the crunch questions:

a) Would you do it??

b) Would you be prepared to work with me to help develop the programme; talk about the people necessary; etc, etc, etc, etc.

c) If I was running the campaign would you see this as the best way to win for NZ (understanding that sponsors will want a say in who is or isn't at the top of the pyramid.) Would you rather work with someone else? Or would you shy away and say thanks but no thanks??

d) And what do you see as a likely fee for your involvement?

The reason for raising this subject at this time is so that I can plan my 80 day project accordingly. And either work it into next years reasonably quiet time ... or say no to the America's Cup altogether. However, if I do say no, Steinlager probably won't get involved for 1995, and some others may follow.

There are many aspects that I would like to discuss with you at length, but understand the tightness of your current work schedule. However, perhaps it would be a good idea if I could schedule in a visit to see you in the next few weeks or so and certainly before I go back to Auckland in early/mid July.

What do you think??

Kind regards,

Peter[36]

June 24, 1992
To: Blake Offshore Ltd
Attention: Peter Blake
Peter,

Thanks for your fax. We would be glad to entertain a visit from you on your way to N.Z. ...

We are still not particularly settled about what to do about future America's Cup involvement, still a bit 'shellshocked' from the last one, but we are happy to have a chat about the potential for another shot from N.Z. ...

Let us know your plans, we look forward to seeing you.

Bruce[37]

June 25, 1992
To: Bruce Farr & Associates
Attention: Bruce Farr
Dear Bruce,

Re my potential visit to see you at the start of July, I have put back my America's Cup visit to Auckland until September, so won't bother you until later in the year. However, if you would have a further think about the matters I raised, I would appreciate it ...

Changing the subject to the Whitbread, if you have any clients who may need some help in preparing the yachts and/or their projects, I am pretty much a free agent at the moment until the 80 day proposal gets further down the track ...

All the best,

Peter Blake[38]

September 5, 1992
Peter Blake OBE
Attention: Blakey
Dear Peter,

With reference to our telephone conversation a week or two back, I set out my thoughts as to your role in a possible NZ Challenge in '95. These are mine and not necessarily the same as BF's.

I believe you were seeking endorsement from Bruce and myself for the idea of you leading the next NZC. I have tried to respond as honestly and directly as possible.

Firstly, there is only one (or perhaps 2 people ...) that could successfully get both NZ corporate and NZ people support for the sort of fiscal contributions required for a Challenge ... That would be your role.

Secondly, you do not belong in a role of campaign manager or leader (by campaign I mean boat, sails, research, design, keels, masts, sailors, rules). I don't believe anybody can *lead* the campaign. Some perceive there is a role where the right single person will provide the direction and leadership — I think this is a dream. Such a role and such a person does not exist because no one has sufficient technical knowledge in all areas to effectively lead. Also, if a *leader* is put in place a large part of the accountability stops with that leader when it should stop at department heads.

The campaign (as defined above) will be successful if competent committed communicating experienced people (CCCE people) are placed in accountable leadership positions (ALP's) in each area of activity. Some co-ordination may be required but leadership will come from that group of leaders ...

You should be involved in or even lead the setup ...

As to a meeting with BF and myself here are the possibilities:-

— Week Sept 28 to Oct 2.

— Oct 19/20

— November.

Regards,

Russ[39]

Blake seemed pleased with what he read.

September 22, 1992

Bruce Farr and Associates

Attention: Russ Bowler

Dear Russell,

Thanks for your America's Cup fax ... and ... for being honest about what you think.

As it happens your ideas are very similar to mine in many ways.

This is very much the way I ran the Steinlager campaign and that worked out well ...

I am off to NZ on the 27th so won't be able to visit you. However, if I get bogged down with discussions while in Auckland, I hope you won't mind if I give you a call.

Regards,

Peter[40]

Russ Bowler replied the same day wishing him well with his meetings. It was not until December 9 that he and Bruce heard from Blake again, by fax from his home in England.

> December 9, 1992
> To: Bruce Farr and Associates
> Attention: Bruce Farr and Russell Bowler
> Dear Bruce and Russell,
> ... I have formed a syndicate — Team New Zealand[41] — we have the backing of the Royal New Zealand Yacht Squadron — and our entry has been accepted by the San Diego Yacht Club.
> I am just back from a week in New Zealand, meeting with potential sponsors, media, and prospective key team members. We have had some preliminary planning sessions and have announced the skipper — Russell Coutts. It will be his role to help develop the direction that the project should take, along with the other key members of the Team ...
> I feel that now is a good time to contact you officially and bring you up to date on our present thinking and in particular to suggest a possible scheme of operation between Team New Zealand and Bruce Farr and Associates.
> We realise that you are extremely busy and probably don't wish to commit at this time, but if you have any strong objections then now would be a good time to raise them.
> Here are the key points:
> 1) Team New Zealand contracts BFA as the yacht designer for the America's Cup.
> 2) We plan to sail and possibly develop NZL20 and to build one or two new yachts ...
> 3) We plan to start by re-evaluating our current position in the light of final racing for the last America's Cup and the new rules and circumstances applying to the next.
> As an aid to this, we plan to solicit input from a variety of designers, informally, on a consulting basis (including BFA) ...
> 6) ... Our instinct is to welcome as much involvement as you may be able to give, but along the lines of your involvement in a Whitbread project at least ...
> 7) In the event that you are successful in separating yourselves into more than one national designer, with more than one America's Cup customer, then we could also work on a non-exclusive basis. However, we would obviously prefer that you be our designers and principal contact and we know that the New Zealand public and our sponsors would prefer that also.
> I would really appreciate it if you have time to give me an indication of your interest.
> Kind regards,
> Peter[42]

It didn't seem to fit with the notion of 'Team' that after his destabilising behaviour in San Diego Coutts could be chosen as skipper. However, Bruce did agree to see him when he turned up in Annapolis, the day Blake's fax arrived from England. They had a general discussion covering such matters as a two-boat budget, BFA as designer, Tom Schnackenberg's role as navigator and some idle speculation on future America's Cup combinations: Cayard-Frers, Dickson-Swarbrick.[43] At no stage did Coutts outline 'our thinking and our objectives'.[44] He did not simply because he could not reveal to Bruce, upon whom he was foisting his sham of a meeting, that he and Blake had, with Tom Schnackenberg's appointment, already discounted BFA as TNZ's designers. Besides, Coutts was at the beginning of his 'education in America's Cup'[45] and his ideas could have been little if any more advanced than those outlined in Russ Bowler's 5 September fax. And contrary to his claims in *Course*, the letter that was sent to Peter Blake following this meeting was neither 'in the form of a 'report card' which gave letter grades'[46] nor was it written by Bruce Farr.

> December 14, 1992
> Team New Zealand
> Attention: Peter Blake
> Dear Peter,
> Thank you for your fax of December 9 updating us with the progress of Team New Zealand.
> As the points of your fax are of a general nature we find it difficult to [make] comments at this time. We would need to have further discussions before we could make any meaningful comments. Your absence over the next few months is probably the most alarming element of the plan ...[47]
> We have been talking to Russell Coutts who has been endeavouring to set up a meeting with us before the end of the year. We assume this had your blessing ...
> We assume these meetings would be a part of the consulting arrangement with designers which you refer to in your letter. Can you confirm that Russell Coutts does have the authority to talk to us on your behalf.
> Please keep us informed of progress.
> Good luck with your Round the World campaign,
> Regards
> Russ[48]

The reply was classic Blake, and the template of the fraud.

> December 15, 1992
> Bruce Farr and Associates
> Attention Bruce Farr and Russell Bowler

Dear Russell,

Thanks for your fax reply of Dec. 14th.

At a meeting that I had whilst in Auckland with people I consider are very important New Zealanders when considering an America's Cup campaign, the unanimous opinion was that we wanted to have you and Bruce as the designers for our new yacht(s).

But what we also wanted to do was to try and get as much worthwhile outside opinion (from some selected quarters) that would help all of us to make the right decisions re style of yacht, and best type of programme considering our budget.

This doesn't mean we don't have the highest regard for your ideas — quite the opposite, in fact — but I imagine it would be an approach that you would also favour.

And we wanted to do as much as possible before the Feb. '93 deadline.

The people attending the meeting, at my request, are people whose opinions I value, but whose names MUST remain confidential at this time. They included the following:

Mike Spanhake

Peter Walker

Tom Schnackenberg

Tom Dodson

Jim Lidgard

Peter Lester

Brad Butterworth

Mike Quilter

Simon Daubney

The three persons we considered for the role of skipper were interviewed earlier in the day, as I did not want them present when it came time to discuss them ...

My role is really the catalyst to get the finance together, to make sure that we are keeping to budget (once set) and to help give the project its overall direction ...

So any thoughts you may have would be appreciated.

The budget figure at the moment is approx NZ$30 million ...

I understand that Russell Coutts has been talking to Bruce and that they plan to meet on December 24th — which would be great if there is time available. The sooner you, Bruce and he get a 'feeling' for each other, the better. I would be disappointed if this meeting has to wait for another month or so.

I know you may have reservations about my being away over the first few months of next year, but I think you will agree that I have been trying to meet with you for a long time, and you have been too busy ...

However, Russell, I would ask that we don't end up in a Dutch auction, as far as your and Bruce's involvement is concerned.

Design 182

The offshore racing yacht of the Eighties: *Propaganda.*

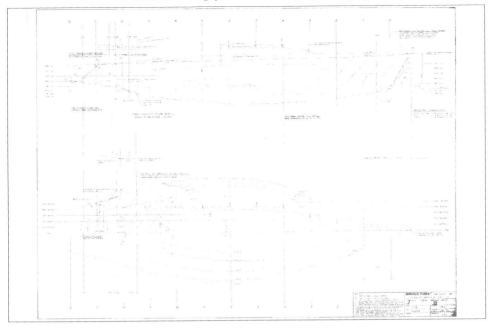

Design 182, *Propaganda*: Lines plans. *(Bruce Farr Yacht Design.)*

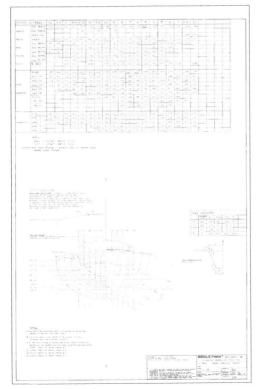

Design 182, *Propaganda*:
Sections and offsets.
(Bruce Farr Yacht Design.)

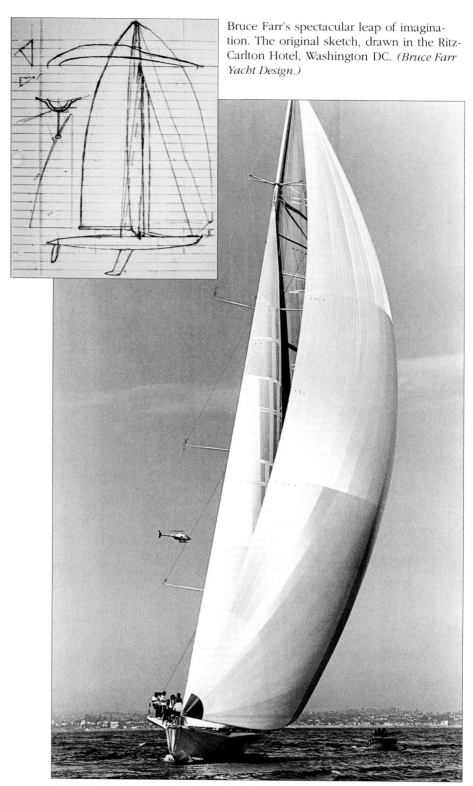

Bruce Farr's spectacular leap of imagination. The original sketch, drawn in the Ritz-Carlton Hotel, Washington DC. *(Bruce Farr Yacht Design.)*

KZ-1/New Zealand — 133 feet of sublime creativity. *(Bruce Farr Yacht Design.)*

KZ-1 under construction at Marten Marine, Pakuranga, Auckland. *(Bruce Farr Yacht Design.)*

With David Barnes on the helm, the world's fastest monohull is all grace and power. *(Bob Grieser; courtesy of Marten Marine.)*

The brilliant Kiwi ketches

Steinlager 2 (Design 190). *(Bruce Farr Yacht Design.)*

Fisher & Paykel (Design 191). *(Bruce Farr Yacht Design.)*

Put the best in the world together and this is what you get:
Beneteau's First 45F5 (Design 202).

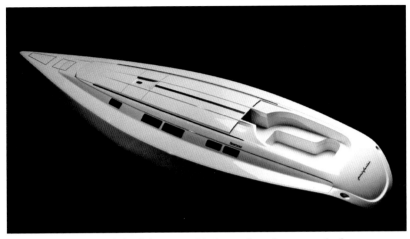

The Pininfarina model of the First 45F5, produced as a study for
Beneteau. *(Bruce Farr Yacht Design.)*

The First 45F5 undergoing sea
trials. *(Bruce Farr Yacht
Design.)*

Bruce Farr with Beneteau's Mme Roux and Francois Chalain.
(Bruce Farr Yacht Design.)

Longobarda (Design 207), Maxi World Champion. *(PPL.)*

World Two Ton World Champion, *Larouge* (Design 242). *(Rick Tomlinson; Bruce Farr Yacht Design.)*

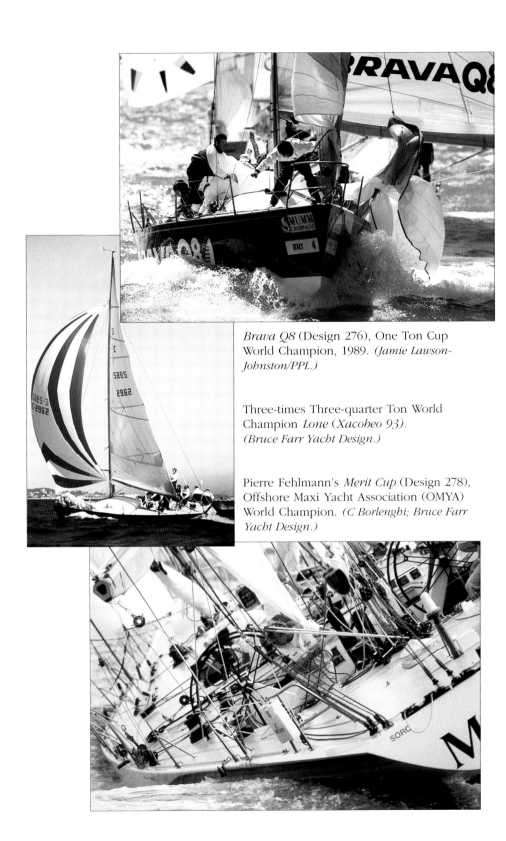

Brava Q8 (Design 276), One Ton Cup
World Champion, 1989. *(Jamie Lawson-
Johnston/PPL.)*

Three-times Three-quarter Ton World
Champion *Lone* (*Xacobeo 93*).
(Bruce Farr Yacht Design.)

Pierre Fehlmann's *Merit Cup* (Design 278),
Offshore Maxi Yacht Association (OMYA)
World Champion. *(C Borlenghi; Bruce Farr
Yacht Design.)*

Hoping to halt sky-rocketing campaign costs and to provide real-time (no handicapping) racing for the sailors, Pierre Fehlmann's dream was for a round-the-world race in identical Maxi-Monotypes rented from the organisers. Although insufficient time to complete the fifteen yachts before the Whitbread 60 class arrived on the scene brought a temporary halt to the grand plan, it still hasn't stopped competitive racing in Europe in the Farr-designed Grand Mistral (Design 309), now renamed the Maxi One Design.

Grand Mistral Number 13 sailing off the coast of La Ciotat, France.
(C Borlenghi; Bruce Farr Yacht Design.)

NZL20 (Design 237). New Zealand 'snatched defeat from the jaws of victory'. *(Alan Sefton; Bruce Farr Yacht Design.)*

'Breaking through the IMS's equivalent of the sound barrier' during Key West Race Week, Florida, 1992. *Gaucho* (Design 266). *(Chris Witzgall; Bruce Farr Yacht Design.)*

The Whitbread.

Hull and deck
layout of
Yamaha
(Design 293.)

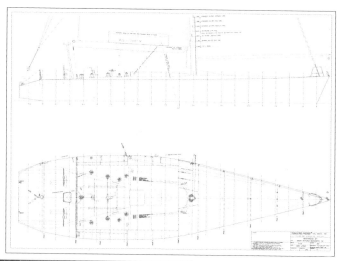

Yamaha and *New
Zealand Endeavour* at
the start of the
Auckland–Punta del
Este leg of the 1993-94
Whitbread. (*Ivor
Wilkins.*)

Tokio (Design 286)
flying towards
Southampton to win the
last leg of the 1993-94
Whitbread RTWR.
(*Thomas Lundberg, PPL.*)

New Zealand Endeavour (Design 274) winning the 1993-1994 Whitbread Round the World Race. *(Jim Lidgard.)*

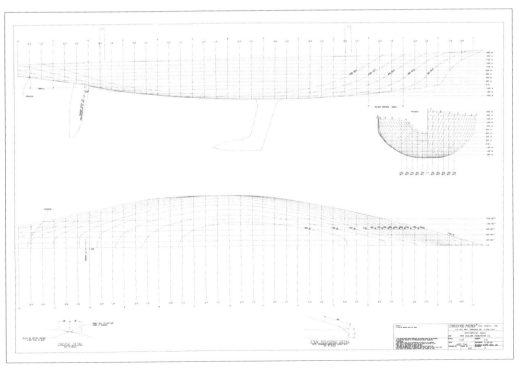

Lines plans: Design 274.*(Bruce Farr Yacht Design.)*

The start of the Auckland–Punta del Este leg of the last Whitbread Round the World Race (1997-98) before its sponsorship was taken over by Volvo. *(Ivor Wilkins.)*

The Whitbread 60 *Merit Cup* (Design 355) charging to the front of the fleet, Auckland 1998. *(Ivor Wilkins.)*

Ian Gibbs' *Swuzzlebubble 9* (Design 299). *(Franco Pace.)*

The Team

The staff of Farr Yacht Design. From left to right: Britton Ward, Graham Williams, Bobbi Hobson, Jennifer Charnesky, Russell Bowler, Harry Dunning, Jim Schmicker, Bruce Farr, Patrick Shaughnessy, Michael Price, Steve Morris.
Insets, from top to bottom: Amy Fazekas and David Fornaro of Farr Yacht Design; Geoff Stagg and Tink Chambers of Farr International.

Premises of Bruce Farr Yacht Design and Farr International, Eastport, Annapolis, Maryland. *(John Bevan-Smith.)*

Bruce and Gail Farr sailing *Farrocious,* a Farr 33, Chesapeake Bay, 1991. *(Bruce Farr Yacht Design.)*

Karli, Lynda and Gareth Bowler, Severna Park, Annapolis, New Year's Eve, 1998. *(Russell Bowler.)*

Russ Bowler during the office Laser match-racing regatta, Severn River, Annapolis, 1995. *(John Bevan-Smith.)*

Mary and Geoff Stagg, Severn River, Annapolis, 1995. *(John Bevan-Smith.)*

Larry Ellison's 79-foot *Sayonara* (Design 323), ILC Maxi World Champion. *(Ivor Wilkins.)*

I know that the Tutukaka yacht club is an accepted entry for the A.C. And I also know that C.D. [Chris Dickson] put up the money for their entry. But he has stated that he wants his Australian designer to design his A.C. yacht, and that's something I won't even consider. I think it vitally important to New Zealand that we 'do it' with New Zealand manpower where possible ...

I am approaching you (as I have been for the last 6 months) wanting you to have a major role to play as designers of the Team New Zealand yacht(s) and also to be a part of the 'think tank' of people that I feel are vital if we are going to have a chance of winning.

You may disagree, but I still think that a very close association between the design team and the yachtsmen (who may not be as technical but who often do have very good ideas) is vitally important right from the start of the project.

For designers to 'go it alone' I don't think is right ...

Sorry to go on so long ... but I am extremely keen to see you on our side. We have a great bunch of people assembled who want to win. And we all want you with us. We have good promised sponsorship backing. We have a first class skipper/helmsman who really wants to win. We have excellent crew potential, probably second to none. We are prepared to listen, to take advise, to modify our plan. We have a country that wants what we want ... We want you to be part of our Team.

What more can I say? except

Happy Christmas.

Kind regards,

Peter Blake.[49]

It was the contents of this fax — and not their 9 December meeting with Russell Coutts — that first elicited indignant replies from both Bruce and Russell.

Date: December 15, 1992

To: Team New Zealand

Attn: Peter Blake O.B.E.

Dear Peter,

I received your Fax this morning and I feel a need to reply personally. I imagine that Russell will also want to reply himself.

Firstly, thank you for a more detailed outline of your plans and your actions to date. Generally, it seems that you are off to a good start and I agree with a lot of what you say and the way that you are going about your preparation work.

Frankly, until we received your fax of December 9 we had been very much in the dark as to Team N.Z.'s activities, plans and progress, other than that contained in Press releases ...

I am not sure what is the significance of 'the Feb. 93 deadline'. As we are not involved in any syndicate, we also have no accurate idea of what is going on [on] the Challenger/Defender front ...

From your correspondence it is very difficult to ascertain exactly what role you want us to play in the short term until the Team becomes fully functional — act as consultants, sign up as designers, negotiate a deal, just help as friendly New Zealanders, jump in and get involved, or wait until something more formal is set up? Some clarification on your desires for the short term would be useful. I presume from your letter that Russell Coutts is planning to see us on behalf of the 'Team'. Could you please confirm on what basis we should deal with him ...

I was somewhat offended by some of your remarks ... Frankly, I see them as an attempt to rewrite history with your spin on it and to 'presume' our thoughts for your version of the record, and I feel compelled to comment from my perspective.

It is simply not true that we have been too busy to meet with you. We are busy, but have made time available for you. Several meetings have been tentatively set up, but YOU have cancelled them. We have made time available for meetings and even offered to meet you in England. During the same period I have also personally given you, without any request for payment, a substantial amount of my time in relation to your Round the World in 80 days project. From our perspective I see that YOU HAVE BEEN AVOIDING MEETING US.

We are interested in having a 'major role ... as designers' if the conditions look good for victory and the arrangement is workable.

I have not been invited to participate in, or even been made aware of any 'think tank' so it is unfair and inaccurate to criticize me for not participating.

We strongly support (and always have) a 'very close association between the design team and the Yachtsmen'. Please don't suggest otherwise.

I do not understand your comments about 'designer to 'go it alone'', but there seems to be some inappropriate negative inference in that statement.

Your suggestion of a 'Dutch auction' is offensive. You need to understand that we must make decisions that are commercially correct as well as ones that fulfil our emotional desires and any loyalty to countries or people.

Your intention to have no non New Zealanders involved is potentially commercial suicide for our company as 80% of our staff are not Kiwis. To pull Russell, Steven [Morris] and myself out of the company has a huge cost to our company's other income and makes it very difficult for us to do your programme without a huge net loss unless our involvement is limited so that we can continue 'normal activities', or we are appropriately compensated ... I have personally put an unreasonable contribution and commitment, much of it not compensated, into 3 previous N.Z. challenges, with consequential huge compromise and sacrifice to my personal and family life and it is pretty tempting to consider approaches that allow for a more normal life. I think it would be grossly unfair of you ... to expect us to commit to a commercially unreasonable arrangement ...

If you wish to avoid competing for our services, then you can either decide not to use our services or make us a serious proposal that we could consider. The decision is yours to make, not ours, and we don't wish to see you blaming us if you choose to not involve us. We would very much welcome some indication from you as to how you want to 'employ' our company, or us individually, and what sort of budget you consider appropriate for the design development side of the Challenge.

I hope this gives you some insight into where I 'am coming from' and I am sure Russell will also want to comment ...

Regards,

Bruce[50]

Russell did. Which wasn't surprising given that despite Blake's lengthy fax he and Bruce were still completely in the dark about Schnackenberg's appointment and the Clayton's invitation they were seriously considering in ignorance of already having been excluded from the TNZ team they were still being asked to join.

December 15, 1992

To: Blake Offshore

Attention: Peter Blake

Dear Peter,

Thank you for your fax's today ... but I find myself getting hot under the collar at a number of issues ...

Management is simple when things are done in the right order, and all decisions are made on clear accurate information ...

Your efforts so far are enthusiastic and commendable and may be the right way to get Team NZ to be a reality, but from the 'win the campaign point of view' they are floored.

The first task on your list should have been recruiting the right people for the multitude of rolls ...

If you get this bit right the rest of the program will be a dream ... The group you name in your fax are a fine body of men and I mean them no disrespect, but they should all be put under close scrutiny, compared with alternative candidates, selected or ejected, and placed in some form of structure before they are let loose. To let them loose in unstructured planning sessions is not only a waste of time, but it is setting the freight train off at good speed in an unknown direction.

If I was a lecturer in management I would be giving you a D- for the work so far. It may be the NZ way, and maybe I've been here too long but it is not the way to win the campaign ...

At this stage I should terminate this fax. Perhaps we should talk on the phone. At least I feel our relationship is such that I can send you these thoughts knowing they will not cause a spin out.

I just want to be involved in a Challenge done right.

Regards,

Russ[51]

Blake's reply was nice as apple pie (wrapped in a poisonous pastry). Having discounted Bruce and Russell as designers at least a month before, he now suggested they be 'absolutely honest with each other' although he hadn't told them of their exclusion from the TNZ team.

> December 15, 1992
>
> Bruce Farr and Associates
>
> Attention: Bruce Farr
>
> Dear Bruce,
>
> Thanks for your fax.
>
> I'm sorry if you viewed my comments in the wrong light — nothing such as you are suggesting was ever intended.
>
> I think it very important at the outset of such a large project as the America's Cup that any potential participants are absolutely honest with each other ...
>
> As far as your role is concerned, we are in the formative stages, and I am keen to get as much sound advise as possible, particularly from yourselves — on a consulting basis of course. An America's Cup challenge from New Zealand would not be the same without your involvement ...
>
> What really needs to happen is for Russell Coutts to sit down with you at a suitable time and run through our initial ideas and budgets ... After this discussion it may be that we change/refine the concept of the project ...
>
> From the tone of your note, I will assume that you are very tired and need a break — as do I. But nothing offensive was ever intended from me. I think we have had a very good association in the past — I hope you guys realise that — and I want it to continue.
>
> All the best,
>
> Peter[52]

The designers met Russell Coutts in Annapolis on Christmas Eve. Not even during this their second meeting did Coutts outline how he 'saw the management and administration ... [or] explained what the consensus view was'.[53] But although they only planned an agenda for their first formal meeting six weeks hence, the TNZ skipper scored more than he could have hoped for.

Russ Bowler not only told him about The Chesapeake Sailing Yacht Symposium at which papers on past America's Cups are presented and promised to send invitations. Exasperated by the lack of information and apparent lack of progress, Bowler penned him a message at the end of which he added his business partner's name.

December 28, 1992
To: Team New Zealand
Attention: Russell Coutts
    cc Peter Blake
Dear Russell,

Thank you for stopping by last Thursday. My attempt to summarize the discussions is as follows:-

1. You requested a meeting with BFA ... for February 8 ...

2. The purpose of the meeting is to go over the '92 event with an in depth review of the NZC [New Zealand Challenge] effort, and other syndicates efforts, with respect to the boats and related technology. With this information you intend to formulate the TNZ plan with respect to the existing NZC equipment and new construction projects.

3. We (BFA) agreed that an accurate and in depth assessment of the boats and associated technology at last regatta is an extremely useful ingredient in the forward planning process ...

4. TNZ would need to provide documentation confirming NZC's agreement to use the data, and confirming that the attenders are committed to TNZ through the '95 Challenge ...

5. For the exercise to be done properly and have any meaning BFA advised that the work ... needs to [be] planned into their schedule and that other commitments through mid February would prevent much work being done on this until then ...

We see little value in an off the cuff, rambling, shoot the breeze style of meeting, much as we would enjoy it and the company of yourself and Tom [Schnackenberg]. If you still want to do this we suggest that it be done in a long evening or two with a few very good bottles of wine so that at least it is enjoyable. If you got any high quality decisions from a meeting like that it would be a fluke. Any more of these and the D management rating handed out last week will drop to an E.

Quite honestly we got no good feelings from our session other than from the red wine. TNZ seems lost in start up euphoria, without a plan for a plan, and I see no evidence of the real start up tasks being addressed.

Could you both please get the act together.

Regards,

Bruce & Russ[54]

Coutts replied on January 5 suggesting the designers 'apply the same assessment procedures and standards to ... [their] own people management and communication skills'.[55] He concluded: 'Suffice to say ... that Team New

Zealand wants to be involved with winners too — that is why we are talking to BFA ... so, the sooner we start working as a TEAM the better.'[56]

It seemed little more than boys' edgy banter. But on January 30, when the designers bumped into Coutts and Laurie Davidson at the Chesapeake Symposium, they knew something wasn't right. If everyone was following Blake's dictum to be 'absolutely honest with each other', why hadn't they been told about Davidson's involvement? And why, when Bruce suggested a meeting, did the pair leave Annapolis without even a phone call or a thank you for the Symposium invitations?

That BFA was not under serious consideration as designer became even more apparent when Bruce visited Auckland at the end of February 1993. There for nine days on BFA business, he finally managed to arrange a meeting at the Royal New Zealand Yacht Squadron (RNZYS) on the day before he left. The meeting with Simon Daubney, Mike Quilter and Alan Sefton on 27 February was unsatisfactory, at least from Bruce's point of view; he spent the entire time answering questions without being offered explanations in return. But that didn't stop him sharing his ideas for the future challenge, based on what he considered were the shortcomings of the previous one. Together with Russ Bowler's advice to Peter Blake, it would become the blueprint for the team approach used by TNZ.

> First, you've got to put the right people in the right places. You must have the designers designing, the sailors sailing. The most important thing you need to do is to appoint a design coordinator to bring all these people together so they can operate as a team. It is preferable that you are based in one place geographically and that you meet as often as possible ...[57]

Alan Sefton phoned Bruce that evening. He had, he said, not been happy with the discussion and wanted to reassure him of TNZ's best intentions. A few days later Bruce sent him a fax from his holiday home in Aspen.

> March 3, 1993
> To: Team New Zealand
> Attention: Alan Sefton
> Dear Alan,
> It was good to have a chat with you on the way to the airport. Thanks.
> It struck me that we actually know very little about how Team NZ currently plans to operate ...
> TNZ seems to have changed it's approach since the info we got from Peter last year and I have probably gleaned more details of the present scheme from the Press and the questions that they ask (they having obviously been briefed by you guys) than I have learned directly! There seems to have been a quite negative 'spin' given to the press on our relationship with TNZ. This is not healthy ...

It seems that TNZ want us to make a commitment to them but are unwilling to tell us much about how they intend to operate or want us to work. I appreciate that you undertook to get something to us and I have put together the following 'list' of some subjects that might be covered ...

Who will elect/control/lead the Design 'Team'.

"   "   " / "   "/ "   the Research team ...

Who do you currently see ... in those teams and who do you see managing them? ...

How would you plan to have us involved? ...

Who/what organisation will be responsible for producing the working drawings needed to build real boats? ...

Can you give us any indication of what you have in mind for a compensation package for our (Russ and my) and BFA's involvement? ...

What is the anticipated budget for Research?

"   "   "   "   "   " Design?

"   "   "   "   "   " Designer involvement in Sailing/testing and other activities?

How much involvement would you want from us in areas that are ... outside of design and research, such as sailing, performance development, project direction, syndicate direction? I think it is interesting to note that whilst my involvement in the previous NZ effort became too extreme, it was actually my additional presence in San Diego that was demanded, perhaps pleaded for, by the management AND SAILING teams that actually made my life nearly impossible. This requirement was so great that it actually had a significant NEGATIVE impact on the research, design and development program in the final 9 months or so.

Can you give us any sort of indication of the overall management structure as well as the operational structure of the entire organization.

Hope this helps you to understand what we might like to know. Please add any issues on your minds so that we can clarify them also. I gather that you are anxious to know where we stand, but I think that we need to know a lot more about your ideas before we could hope to make any sort of preliminary decision ...

Regards,

Bruce[58]

In Cape Town to assist *Enza* after its withdrawal from the Jules Verne race, Sefton sent a detailed fax to Annapolis in which he 'stressed our earnest desire that Bruce Farr should be a key part of ... [our] team'.[59] But in spite of its verbosity, his 10 March message still contained no concrete proposal with which Bruce and Russell could make a decision. While Blake kept in touch with his 'team' and 'potential sponsors'[60], he didn't with Farr and Bowler. It would be nearly eight weeks before they heard the news.

3 May 1993
To: Bruce Farr/Russell Bowler
From: Peter Blake
Dear Bruce and Russell,
Since I returned to New Zealand after the Jules Verne attempt in the big catamaran, we now have the funding in place for Team New Zealand's 1995 America's Cup challenge.
In strict confidence, I wish to inform you that we have engaged the services of Doug Peterson and Laurie Davidson as two members of our design team, which is being headed up by Tom Schnackenberg.
We are still keen to have you as an important part of our group and will keep you advised of our progress while you make up your minds on who you decide to design for ...
Hope all is well...
Kind regards,
PETER BLAKE[61]

After the convoluted messages, the lack of information, and the going-nowhere meetings, the engineer-designer had had it up to here.

May 3, 1993
To: Team New Zealand
Attention: Peter Blake
Dear Peter,
Thank you for your fax informing us that you have engaged Doug and Laurie as two members of your design team headed by Tom.
Clearly you have been working hard at forming a team, but we have had no communication from TNZ since mid March that would assist us in assessing our possible involvement. Naturally we are disappointed that Tom has not communicated with us.
At this time it is difficult for us to feel welcome in your design team. We see your inviting us 'to make up our mind' nothing more than a gesture in support of an anticipated future PR posture.
Now that the May 6 nationality date is effectively past, and the selection of your design team is now limited by Nationality, we would appreciate some honest indication of TNZ's desire to have us involved ...
We were sad to learn of ENZA's mishap in the Southern Ocean ...
Kind regards,
Russ[62]

Given these latest developments, Bruce and Russell were surprised to receive a fax from Alan Sefton asking them to meet with Peter Blake on his return to the UK from a New Zealand fundraising trip. They replied offering to make the hotel booking, to collect him from the airport, to meet for dinner and to have a more formal meeting the following day, 28 May. Still hoping they'd got it wrong, that it was nothing more than a syndicate stumbling in start-up mode, Bruce asked for a faxed proposal that he and Russell could consider before Peter arrived. Needless to say, they got neither a faxed nor a hand-delivered one — just more of the same as Blake accepted their hospitality while squeezing every last shadow from the charade.

> 8th June 1993
> To: Bruce Farr & Russell Bowler   CONFIDENTIAL
> Dear Bruce and Russell,
> Thanks for making time to talk with me when I was in Annapolis.
> I have spent a lot of time thinking on our discussions and am convinced we could work together for mutual benefit.
> I feel that the best association we could have would initially be with yourselves as consultants to Team New Zealand, where we could seek your advise on any of the number of matters that will arise ...
> I would be very keen if this association could be an exclusive one and am sure that we can come to appropriate financial arrangements to secure this....
> Kind regards,
> Peter[63]

It would take another five weeks — thirteen months after their first America's Cup discussions — before Blake officially notified the designers that their services would not be required.

> 14 July 1993
> To: Bruce Farr/Russell Bowler
>   Bruce Farr & Associates
> Dear Bruce and Russell,
> My apologies for not responding earlier following our last communications, but our deliberations on Team New Zealand's design team and programme have been exhaustive and time consuming.
> We regret that we are presently unable to enter into any agreement with the Farr Office for its exclusive involvement with Team New Zealand ...
> Best regards,
> Peter Blake[64]

Ten months later — after the charade had apparently run its course — Blake repaid Bruce and Russell the best way he knew how; by flying his favourite fraud in the *Sunday Star-Times*: 'I told Bruce my concept was having a lot of people have a say, including in the style of boat we sail ... He wrote back saying: 'Well Peter, we'll give you a D, or maybe an E for management concept'. I thought 'What a cheeky bugger'. But I kept at him to join. He just didn't want in under our terms.'[65]

But not even this knife-deep defamation could stop the two designers offering their support after *NZL32* had beaten their design, *TAG Heuer* (NZL39), and was racing towards a date with Conner, Cayard and history.

April 11, 1995
To: Team New Zealand
Attention: Russell Coutts
    Peter Blake et al
Gentlemen,
Best of luck for the Finals from all of us at BFA.
Your performance to date has been phenomenal and we hope it will continue through the next round.
May the wind gods of the left be with you.
Regards,
Bruce and Russ.[66]

April 25, 1995
To: Team NZ
Attention: Mike Drummond
Ref Observations Our Telephone Discussion Yesterday
Dear Mike,
Further to our discussion yesterday we have another observation.
The defenders seem to have developed some big asymmetricals that seem to set at surprisingly low wind strengths. We have not seen the likes of these in the challenger fleet.
If you want further discussion please give me a call.
Regards,
Russ[67]

April 25, 1995
To: Team New Zealand
Attention: Peter Blake
    Russell Coutts et al
Gentlemen,
From all of us at BFA we would like to congratulate your team on another magnificent performance in the finals.
We wish the team all the best of luck in the Challenge itself. We will be rooting for you all the way.
Our observations are being passed on to your team, and if there is any other way we can help please let us know.

We believe your team can do it. Our only concern is that Sefton will be unable to produce poetry of the quality befitting the occasion.

Regards,

Bruce and Russ[68]

Of 'the letter [that] gave us a D',[69] Coutts still had more to say: 'Blake ... was outraged ... The obvious slight burned inside him for more than two years ... [and] in the official team video produced after the 1995 Cup, Peter could contain himself no longer'.[70, 71] But not content with hyping up this Hollywood-worthy performance, Coutts doesn't stop there. Like a serial sponge he soaks up every last drop of sympathy by rounding out 'Schnack Attack' with a further embellishment of the fraud. 'Why that letter was written will forever remain a mystery to me. After we won the Cup the team travelled to Washington for a reception and, to their credit, Bruce and Russell were there to congratulate us. I couldn't help but ponder what it would have meant to them had they not taken themselves out of contention. It seemed to me that a huge design talent had been wasted.'[72]

Indeed there was a lot to ponder. And while Bruce was away on holiday he wrote to the RNZYS, now the official holders of the America's Cup, offering his and Russell's services for the Squadron's first defence. BFA would, he said, be pleased to play a part in a single defending unit or take a role in a defender's series. It was a small but significant gesture. Although he now dismissed the Blake-Coutts hocus-pocus as New Zealanders playing the political game they'd publicly pledged they wouldn't, Bruce still hoped he could get the relationship out of jail. As did Russ when he organised a job for Russell Coutts and Simon Daubney after their America's Cup win sailing Steve Kaminer's new ML 39[73] *Predator* in the June Block Island regatta, along with himself and his nephew Alan McGlashan. But what they couldn't escape, with the Squadron's non-reply, was the now sure knowledge that New Zealand no longer wanted them.

Chapter 31
The Shape of Speed

Somebody will use him for a design [in the next America's Cup], probably the Americans, and they'll end up with the quickest boat.
David Barnes, New Zealand yachtsman[1]

Someone who has never copied is Bruce Farr. He developed all the philosophies we use today: fractional rig, light-displacement, weight concentration ... Monohull design is the hardest area in which to dominate ... and if you succeed there, everywhere else will be easy.
Philippe Briand, French yacht designer[2]

Perhaps, if you have time for 'the dog work needed to make things work'.[3] But when it came to designing *TAG Heuer* Bruce and Russell had even less than that. When their funding coughed to a stop in May so did their R&D; the Pacific Challenge float had been muscled right off the market, leaving Chris Dickson strapped for cash. But thanks to the Swiss watchmaker's funds the young *TAG* crew[4] eventually made it to San Diego. That they finished third in the Louis Vuitton Cup on their minuscule budget — and miles ahead of the monied might of Japan and France, and Syd Fischer's Sydney 95 — was a major achievement. It was even more remarkable given the extremely limited time the Farr team had had to design and engineer *TAG*.

But Blake had won off the water, Coutts emphatically on it. That their win was comprehensive is now a matter of record. What isn't is just how lucky those 'lucky red socks' really were. US money, morale and interest were at their lowest ebb ever in the Auld Mug's history thanks to San Diego's shenanigans in the previous two defences. In fact, so weak was *Stars & Stripes* that any of the top three challengers might well have produced a similar whitewash. Certainly if there'd been even the hint of a win

Conner would have been on the helm at the end, not Cayard. Nor if he had had a strong defence would he have had to jimmy the rules to beat Leslie Egnot's *Mighty Mary* in the defender series or jump on the mermaid boat at such short notice.

It is also a matter of record that having promised a level playing field Peter Blake promptly moved the goalposts.

'[I want] identical rules for both sides. It will be like going to any ordinary regatta. There'll be no difference to whether you are a challenger or a defender.'[5]

The Challenger of Record might have reined Blake in after acceding to an inordinately long wait for the next America's Cup Match. But that didn't stop him denying his fellow New Zealand sailors the right to race a defenders' series. Not only does this leave his single defending unit unable to test their boats or hone their skills against serious opposition. It has also meant some 40 to 50 Kiwis have been denied the chance to sail for their country in an America's Cup regatta. And as a result of the defenders altering the period of residency from two to three years, even some of those have been forced to choose between joining a foreign syndicate to compete in the Cup or remaining at home to seek selection for the Sydney 2000 Olympics. It has also meant that all but a few of New Zealand's boat-building, spar- and sail-making industries have likewise been denied the opportunity such a series provides; as has New Zealand's yachting-mad public been denied the spectacle and excitement. But arguably the most disturbing feature of this very un-Kiwi behaviour is the openly gang-like tactics adopted by the defenders. Not only do they suggest the Challenger of Record is spying on their boats.[6] They hurl insults at fellow New Zealanders forced to work for foreign syndicates, and have even been known to ram them out on the Hauraki Gulf.[7] Blake's main man might laud his tactics with sponsor-speak.[8] But in reality they are just more of those policies that isolate and divide. They also leave the claim of that top-shelf beer — 'the force behind New Zealand yachting'[9] — with a rather stale head.

Only in a growing celebrity culture in a sponsorship-driven sport in a market-driven economy, could a person who had done so little receive so much. It is an irony that will not be lost on those competing for the ewer in Auckland's AC2000, especially those New Zealanders who have felt the heat of Peter Blake's pogrom. For if all those who have paid their money make it to the start line, this will be the largest and most technologically advanced fleet ever to challenge for the America's Cup. The rubber ducky rammers may have started with the fastest boat from 1995. But having lost a key designer[10] they could have rather more on their minds than intimidating fellow New Zealanders[11] when February 2000 arrives.

***

For Immediate Release
WENHAM, MA (August 5, 1996) — PACT2000,[12] the New York Yacht Club's America's Cup Challenge today announced that it had selected Bruce Farr & Associates, Inc. of Annapolis, Maryland as principal designer for its America's Cup challenge.

'Bruce Farr & Associates is the pre-eminent designer of large racing yachts in the world. Over more than two decades, boats by Bruce and his partner Russell Bowler, a genius in composite technology, have earned the most remarkable list of international and national championships and first-place wins ever compiled by yachts of a single design group ... We are thrilled to have brought on board the world's best yacht designers ... Bruce Farr brings two vital elements to our program beyond his extraordinary track record of winning designs. First, he and his partner Russell Bowler grew up in New Zealand and have vast experience sailing in Auckland. Second, Bruce is an America's Cup veteran ...'

'We are more optimistic than ever that we will bring the Cup home to America in 2000.'[13]

Having been rejected by Gang New Zealand because it was 'believed unanimously'[14] that they would not want 'to justify their decisions to someone'[15] like Tom Schnackenberg, Bruce and Russell and their team had just been invited to join the most prestigious team of designers ever assembled in the USA. It was an honour no New Zealanders had ever been accorded.[16] As for the NYYC's PACT2000 team, it bristles with professors and PhDs, with aerospace engineers and computational fluid dynamic specialists, with tank-test and wind-tunnel experts, with software inventors and structural engineers, and with veterans in design and management from past America's Cups.[17]

Although it wasn't needed, the NYYC's choice of principal designer received further endorsement from the continuing success of Farr-Bowler boats all around the world. Larry Ellison's 79-foot Maxi *Sayonara*, often with Chris Dickson on the helm, dominated everything it entered, including the tragic Sydney–Hobart of 1998–99.[18] Karl Kwok's *Beau Geste*, Syd Fischer's *Ragamuffin* and Giorgio Gjergja's *Ausmaid* finished 1-2-3 in the previous year's Sydney–Hobart, with *Beau Geste* and *Ragamuffin* the top two boats in that season's Southern Cross Cup. The Farr-designed Beneteau First 40.7 was soon winning events all around Europe after its unveiling at the 1997 Paris Boat Show. The Farr 40 One Design was named Overall Boat of the Year in 1997 by *Sailing World* magazine. Clients returning for new designs included Norway's H.M. King Harald V (Design 412), Italy's Bruno Finzi (Design 415) and USA's Helmut Jahn (Design 416). In 1997, with two of its three boats designed by BFA,[19] the USA won the Admiral's Cup for the first time since 1969. In the 1998–99 Whitbread, eight of the nine finishers were Farr W60s,[20] a statistic that not only underscored BFA's dominance in this race but how much it continues to provide a platform for New Zealand's offshore yachtsmen.

> When the 10 entries crossed the start line in Southampton, there were 36 New Zealanders sailing in the fleet — enough to crew more than a third of the boats. Four of the entries ... were skippered by New Zealanders. Three of the [nine] new generation Whitbread 60s [in the race] — two *Merit Cups* and *Swedish Match* — were built in New Zealand by Marten Marine and Cookson Boats/Yachting developments respectively, more than any other single country produced for the 1997 event. Six of the 10 yachts carried masts built by Southern Spars in New Zealand. Several wore sails designed and built at New Zealand's North Sails, Halsey Lidgard Sails and Doyle Bouzaid lofts. All this, despite the fact that there was no official New Zealand entry for the first time since 1981.[21]

But the name behind New Zealand's rise to offshore racing supremacy has all but been forgotten by the New Zealand media. When it won the Kenwood Cup in 1998 no mention was made in two *New Zealand Herald* articles that the three boats used were Farrs,[22] just as they had been when New Zealand last won in 1986.[23] Where once it would have been acknowledged for its excellence and proudly owned by New Zealand, the name that dominated this regatta was now hard to find. In fact, so far has the F word been removed from Kiwi sporting consciousness that when an interview with the designer was screened on New Zealand television in November 1998, Murray Deaker, an award-winning radio sports broadcaster, was later heard to ask: 'Who is this guy Farr? Is he a mercenary or something?'

Perhaps it is like Farr-Bowler structures and shapes, now so universally accepted and copied that where they came from has almost been forgotten. Perhaps it is like their business, now such an integral part of the international marine industry that their contribution to national economies is overlooked by talk-show hosts. Still, even that is hard to believe when every year Farr-Bowler designs produce an estimated one hundred million dollars worth of product worldwide, with a significant percentage of that manufactured in New Zealand.[24]

Fortunately there are those still happy to acknowledge the ongoing contribution to New Zealand's industry and sporting culture by the two expatriate designers.

Mick Cookson, boat-builder, New Zealand:

> Bruce has got a nice blend of both the art form and scientific aspect of yacht design, and the absolute attention to detail that goes with it. He has a certain style and look that's peculiar to him. And from what I've seen he's never

copied anybody; everyone's copied him! His drawings and Russell's engineering are second to none. They're unbelievably professional, but they've also pushed the envelope.

I remember when I first started an apprenticeship, I was on a dollar an hour, working with Andy Ball. Bruce had designed and built him a Javelin for something like 40 cents an hour. But it was such a work of art Andy was too scared to take it sailing at first. It's that level of professionalism, that attention to detail, and that almost unrealistic approach to boat-builders making money — their passion should be to do a better job! — that's been with Bruce from the beginning. It's something he's never backed down from. And that's how he is today. He understands fully what an owner wants from a custom composite race boat, as well as understanding rating rules in the most intimate detail.

The sport will never match Bruce's ideal — it can't! Some might say his level of expectation, his things-must-be-perfect approach, is unrealistic. But he is, in my opinion, nothing short of a genius who's led the yachting world, and dragged it and the rulemakers, screaming, to where it is now.[25]

Michael Fay, merchant banker, Switzerland:

[They're at] the cutting edge of revolutionising things they've always worked with. And ... always revolutionising them — not once, not twice, [but] every time. I mean, that's an interesting person who has a frame of mind that's always going to be new, always going to be different, always going ... forward, that's relentless in the pursuit of excellence. They're words that are easy to say ... but it's hard to find people who ... [live] them as much as Farr and Bowler do. All the time. [It's] tremendous, exciting and stimulating ... to be around them, and they deserve all the rewards, all the accolades and all the success that they get. And they should be a source of inspiration to a lot of Kiwis ... But, you know, sometimes New Zealanders aren't prepared to necessarily see how fantastic some of this is.'[26]

Ian Franklin, boat-builder, New Zealand:

As well as being exceptional designers, Bruce and Russell also possess great skills as businessmen. Not only do other designers copy their shapes and structures. They have forced the industry to become totally professional. It is their influence that has made our marine industry the envy of the world ... I have always said the industry should take out a large insurance policy on Bruce's life. If he got knocked down tomorrow the industry would suffer terribly.'[27]

Kim McDell, boat-builder, New Zealand:

We made a commitment to them back in 1976. From our hands-on experience in the 18-footers, I believed totally in Bruce as a yacht designer. He's brilliant. He's a genius. He's one of the few guys I know who can step on a boat and feel what's needed through his bones. He has incredible intuition. That's why we've stayed with him for twenty years.

And Russell not only brought his own great talents as a designer, helmsman and engineer to the business but that whole experimental side as well. And here we are still going strong with the Platu 25 the Mumm 30 and other new Farr projects.[28]

Steve Marten, boat-builder, Auckland:

Bruce and Russell are the most successful yacht designers we've ever had. They are probably only rivalled by Nathanael Herreshoff. He too achieved almost total domination in his day and in a very similar way ... Their contribution to New Zealand? Well, it's been enormous. Of course it has worked both ways — they sold New Zealand-made boats to clients at a better price and quality than they could get elsewhere. But Bruce and Russell have created links and opportunities for us here as well as feeding an awful lot of work back into our economy.[29]

The accolades, though, have not just come from those who have been ben-eficiaries of Farr-Bowler boats. They have also come from those with whom Bruce and Russell have had to compete for survival in the highly competi-tive world of monohull design.

German Frers, yacht designer, Argentina:

In the mid-Seventies ... [Bruce Farr] was the king with his light-displacement, centreboard designs; then after the rules were changed against him he came back and re-affirmed his status as the top designer.'[30]

Ed DuBois, yacht designer, England:

Bruce Farr's ... work in the mid-Seventies was innovative and extreme ... almost victimised. He had a few wilderness years and then ... came back to the IOR scene with a very single-minded attitude which resulted in boats with good upwind ability and no ... real weaknesses.'[31]

Rob Humphreys, yacht designer, England:

In racing yacht design, the man I have to take my hat off to is Bruce Farr who has managed to stay at the top of the IOR racing scene for nearly fifteen years.'[32]

Olin Stephens, yacht designer, USA:

> Bruce and Russell have dominated offshore racing — there's no doubt about that! — [and] they certainly have contributed materially to making boats go faster than they ever used to. I think that's a big accomplishment.[33]

Rolf Vrolijk, yacht designer, Germany:

> I think Bruce Farr does an excellent job in evaluating the correct parameters the first time around for new design types.[34]

Then there are those standing back, who, with no vested interest, have seen fit to acknowledge BFA's achievements in making our communion with the wind and sea safer and more enjoyable than it ever was before.

> Bruce Farr, Sir Robin Irvine, Friends ...
> At this time of year in the nation's capital, the idea of boating for pleasure defies the imagination. Yet we have only to recall the epic adventure of the Whitbread Round the World Race to realise that Bruce's boats are perfectly at home racing at high speed through heavy seas and gale-force winds in sub-Antarctic conditions. The models that grace the Embassy Library tonight are slender thoroughbreds, but they are tough with it.
>
> Yachting has been defined as the art of going slowly in circles in exquisite discomfort and at great expense. In Bruce's case, the adverb 'slowly' hardly seems to apply. But in any case, I would prefer to think of the design and building of high-performance yachts as a blend of creative flair, rocket science, and business acumen.
>
> Bruce Farr excels in all of these. He has salt water in his blood, and in his younger days was a champion sailor in his own right. His designing prowess also dates from early on, as he reacted against the limitations of the state of the art as it was then. He is a sailor's designer who understands the needs of the people out on the water. He has not been afraid to push the envelope and think the unthinkable — witness the work he did at the time of the 1992 America's Cup campaign to explore the possibility of an asymmetrical hull.
>
> If yacht design is ultimately an art form, it has to have a solid scientific underpinning. The fact that the Royal Society of New Zealand has chosen to confer a prestigious medal for scientific achievement speaks for itself. In Bruce Farr creative brilliance is underpinned by a powerful capacity to apply scientific principles, seeing further than his peers into the practical implications of theory. Scientific endeavour is partly about trial and error, where the consequences of error can be not only expensive but fatal. But in Bruce's case the experimental process simply fine-tunes what he has already predicted from first principles.
>
> Bruce is also a successful businessman. He has built an outstanding team to

work with him, and he has stayed for an extended period at the top of a mercilessly competitive industry. He is subjected to constant scrutiny from clients and rivals; Bruce stays ahead of the pack not simply through the excellence of his design work, but by taking an entrepreneurial, proactive, and visionary approach to developing business strategies. I know that he is the subject of case studies in management courses in fields far removed from yacht racing.

Bruce, we are proud of your achievements as a New Zealander. We are a fiercely patriotic nation, especially when it comes to sporting endeavour. We can also be harsh in cutting tall poppies down to size. But there is no mistaking the goodwill New Zealanders extend to you — not only among yachties, but right across the community. We like people who have proved themselves on the world's stage. And the way you have achieved your success — through intelligent innovation, hands-on effort, and letting your boats walk the talk — makes you the sort of role model New Zealanders love to identify with.

It is fitting, therefore, that the Royal Society — our nation's pre-eminent learned body — has chosen to honour you for your exceptional career, and the way in which you have stamped your personal mark on what has become to us in New Zealand a glamorous and economically important industry.[35]

In a few moments I will be inviting representatives of New Zealand's scientific and boat-building communities to say something about the significance of this award from their perspective. But first ... I have the honour to present to you the Silver Medal of the Royal Society of New Zealand and citation in recognition of excellence, your significant contribution to racing yacht design, and the impetus given to the development of this technology in New Zealand.[36]

But above all it is their world-championship-winning designs, in which they have invested their minds and hearts and their undying passion for sailing, that speak the most eloquently of Bruce and Russell's achievements.

1970	*Jennifer Julian*	Cherub Worlds
1972	*Smirnoff*	18-Footer Worlds
1973	*TraveLodge*	18-Footer Worlds
1974	*TraveLodge*	18-Footer Worlds
1975	*45° South*	Quarter Ton Worlds
1975	*KB*	18-Footer Worlds
1977	*Joe Louis*	Three-quarter Ton Worlds
1977	*Gunboat Rangiriri*	Half Ton Worlds
1977	*The Red Lion*	One Ton Worlds
1987	*Fram X*	One Ton Worlds
1987	*KZ-7*	Twelve-Metre Worlds
1988	*Propaganda*	One Ton Worlds
1989	*Brava*	One Ton Worlds
1989	*Carat VII*	50-Foot Worlds

1989	*Longobarda*	ICAYA Worlds
1990	*Lone*	Three-quarter Ton Worlds
1990	*Merit*	OMYA Worlds
1991	*Lone*	Three-quarter Ton Worlds
1991	*Vibes*	One Ton Worlds
1991	*Larouge*	Two Ton Worlds
1991	*Merit*	OMYA Worlds
1992	*Xacobeo 93 (Lone)*	Three-quarter Ton Worlds
1992	*Brava*	One Ton Worlds
1992	*Shockwave*	Two Ton Worlds
1992	*Merit Cup*	OMYA Worlds
1993	*Larouge*	Two Ton Worlds
1993	*Merit Cup*	OMYA Worlds
1993	*Carat VII*	50-Foot Worlds
1995	*Brava Q8*	ILC 40 Worlds
1996	*Sayonara*	ILC Maxi Worlds
1996	*Brava Q8*	ILC 40 Worlds
1997	*Sayonara*	ILC Maxi Worlds
1998	*Sayonara*	ILC Maxi Worlds
1999	*KZ-7*	Twelve-Metre Worlds

They had come a long way, the shy lad from Leigh and the cool kid from Kohi, this list proof of the distance they had travelled. It is of course a staggering list, a dream come true for any yacht designer. But it is also a dream that had been a long time in the making. From the mullet boats of Auckland to the Patikis of Napier, from the Jack Logan skimmers to the John Spencer Cherubs, from the Stewart 34s to the Jim Young 37s, it had taken a hundred years and an extraordinary range of gifts meeting in two young men before the dream became invention and the breakthrough commonplace. You see it in the minimal wetted surface of their light-displacement hulls. You see it in the plumb bows, fine entries and long flat runs aft to powerfully wide sterns. You see it in uncluttered decks, the wedge-shaped deckhouse. It's in the high-flared freeboard and the knife-edged appendages. You see it in the fractional rigs with their swept-back spreaders. You sense it in the balance. You feel it in the helm. And you know it when a yacht leaps to a plane at the slightest provocation. But above all you see it in the faces of the thousands around the world who race and cruise this new breed of boat. And whether they are Farr designs or Farr simulacra, they are fast and fun to sail. For in freeing yacht design from its superstitious past, Bruce Farr and Russell Bowler have not only drawn new horizons and designs to get you there much faster than before. They have delivered to the world from their science-based art what the New Zealand-nurtured dream had promised all along: the shape of speed.

Afterwords
Bruce Farr

It seems that my life forever moves fast forward and I seldom have the time or inclination to look back or dwell on the past. John has quite inadvertently done me a truly great service by writing this book. He has very thoroughly recorded my accomplishments and those of our design businesses as they have morphed from Bruce Farr Yacht Design to Bruce Farr and Associates Inc., and just recently to Farr Yacht Design Ltd. This latest change is intended to reflect that we are now a group of multi-talented designers and to make our mission clear from our name.

The process of confirming and checking facts for this book has been both interesting and rewarding, forcing me to spend a lot more time with my past than I would otherwise and producing some revelations. It has been quite startling and pleasurable to review the degree of success that we have had over so many years and to relive some experiences I had completely forgotten in the intervening time.

It also struck me that we have been profoundly fortunate to be supported by so many very kind and capable people, sailors, sailing industry folks and, in particular, boat owners prepared to put their faith and monetary risk in our design work. So many people from all over the world have touched our career in so many different ways that it is truly difficult to know where to start to thank them. I think they know who they are and I hope they understand that I am eternally grateful to them all.

Some of the most important were those there at the beginning who helped get the show on the road.

My parents, Jim and Ileene, introduced me to sailing before I knew anything and created an interest that I took for granted but from which I have never wavered. They totally supported my early sailing exploits and also encouraged me to create my own boats and make my own way rather than follow others. My father's interest and involvement in construction of his own boats and boats for me, and his encouraging me to participate in that process, was my inspiration from a very young age. This ultimately led to my choosing a career in building and designing boats — a decision that perhaps caused concern to my parents when academia would have had me become an engineer or architect. But they understood my choice and were always fully supportive. To them I owe a huge debt for allowing me and encouraging me to follow my dreams.

My older brother Alan, who was much more outgoing than I, gave me a huge amount of support in my earlier years when I was very shy and had some difficulty finding my way in the world. I regret that I never properly made him aware of just how grateful I was for that. He also was my first client!

In New Zealand in the sixties there always seemed to be people ready to support someone with an indication of talent or promise. Perhaps this was the lifeblood which produced the successes which New Zealand has enjoyed in the yachting business. It was certainly a great confidence booster for me when, as a teenager, I began designing and building boats for people outside my immediate family. Vern Smith, the Hooks, the Torkingtons, Lindsay Lovegrove, Des Matheson and other Whangateau area residents were all early clients or great sailing friends as I began designing, building and sailing International Moths at about 15 years of age. When I dropped out of school Vern also took an interest in my continuing education, helping me to develop further skills in mathematics and basic engineering. This relationship was to be rekindled in the mid-1970s when he looked after programming for me as the age of programmable calculators and rudimentary computers arrived. Merv Elliott, who had a passion for building small boats in his spare time, and whose son Greg is now a top New Zealand designer, provided me with opportunities to produce designs for 12- and 18-foot skiffs. He introduced me to Gary Banks for whom I designed and built my first 18-footer. Other local people provided inspiration and open doors for advice, with Len Naggs and Joe Joyce being notable.

When I was casting around trying to figure out how to best follow my dreams, people in the industry also helped in a very unselfish way. I especially recall John Spencer and Des Townson taking the time to give me the wisdom of their life experiences even though I was in some small way at the time, and could in the future become, their competition.

Jim Young took me into his design and boat-building business at a time when jobs were hard to get and while I worked for him he gave me unending encouragement and knowledge as I set out on various of my own design and building exploits in my spare time. He, his brother Alan and his staff helped me develop experience in the design and building of keel-yachts and powerboats, and some insight into how to manage a business. I was very fortunate to have a job involving my primary interest and also spare time to follow my own sailing and design career. During this period I honed my sailing and designing skills and met many of Auckland's yachting industry and sailing characters, some of whom would become clients later in my life. I established relationships with industry people such as Chris and Tony Bouzaid, Jim Lidgard, Helmer Pedersen, Murray Ross, Max Carter, and

Trevor Geldard, all of whom would be part of the New Zealand support group that would help me and the New Zealand boat-building industry in the future. Through Jim's business I also met Graham Painter who introduced me to the wild world of international commerce and later would shepherd me through the expansion of my own business into the international arena. These were key people who, along with many others of that era, established the environment in which I, and many other youngsters, flourished.

It was in this period that I also established a sailing relationship in 12-footers with Rob Blackburn who would later provide me with perhaps the single most significant opportunity of my career — my first commission for a keel-yacht, *Titus Canby*, that would launch my big boat design business. Looking back, it is hard to believe that Rob, and George Knightly with *Moonshine,* were brave enough to take the risk investing most of their assets in the hope that I would produce competitive designs at my first real attempts. A belated but HUGE thank you to you two in particular. At this time, following a reasonably successful second 18-footer design for Gary Banks, armed with these keel-yacht design commitments and an 18-footer construction job for Noel and Wayne Fleet, I left Jim Young and set up a boat-building and design business. The 18-footer worked, and my 18-footer career was launched conclusively when Don Ligard's *Smirnoff* won my first world title. *Titus Canby* worked, and so began my nonstop keel-boat design career.

My design business then flourished, and it becomes difficult to separate clients, sailors, members of the press, friends and supporters in terms of their significance. They were all significant! All these clients had a certain level of commitment to a new wave of yacht design and their results do not necessarily represent a true measure of their value or my gratitude. I thank them all. Some stick out because of particular deeds, support, the timing of their commissions, or because they took me sailing with them: Terry Harris (first Quarter Tonner); John Senior and Graham Eder (first really big boat at 42 feet); Murray Crockett (built and campaigned the Farr 727 Class, my first production boat); Noel Angus (*Prospect of Ponsonby*); Graeme Woodroffe and Roy Dickson (our first offshore World Championship); Stu Brentnall (*Jiminy Cricket*); Ian Gibbs (*Tohe Candu* and *Swuzzlebubble*); Eric Simian (our first European-based world champion); Kim McDell (our most successful early production boat-building relationship). Gil Galley, the 'godfather' of Devonport, provided endless personal support during this period.

Many of these people helped take my business global and should take a good deal of the credit for where we are today. We were then established on the world scene and it was up to us to make something of it. To this end Peter Walker, Roger Hill, Cathie Cochran, Robyn Curtain, Sally Gray and Russell Bowler joined me in our expanding endeavours, providing huge effort, support and enthusiasm to our task. Russell, Roger, Cathie and Sally were even prepared to uproot and follow the business to the USA in 1981. I have always felt blessed with terrific associates and staff.

In our modern post-1981 form we have developed many, many successful projects and a huge client base. All of these people deserve our thanks for being there at the right time. As do those who have gone out of their way to assist us in connecting and communicating with clients in foreign lands: Harunobu Takeda (Japan), Marco Holm (Italy), Bob and Sue Farrell (New Zealand), Kevin Shephard (Australia), Luciano and Silvana Lievi (Italy), Clas Kruger (Spain), Lorenzo Bortolotti (Italy), Peter Morton (England). We have also had the pleasure of working with some extraordinary clients who deserve special mention — Pierre Fehlmann, Sir Michael Fay, Pasquale Landolfi, Grant Dalton and Helmut Jahn — all of whom have been unswerving in their support for us and successful in their ventures. In 1982 Geoff Stagg came aboard and through Farr International developed a market in selling and brokering Farr yachts, supporting racing and other yachting projects. This made a big contribution to our design efforts, producing some high-yield One Design and Production Projects and many opportunities to get the best teams into our custom race- boat designs.

Our two biggest production builders are Beneteau, who have built a staggering number of our cruising yacht designs along with a few racing yachts, and Carroll Marine, who have built many production One Design racing yachts and a few cruising yachts. Mick and Terry at Cookson Boats have been our most prodigious builders of high quality custom race boats for two decades.

None of this would have been possible without a superb yacht design team. Over the last 15 years we have developed a very strong team with tremendous depth of talent, expertise and commitment. Over these years the quality of the team has developed to a level where it can produce successful designs of its own authorship. They are not people working for Bruce Farr and Russell Bowler, or for Farr Yacht Design Ltd. Jennifer Charnesky, Harry Dunning, Amy Fazekas, Dave Fornaro, Bobbi Hobson, Steve Morris, Mick Price, Jim Schmicker, Pat Shaughnessy, Brit Ward and Graham Williams *are* Farr Yacht Design. The shapers of speed!

Lastly, I must thank my wife Gail, who has always been very patient with my habit of putting boats and business before most else. She has given me support and understanding through dark and discouraging periods and provided so many other aspects of life which I would have missed without her.

Russell Bowler

Life is an unpredictable flow of events. John Bevan-Smith, through his tireless research in putting this book together, and his insatiable desire to get the issues presented accurately, has produced a chronicle that thoroughly endorses this statement.

While I knew from an early age that the physics of sail boats held a particular fascination for me, until my early thirties it remained just an enjoyable pastime. Not in my wildest dreams would I have guessed that it would become my trade, the reason to take up residence and raise my family in the USA and the means for travelling and meeting people at the four corners of the globe. Nor did I anticipate that two of the sailors I met early along the way would become my business partners united in a successful endeavour.

Before I moved to the seaside suburb of Kohimarama, we lived in the inland suburb of Mt Eden. My family was Presbyterian and the normal Sunday involved attending church and dispatching me to Sunday School beforehand. I did not enjoy this intrusion on my free time, not because I didn't think their teachings had anything for me, or that they weren't a darn fine group of people to be with. My concern was that the time it took to attend Sunday School robbed me of those valuable hours in the weekend needed to complete projects, climb trees, ride bikes and go on adventures. My reluctance to attend, however, was overcome by my mother's insistence that the church was a valuable part of our lives and something that I must not miss. So when we moved to Kohimarama and I purchased my first competition P Class I could see a conflict raising its ugly head. Competing at the Kohimarama Yacht Club (KYC) meant racing on Sundays.

I sat down with my parents and we talked about the problem. My mother resolved that sailing was an activity on the sea, with the wind and waves, and had a closeness to nature and our creator that made it an almost sublime activity. While sailing would not be considered quite as lofty a form of worship as attending church, it was close. This was my mother's genuine belief as she used to talk in a similar manner about her younger days rowing dinghies around Castor Bay where her family had a bach.

At the time I went right along with this explanation, for obvious reasons — I was now free to sail at KYC events. However, the reasoning has left me with the permanent belief that sailing is a sublime activity and a way of appreciating the beauty and physics of the world we have been given. I also consider all those captivated by the charm of the sport as being of the same conviction.

While we did not know it at the time, the design, building and sailing that was taking place in Auckland through the middle half of this century has had an enormous impact on the design of sail boats worldwide. I feel extremely privileged to have been involved in some of the action during this period and to have played a small part in the adventure. Limitations of space and a touch of amnesia prevent me from naming and thanking all those who were a part of that powerful and formative experience for me. They were true amateurs driven only by genuine interest and the sheer joy they derived from designing, building and sailing boats. Could anyone in the 60s have guessed that sailing would develop into the international sport it is today; that international air travel would shrink the globe and make it possible for New Zealand sailors to participate in events around the world as if it were all happening in Auckland and surrounding areas; that relative world peace and prosperity would expand the time and money available for leisure activities such as sailing and develop it into the industry it is today.

I feel deeply honoured to be part of today's worldwide sailing fraternity. Its membership is made up of people from all walks of life, from royalty to Australians, from many nations, races and languages, all united by the fascination of sail-boat racing. I admire their willingness to put up with severe discomfort, their energy and the enthusiasm they dedicate to the sport in a day and age where the easy life and immediate satisfaction are growing trends. The racing fraternity practise hard, play hard, have strong camaraderie and risk their lives in this glorious endeavour. After all the effort that goes into preparing for a race they literally cast their fate to the wind and accept the sometimes frustrating, sometimes exhilarating consequences that are dealt to them. They are just simply a wonderful group of people and I salute them all.

In contrast it is a sad observation that the administration of our sport as a whole, by individuals who are sailors themselves, has so many blemishes on its record. It may be that the utopia we would all like to see on the administration front is simply unattainable due to some of mankind's basic weaknesses.

New Zealand sailors have been among the very best of New Zealand's overseas ambassadors. From the early exploits of Peter Mander, Geoff Smale and Ralph Roberts, through Chris Bouzaid's One Ton Cup success to Grant Dalton's Whitbread crews and everything in between, I have felt very proud of the men and women who have done a splendid job of representing our country, as have many others in a wide range of sporting and other endeavours including mountaineering, soldiering and academic life. Some events in recent years have taken the shine off New Zealand sailing's glowing reputation. But I am sure with the passing of time, the talent and integrity of those sailors who will rise to the top of New Zealand sailing and extend their activities overseas, will resume the good work to restore and embellish that reputation. Only the Olympic games could hope to claim the same sort of ambassadorial penetration that sailing reaches, and then only once in four years in contrast to the continuous impact of sailing. I'm sure I am not alone among New Zealanders living outside the country who, despite being physically removed from Aotearoa, still consider themselves 100 percent Kiwi. I will always be grateful to our magic little country for the spirit that intoxicates its people, and yearn that some day my family will return to enjoy that spirit and settle where our roots lie.

The opportunity to add something to the end of this book cannot be passed without a sincere expression of gratitude to those who were a part of the story. To my mother and father for their support and setting the standard, my sisters and their husbands who were strong catalysts along the way. To Lynda for the sacrifices over the years, and Gareth and Karli for coping with my absences. To Bruce, whose focus, honesty and dedication make him the best possible business partner one could hope for. To Geoff, whose drive and effervescent enthusiasm adds another dimension to our business. To Peter Walker, who kicked me into the programme. To the staff at FYD who have been part of our success. To the crews that I sailed with over the years. To the journalists and commentators. To sailing and racing officials throughout the world who unselfishly dedicate their spare time to the sport. To the builders who went along with it all, and the clients who shared their dreams. And many more who shared the journey.

It has been five years since John first phoned with the thought that writing a book about Bruce and myself is something he wanted to do. He warned it would take two years! At that time we had no idea of the form it would take and probably neither did he. I first met John when we were both enjoying those carefree years of discovery, 19 through 22. Following that period we did not have any real contact with each other until that phone call five years ago. In the intervening 26 years the John I knew as a talented musician, ace cricket spin-bowler and athlete with the enviable knack of being able to date the best-looking girl in the village had put together a fascinating life that covered raising a family, business success, missionary work and more. It seemed to me that a book about him would make far more interesting reading than anything he could ever write about me. After his years of work interviewing, researching, validating, checking, rewriting, I need to offer him my eternal thanks for putting together a chronicle remarkable in its depth of perception and for expanding its scope to those around us who are part of who we are today. A narrative like this can only be composed by someone with talent, a broad experience of life and a thorough commitment to the project. Bruce and I are indeed fortunate that, for whatever reason, John woke up one morning with this mission tattooed on his mind.

Where to from here? Despite having achieved more than we set out to do even in our most optimistic business plans I do not feel any sense of having finished. There is still a lot of satisfaction and some ecstasy in receiving a brief from a client, seeing Bruce bring his instinctive powers to bear, adding a little of my own, having our team pour on the technology and watching something new, unique, fast and exciting grow from the process. Until that satisfaction and excitement fades, or sanity fails me, whichever comes first, it will be business as usual.

Appendix 1

Farr Yacht Design Ltd Honours and Awards

Year	Award	Awarded by
1999	Boat of the Year, Cruiser/Racer Category: Beneteau First 40.7	*Cruising World* Magazine
1998	Boat of the Year: Farr 40 One Design	*Sailing World* Magazine
1997	Alan Payne Memorial Trophy; Designer of 1997 Sydney-Hobart Overall Winner: Farr IMS 49, *Beau Geste*	Cruising Yacht Club of Australia
1996	New Zealand Science & Technology Silver Medal	Royal Society of New Zealand
1995	RORC Yacht of the Year: Farr ILC 40, *Brava Q8*	Royal Offshore Racing Club
1995	Citation of Merit	City of Annapolis
1993	Best Yacht Designer	*Yachting* Magazine
1993	Most Successful Boat: IMS: Mumm 36, *Pigs In Space*	*Yachting* Magazine
1993	Most Successful Boat: IOR: Farr Two Ton, *La rouge*	*Yachting* Magazine
1992	Best Yacht Designer	*Yachting* Magazine
1992	Most Successful Boat: IMS: Farr IMS 44, *Gaucho*	*Yachting* Magazine
1992	Offshore Medal of Achievement: Farr IMS 44, *Gaucho*; Honorable Mention: Farr IMS 40, *Cookson's High 5*	*Sailing World* Magazine
1992	Elected to the Sailing World Hall of Fame	*Sailing World* Magazine
1991	Named the Expert's Expert: The Racing Yacht Designer's Racing Yacht	*Observer* Designer
1990	Yacht of the Year: *Steinlager 2*	*Yachting* Magazine
1990	Best Designer	*Yachting* Magazine
1990	Inducted into the NZ Sports Hall of Fame	New Zealand
1990	Order of the British Empire (OBE)	HRH Queen Elizabeth II
1989	Architecte de L'Annee	*Mer & Bateaux* Magazine
1988	Designer of the Year	*Yachting* Magazine
1988	Distinguished Business Award	Marine Trades Assoc. of Maryland
1988	Distinguished Business Award	City of Annapolis

1988	Admiral of the Chesapeake Bay	State of Maryland
1987	Architecte de L'Annee	*Bateaux* Magazine
1987	Annapolis Sailor of the Year	*The Capital* Newspaper
1987	Governor's Citation	Wm. Donald Schaefer, Governor, State of Maryland
1987	Boat of the Year: Farr One Ton	*Sailing World* Magazine
1976	Yachtsman of the Year New Zealand	

Appendix 2

First Place Victories Won in Farr Designs 1964–1999

Year	Event	Class	Yacht	Design
1999	Block Island Race Week	1st IMS	*Rima*	414 — CM60
		1st PHRF	*Fatal Attraction*	336 — Farr 40
		1st Farr 40 Class	*Solution*	374 — Farr 40 OD
	Mumm 36 World Championship	1st Overall	*Thomas L Punkt*	299 — Mumm 36
	New York Yacht Club 145th Annual Regatta	1st IMS 2	*Passage*	374 — Farr 40
	Rolex 12-Metre World Championship	1st Overall	*Kiwi Magic*	196 — 12 Metre
	European IMS Championship	1st Racer	*Bribon*	411 — Farr IMS 49
	RORC Red Funnel Easter Challenge	1st IRC Class 0	*Warlord VI*	374 — Farr 40 OD
	Sydney-Mooloolaba	1st IMS Division C	*Sagacious V*	374 — Farr 40
	GMC Yukon Yachting	1st IMS Fleet and IMS 1	*Hi Fling*	414 — CM60
	Key West Race Week	1st Team Italy	*Breeze*	415 — Farr 49
		1st Team Italy & 2nd IMS 3 Class B	*Malinda/Invicta*	338 — Mumm 30
		1st IMS 1 Class B	*Brava Q8*	330 — Farr ILC 40
		1st IMS 1 Class C	*Hissar*	374 — Farr 40 OD
		1st IMS 3 Class B	*Sector*	338 — Mumm 30
1998	Telstra Sydney to Hobart	Line Honors & 1st IMS A	*Sayonara*	323 — ILC Maxi
		1st IMS B & 2nd Overall	*Ausmaid*	328 — Farr 47

Year	Event	Class	Yacht	Design
1998		1st IMS C	*Yendys*	219 — Beneteau 53
		1st Div. 2	*Jubilation*	72/2 — Farr 40
	St. Francis YC Big Boat Series	1st IMS		
		Grand Prix	*Flash Gordon III*	416 — Farr 49
	Kenwood Cup, Hawaii	1st Richard Rheems	*Samba Pa Ti*	374 — Farr 40
		1st IMS Class C, Winning Team (NZL) &		
		Top IMS Racing	*Big Apple III*	418 — Farr IMS 45
		1st IMS Class D, Winning Team (NZL) & Top		
		IMS Cruiser/Racer	*White Cloud*	336 — Cookson 12MT
		1st IMS Class A, 2nd Team	*Ragamuffin*	324 — Farr IMS 50
		1st IMS Class E	*Santa Red*	299 — Mumm 36
	Copa del Rey, Spain	Top IMS Cruiser/ Racer Overall	*Estrella Damm*	354 — Beneteau First 40.7
		1st IMS Racing, 2nd Overall	*Breeze*	415 — Farr IMS 49
	Rolex Commodore's Cup, Cowes, England	1st Rolex Class B Winning Team (Germany)	*Sequana* *Red*	354 — Beneteau First 40.7
	Breitling Regatta, Palma, Spain	1st Cruising	*Sebago*	354 — Beneteau First 40.7 mod.

Year	Event	Class	Yacht	Design
1998	IMS Nationals — Newport Race Week, USA	IMS National Champion — Racing Division &		
		1st IMS Class D	Rima	414 — Farr CM 60
		1st IMS Class E	Orient Express	374 — Farr 40 OD
	Volvo Distance Race — Newport Race Week, USA	1st IMS Class 1	Orient Express	374 — Farr 40 OD
		1st PHRF Class 5	Fatal Attraction	336 — Farr 39 ML
	Chicago — Mackinac, USA	Line Honors &		
		1st PHRF Class 1	Sayonara	323 — ILC Maxi
		1st IMS Class E	Nitemare	369 — Corel 45
	ILC Maxi Worlds, Newport	1st ILC Maxi	Sayonara	323 — ILC Maxi
	Block Island Race	1st IMS Overall	Blue Yankee	326 — Farr IMS 47
	Whitbread Round the World Race	1st Whitbread 60	EF Language	378 — Farr W60
	Squadron Cup	1st IMS Overall	Beau Geste	420 — Farr IMS 49
		1st PHRF	Fatal Attraction	336 — Farr 39ML
	SORC	1st IMS Class 1	Flash Gordon 3	416 — Farr IMS 49
		1st PHRF 1	Fatal Attraction	336 — Farr 39ML
1997	Telstra Sydney-Hobart Yacht Race	1st IMS Overall	Beau Geste	420 — Farr IMS 49
	Manhasset Bay Y.C. Fall Series	1st IMS Class B	Rampant	336 — Farr 39ML

Year	Event	Class	Yacht	Design
1997	Japan Cup	1st IMS Overall	*Aoba-Escape 1*	369 — Corel 45
	Maxi Yacht Rolex Cup	World Championship & 1st ILC Maxi	*Sayonara*	323 — Farr ILC Maxi
	PHRF New England Championship	1st PHRF Class 2	*Wired*	374 — Farr 40 OD
	Copa Del Rey Regatta	1st Class CR1	*Chariad*	128 — Farr 37
		1st Maxi Yacht Class	*Sayonara*	323 — Farr ILC Maxi
	Champagne Mumm Admiral's Cup	1st IMS Overall (Top Big Boat) & 2nd IMS (Fastnet Race)	*Flash Gordon 3*	416 — Farr IMS 49
		1st IMS (Fastnet Race) & 2nd IMS Overall	*Madina Milano*	415 — Farr IMS 49
		Winning Team (USA)	*Flash Gordon 3*	416 — Farr IMS 49
			Jameson	299 — Mumm 36
	Berthon/Source Regatta	1st IMS Overall	*Flash Gordon 3*	416 — Farr IMS 49
	Breitling Regatta	1st Overall & Racing Div.	*Bribon*	411 — Farr IMS 49
	Microsoft Round Gotland Race	1st IMS Class 1	*Investor*	369 — Corel 45
		1st Open Class	*Swedish Match*	394 — Whitbread 60
	ILC Maxi Worlds	1st ILC Maxi	*Sayonara*	323 — ILC Maxi
	Block Island Race Week	1st IMS Class 1	*Swing*	368 — Farr ILC 46
		1st IMS Class 2	*Wired*	374 — Farr 40 One Design
	NYYC 143rd Annual Regatta	1st IMS C/R Division	*Rampant*	336 — Farr 39ML
	European IMS Championship	1st Overall	*Evolution*	362 — Farr IMS 38
	GMC Yukon/Sailing World NOOD Regatta	1st Grand Prix	*Fast Tango*	299 — Mumm 36

Year	Event	Class	Yacht	Design
1997	Black Seal Rum Cup	1st IMS Class A	Flash Gordon 3	416 — Farr IMS 49
	Brisbane — Gladstone Yacht Race	1st IMS	No Fear	336 — Cookson12MT
	Antigua Sailing Week	1st Overall & Big Boat	Sayonara	323 — Farr ILC Maxi
	Sydney-Mooloolaba	1st IMS C/R Division	No Fear	336 — Cookson12MT
		1st IMS Racer Division	BZW Challenge	369 — Corel 45
	Air New Zealand Regatta	1st IMS Overall	Georgia	360 — Farr 43
	SORC	1st IMS Overall & 1st IMS 2	Esmeralda	344 — Farr ILC 40
	Key West Race Week	1st IMS 2	Esmeralda	344 — Farr ILC 40
1996	Telestra Sydney to Hobart	1st Overall & Div. B	Ausmaid	328 — Farr 47
	Corum Cup	1st Overall	G'Net	330 — Farr ILC 40
	St. Francis Yacht Club Big Maxi Boat Series	1st IMS Maxi	Sayonara	323 — Farr ILC
		1st IMS Grand Prix	Beau Geste	369 — Corel 45
	ILC 40 European Circuit	1st	Brava Q8	330 — Farr ILC 40
	Sardinia Cup	1st ILC 40	Brava Q8	330 — Farr ILC 40
	Kenwood Cup	1st Maxi Class, 1st IMS A	Sayonara	323 — ILC Maxi
		1st IMS C, 1st Overall	Georgia	360 — Farr 43 C/R Div.
		1st IMS D, 2nd Overall	White Cloud	336 — Cookson12MT C/R Div.
	Copa del Rey	1st ILC 40, 2nd Overall	Brava Q8	330 — Farr ILC 40
	ILC 40 World Championship	1st	Brava Q8	330 — Farr ILC 40
	European IMS Championship	1st	Osama Citizen	336 — Cookson 12MT
	ILC Maxi Worlds	1st ILC Maxi	Sayonara	323 — ILC Maxi
	Mount Gay/Block Island Race Week	1st IMS A	X-Rated	326/1 — Farr ILC 46
	Kiel Week	1st IMS Overall	Lina F	297 — Farr IMS 31
	Rolex Cup Regatta	1st Class A	Morgana	243 — Farr 70
	China Sea Race	1st IMS	Beau Geste	317 — Farr ILC 40

Year	Event	Class	Yacht	Design
1996	Sydney-Mooloolaba	1st IMS Div. I	*Ragamuffin*	324 — Farr 50
		1st IMS Div. III	*Son of a Son*	Farr 1104
	Air New Zealand Regatta	1st IMS Overall, 1st IMS B	*Georgia*	360 — Farr 43
		1st PHRF Over.,		
		1st PHRF D	*G'Net*	330 — Farr ILC 40
		1st PHRF E	*Mumm 30*	338 — Mumm 30
		Winning Team (RNZYS)	*High 5*	336 — Cookson12MT
			Stackerlee	336 — Farr 39ML
			Georgia	360 — Farr 43
	SORC	1st IMS Class 3	*Esmeralda*	344 — ILC 40
	Key West Race Week	1st IMS Class D	*Flash Gordon 2*	329 — Farr 43
		1st PHRF II Class B	*Diana*	294 — Bene. 42S7
1995	Sydney–Hobart Race	Line Honors & 1st IMS Division A	*Sayonara*	323 — ILC Maxi
	Southern Cross Cup	Winning Team (AUS)	*Ragamuffin*	324 — Farr 50
			Amp Wild Oats	159 — Farr 43
	Rolex IMS International Championships	1st Offshore & 3rd Inshore Div. II	*Flash Gordon 2*	329 — Farr 43
		1st Inshore Div. II	*Radical Departure*	299 — Mumm 36
	Big Boat Series — USA	1st IMS Grand Prix	*Flash Gordon 2*	329 — Farr 43
	RORC Channel Race	1st IMS 5-A2	*Brava Q8*	330 — ILC 40
	Copa Del Rey	1st IMS	*Banco Atlantico*	336 — IMS 39
	ILC 40 European Circuit	1st	*Brava Q8*	330 — ILC 40
	Admiral's Cup	1st ILC 40 & Italy Team	*Brava Q8*	330 — ILC 40
	Berthon Source IMS Regatta	1st ILC 40 & IMS 1	*Brava Q8*	330 — ILC 30
	Chicago–MacKinac Race	1st Overall & IMS D	*Flash Gordon 2*	329 — IMS 43

Year	Event	Class	Yacht	Design
1995	LC 40 World Championship	1st ILC 40	Brava Q8	330 — ILC 40
	Block Island Race Week	1st IMS 4	Esmeralda	344 — Farr ILC 40
		1st PHRF 2	Christopher Dragon	151 — Farr 43
		1st PHRF 4	Sugar	143 — Farr 33
		1st PHRF 8	Hat Trick	297 — Farr 31
	Kiel Week	1st ILC 40	Brava Q8	330 — ILC 40
		1st AC Big Boats	Capricorno	326I — ILC 46
	SORC	1st IMS 3 and 3rd Overall	Corum Watches	299 — Mumm 36
	Air New Zealand Regatta	1st Overall and IMS B	High 5	299 — Mumm 36
	Key West Race Week	1st IMS B	Flash Gordon 2	329 — Farr 43
	Kodak Sydney–Hobart	Line Honors Winner	Tasmania	274 — Maxi
		1st IMS B, 2nd overall	Ninety-Seven	304 — Farr 46
		1st IMS G	Invincible	51- Farr 1104
	Rothmans Race Week	1st Overall	Bill Rauson	72/2 — Farr 38
	Corum Cup	1st Overall	Corum No Problem	299 — Mumm 36
	Kenwood Cup	1st IMS B and 2nd Overall	Gaucho	266 — IMS 44
		1st IMS D and 4th Overall	Corum No Problem	299 — Mumm 36
	Sesquicentennial Regatta	1st IMS 1 Racing	Full Cry	281 — IMS 50
		1st IMS 3 Racing	Beau Geste	317 — ILC 40
	Rolex Commodore's Cup	1st IMS Overall	Thomas I Punkt	299 — Mumm 36
		1st IMS 2 and 4th Overall	Flash Gordon	279 — IMS 39
	Black Seal Cup	1st IMS 2	Radical Departure	299 — Mumm 36
	Kiel Week	1st IMS Overall	Omen	317 — ILC 40
1994	Block Island Race Week	1st IMS A	Full Cry	281 — IMS 50
		1st IMS C	Beau Geste	317 — ILC 40

Year	Event	Class	Yacht	Design
1994	Swedish IMS Championships	1st IMS	Flawless Lady	Farr 1 Ton
	RORC de Guinguand Bowl	1st IMS	Sodifac	299 — Mumm 36
		1st CHS	Brittany Ferries	202 — First 45F5
	Detroit Nood Regatta	1st Offshore	Sensation	299 — Mumm 36
	Whitbread Round the World Race	1st Maxi	NZ Endeavour	274 — Maxi
		1st W60	Yamaha	293 — Whitbread 60
	Japan Cup	1st IMS Class A	Wind War	299 — Mumm 36
	Orange Cup	1st IMS Overall & IMS II	Bird	297 — IMS 31
		1st IMS I & 3rd Overall	Konakai Baby	299 — Mumm 36
	SORC	1st IMS 2	Gaucho	266 — IMS 44
	Key West Race Week	1st IMS Class B	Flash Gordon	279 — IMS 39
1993	Kodak Sydney–Hobart Race	1st IMS Class A, 2nd IMS Overall & Line Honors	Ninety Seven	304 — IMS 46
		1st IOR Overall / 1st IOR A	Soulbourne Wild Oats	159 — FARR 43
	Southern Cross Cup	1st IOR Fleet	Soulbourne Wild Oats	159 — FARR 43
	Asia Pacific Championships	1st IMS Fleet	Ninety Seven	304 — IMS 46
	OMYA World Cup	1st	Merit Cup	278 — Maxi
	50-Foot World Cup	1st	Carat VII Citroen	203 — IOR Fifty
	Admiral's Cup	1st indiv. yacht scorings	Ragamuffin	260 — IOR Fifty
	Fastnet Race	1st elapsed time	Galicia 93 Pescanova	284 — Whitbread 60
	Royal Ocean Racing Club Yacht of the Year	(Beneteau First IMS 40.7)	Bounder	289 — IMS 40.7 M
	Two Ton World Cup	1st	Larouge	242 — Two Ton
	One Ton Circuit	1st	Brava Q8	276 — One Ton

Year	Event	Class	Yacht	Design
1993	Open UAP Round Europe Race	1st Maxi	*NZ Endeavour*	274 — Maxi
		1st Whitbread 60	*Galicia 93*	284 — Whitbread 60
			Pescanova	287 — Whitbread 60
	Rolex IMS Championship	1st IMS 2	*Flash Gordon*	297 — IMS 39
	San Francisco Big Boat Series	1st IMS GP	*Pigs in Space*	299 — IMS 36
	Gold Cup Transatlantic Race	1st elapsed time record	*Winston*	287 — Whitbread 60
	Yachting's Key West Race Week	1st IMS A	*Full Cry*	281 — IMS 50
		1st IMS C &		
		1st IMS Overall	*Rush*	279 — IMS 39
1992	Rolex Commodores' Cup (IMS)	1st Overall	*Califa 3*	266 — IMS 44
	Kenwood Cup	1st IMS D &		
		1st IMS Overall	*Cookson's High 5*	277 — IMS 40
		1st IOR C & IOR Overall	*Larouge*	242 — Two Ton
	Yachting's Key West Race Week	1st IMS A & Fleet Overall	*Gaucho*	266 — IMS 44
	Kodak Sydney-Hobart Race	1st IMS B & IMS Overall	*Assassin*	277 — IMS 40
		1st IMS A &		
		2nd IMS Overall	*Morning Mist III*	281 — IMS 50
		1st IOR A & IOR Overall	*Ragamuffin*	260 — IOR 50
		Line Honors	*NZ Endeavour*	274 — Maxi
	Kodak Asia Pacific Championships	1st IMS Fleet	*Assassin*	277 — IMS 40
		1 st IOR Fleet	*Ragamuffin*	260 — IOR 50
	San Francisco Big Boat Series	1st IMS Atlantic Fleet	*Cookson's High 5*	277 — IMS 40
	Royal Ocean Racing Club	(Beneteau First 45f5)	*Bounder*	202
	Yacht of the Year	1st IMS	*Dreampic*	266 — IMS 44
	Corum Japan Cup	1st IOR B	*Cha Cha II*	253 — One Ton

Year	Event	Class	Yacht	Design
1992	OMYA World Cup	1st	*Merit Cup*	183 — Maxi
	Two Ton World Cup	1st	*Shockwave*	268 — Two Ton
	One Ton World Cup	1st	*Brava Q8*	276 — One Ton
	3/4 Ton World Cup	1st	*Xacobeo 93*	229 — 3/4 Ton
	Copa Del Rey (King's Cup)	1st Class A	*Publiespana*	191 — Maxi
1991	Champagne Mumm Admiral's Cup	1st One Ton	*Brava*	223 — One Ton
		1st Two Ton	*Bravura*	242 — Two Ton
	Kodak Sydney-Hobart	1st IOR A & 1st Fleet	*Atara*	177 — Farr 44
		Line Honors	*Brindabella*	220 — Farr 65
	Southern Cross (IOR)	1st Overall	*Atara*	177 — Farr 44
	Royal Ocean Racing Club			
	Yacht of the Year	(Beneteau First 53F5)	*Yellow and Blue*	219
	Yamaha Osaka Double-Handed	1st , elapsed time record	*Nakiri Daio*	152 — Farr 55
	Transpac Race	1st IMS B & 2nd Overall	*Kotuku*	165 — Farr 1220
	OMYA World Cup	1st	*Merit*	183 — Maxi
	Two Ton Cup	1st	*Larouge*	242 — Two Ton
	One Ton Cup	1st	*Vibes*	253 — One Ton
	3/4 Ton Cup	1st	*Lone*	229 — 3/4 Ton
	Japan Cup	1st Class A & 1st Overall	*I'm Sorry*	216 — Two Ton
		1st Class B & 2nd Overall	*Cha Cha II*	253 — One Ton
		1st Class C	*Sylphides 4*	201 — 3/4 Ton
1990	Centomiglia	1st Libera Class	*Pleasure*	146M
	Whitbread Round the World Race	1st	*Steinlager 2*	190 — Maxi
	Kenwood Cup	Winning Team (Japan Blue)	*Will*	211 — IOR 5
			Tiger	203M- IOR 50
			Swing	216 — IOR 44

Year	Event	Class	Yacht	Design
1990	Sardinia Cup	Winning Team (Italy)	Brava	223 — One Ton
		1st 50 Foot Class, Overall Fleet Honors, & Winning Team	Mandrake	224 — IOR 50
		Top Boat, 2nd Overall & Winning Team	Larouge	242 — Two Ton
	Japan Cup	1st Overall	Blue Note	223 — One Ton
	Copa del Rey (King's Cup)	1st Class 4-5 and Overall	Lone	229 — 3/4 Ton
		1st Class 2-3	Vento	223 — One Ton
		1st Class 1, 2nd Overall	Larouge	242 — Two Ton
		1st 50 Footer, 2nd Class 1	Mandrake	224 — IOR 50
		1st Class A	Longobarda	207 — ICAYA Maxi
		1st OMYA Class	Merit	183 — Maxi
	Nortel Sydney-Hobart	1st IOR & 1 st Overall	Sagacious	185 — One Ton
		1st Maxi Division & 3rd IMS Class III	Brindabella	220 — ULDB 65
	OMYA World Cup	1st	Longobarda '92	191 — Maxi
	3/4 Ton World Cup	1st	Lone	229 — 3/4 Ton
	Corum China Sea Series	1st Overall	Steadfast	188 — One Ton
	Royal Ocean Racing Club Yacht of the Year	(Beneteau First 45F5)	Cap Sogea	202 — 45' R/C
1989	One Ton Worlds	1st Overall	Brava	223 — One Ton
	Int'l 50 Foot World Cup	1st Overall	Carat	203 — IOR 50
	Admiral's Cup	1st Overall	Jamarella	213 — IOR 50
		1st in class	Librab	216 — IOR 44
		1st in class	Joint Venture	214 — One Ton

Year	Event	Class	Yacht	Design
1989	ICAYA Maxi Worlds	1st Overall	*Longobarda*	207 — Maxi
1989/90	Whitbread Round the World Race	1st Overall	*Steinlager2*	190 — Maxi
	Japan Cup	1st Overall	*Swing*	216 — IOR 44
	Southern Cross	1st Overall	*Heaven Can Wait*	203M — IOR 50
1988	Kenwood Cup	1st Overall & 1st Class E	*Bravura*	182 — One Ton
	Antigua Sailing Week	1st Class R1	*Marlboro*	131 — Maxi
	Japan Cup	1st Overall & 1st Class 4	*Aphrodite*	201 — 3/4 Ton
	One Ton North American Championships	1st	*Rush*	184 — One Ton
	One Ton Worlds	1st	*Propaganda*	182 — One Ton
	Big Boat Series — USA	1st in San Fran Class	*Great News*	155M — IOR 50
	Sydney–Hobart	1st Class A & 4th Overall	*Great News*	155M — IOR 50
		1st Class B & 2nd Overall	*Southern Cross*	One Ton
1987	Route of Discovery	1st	*Merit*	183 — Maxi
	Melbourne Osaka Race	1st	*SDC Nakiri Daro*	152 — 55' Cr
	SORC — USA	1st Class 5 /8th Overall	*Bodacious*	156 — One Ton
	Admiral's Cup	1st Overall & 1st NZ Team	*Propaganda*	182 — One Ton
	Japan Cup	1st Overall	*Will*	211 — One Ton
	One Ton Cup	1st	*Fram X*	185 — One Ton
	Southern Cross Series	1st Overall & Team (AUS)	*Madeline's Daughter*	177 — IOR 44
	12-Metre Worlds	1st	*Kiwi Magic*	12 Metre
	Centomiglia — Italy	1st Libera Class	*Lillo*	Libera
	European Class Liberia Champs.	1st	*Farmeticante*	Libera
1986	Kenwood Cup	1st Class C, 3rd Overall,	*Equity*	43'
		1st Team (NZ)		

Year	Event	Class	Yacht	Design
1985/86	Whitbread Round the World Race	1st	UBS Switzerland	131 — Maxi
1986	Ocean Racing Champs.	1st Overall	Another	128 — One Ton
		1st Class B & 4th Overall	Concubine	One Ton
1985	Sydney — Hobart	1st Class 3, 4th Overall	Paladin	151 — 43'
	Southern Ocean Racing Confer.	1st Class E, 3rd Overall	Snake Oil	136 — One Ton
1984	Pan American Clipper Cup	1st Overall & Div B	Pacific Sundance	One Ton
	Sydney — Hobart	1st Class F	Indian Pacific	124 — IOR 37
1983	Southern Ocean Racing Confer.	1st Overall & team (NZ)	Migizi	136 — One Ton
	Southern Cross	2nd Overall &	Pacific Sundance	
		1st team (NZ)	Geronimo	136 — One Ton
1982	Sydney–Hobart	1st Division C	Scallywag	62 — One Ton
1981	Centomiglia Regatta	1st Open Class	Grifo	97 — Libera
	Transpac — USA	1st Overall & Class D	Sweet Okole	51 — 36'
		1st Class A	Zamazaan	60 — 52'
	Montagu Island Race — Australia	1st Overall & Div 2	Salamander II	36'
	Auckland–Suva Race	1st Overall & Div 1	Kaihua	39 — 42'
	Auckland Trailer Yacht Championships	1st	Hawkeye	101 — 740 Sport
	Donald Hay–Travelodge Race	1st Line & Handicap	Head Office	7.5M
	Trailer Yacht Nationals — NZ	1st Overall	Farr 740 Sport	101 — 740 Sport
	Argentario Winter Championships	1st in class	Roba Da Ppazzi	1/2 Ton
	Cutty Sark — Spain	1st Class III	Alcaravan III	64 — One Ton
	Regata Int'l del Jerez — Spain	1st Class III	Alcaravan III	64 — One Ton
	Rias Bajas — Spain	1st Class I, II & III	Alcaravan III	64 — One Ton
	Finisterre — Spain	Line Honors /2nd hand.	Alcaravan III	64 — One Ton
	Rias Altas — Spain	1st Class B & 1 st gen.	Alcaravan III	64 — One Ton

Year	Event	Class	Yacht	Design
1980	San Francisco–Kauai, HI — USA	Line Honors /2nd hand.	Timberwolf	38'
	Big Boat Series — USA	1st City of San Fran.	Zamazaan	60 — 52'
	Sydney–Hobart	Line Honors & 1st IOR	Ceramco	90 — Maxi
1980	Pan Am Clipper Cup Series — USA	1st Class A	Zamazaan	60 — 52'
	National Quarter Ton Series	1st	Anchor Challenge	85/5 — 1/4 Ton
1979	Transpac — USA	1st Class B	Zamazaan	60 — 52'
1978	Pan Am Clipper Cup — USA	1st Overall & Div 2	Monique	67 — IOR 39
	N.A. One Ton Champ. — USA	1st	Scallywag	62 — One Ton
	China Sea Race	1st	Uin-na-mara	67 — IOR 39
	Southern Ocean Racing Circuit	1st Div D	Mr. Jumpa	66
	New Zealand Eighteen-footer Champ.	1st	Benson and Hedges Designed by Russell Bowler	
1977-78	Southern Cross Cup	1st	Jenny H	62 — One Ton
1977	Dunhill Int'l Series — Australia	1st	Gunboat Rangiriri	65 — 1/2 Ton
	World Half Ton Championships	1st	Gunboat Rangiriri	65 — 1/2 Ton
	World One Ton Championships	1st	Red Lion	64 — One Ton
	World 3/4 Ton Championships	1st	Joe Louis	56 — 3/4 Ton
	French 3/4 Ton Series	2nd	Joe Louis	56 — 3/4 Ton
	Southern Ocean Racing Circuit	1st Class III	Sueet Okole	51 — One ton
	Auckland–Suva Race — Fiji	1st	Country Boy	
	National One Ton — NZ	1st	Country Boy	
1976-77	Sydney–Hobart	1st Overall	Piccolo	One Ton
1976	British Yacht of the Year		Solent Saracen	One Ton
	North American Quarter Ton	1st	Why Why	37 — Farr 727
	National One Ton Series	1st	Jiminy Cricket	51 — One Ton
1975-76	Southern Cross Series	Top points scorer	Prospect of Ponsonby	51 — One Ton

Year	Event	Class	Yacht	Design
1975	Admiral's Cup	Top scoring NZ boat	*Gerontius*	39 — 42'
	World Quarter Ton — France	1st	*45° South*	37M — Farr 727
	National Quarter Ton — NZ	1st	*727*	37M — Farr 727
1974	South Pacific Half Ton Cup	1st	*Titus Canby*	27 — 1/2 Ton
1973	National Quarter Ton Champs	1st	*Fantzipantz*	
1972	South Pacific Half Ton Cup	1st	*Titus Canby*	27 — 1/2 Ton
1975	18 Footer Worlds — Australia	1st	*K.B.*	
1974	18 Footer Worlds — NZ	1st	*TraveLodge NZ*	
1973	18 Footer Worlds — Australia	1st	*TraveLodge NSW*	
1972	18 Footer Worlds — Australia	1st	*Smirnoff*	
1970	12 Foot Interdominion Champs	1st	*Miss Beazley Homes*	
1969	Q Class National — NZ	1st	*Miss Beazley Homes*	
1969	Q Class Interdominion Champs	1st	*Jennifer Julian*	Designed by Russ Bowler
1966	NZ Moth Championship	1st	*Mammoth*	

Appendix 3

Design Team — Farr Yacht Design Ltd

Bruce Farr — president and founder of Bruce Farr Yacht Design Ltd.

Russell Bowler— vice president of Bruce Farr Yacht Design Ltd.

Graham Williams — senior designer. Prior to joining Farr Yacht Design as a designer in 1984, Graham worked for Bruce Robert's design firm and as a consultant to Bruce Farr and Russ Bowler. He had established a career as a boat-builder with experience in the construction of wood and fibreglass hulls up to 130 feet in length. At Farr Yacht Design, Graham works closely with the builders of Farr designs. As the staff has grown, Graham's role has expanded to include the oversight and coordination of projects at the FYD office.

Graham learned to sail at a very young age and built his first boat in Australia at the age of 13. Today, he enjoys sailing his Farr 10^{20}, *Fadoodle*, on the Chesapeake Bay.

Jim Schmicker — design/naval architect. Jim holds a Master of Science Degree in Naval Architecture and Marine Engineering from the Massachusetts Institute of Technology, as well as an undergraduate degree from the Webb Institute of Naval Architecture. Prior to joining Farr Yacht Design, Jim was the Offshore Technical Assistant for the United States Yacht Racing Union. He has also worked for Newport Offshore and with the Heart of America syndicate in 1986/87 as performance analyst and engineer. This education and several years of engineering and technical experience are the qualities that influenced Bruce and Russ to hire Jim in 1988 as a Naval Architect. His primary responsibilities include preliminary design and concept development, rating analysis, optimisation, and managing post-design consulting work. Additionally, Jim was recently appointed a member of both the International Technical Committee of the ORC and US Sailing IMS committee.

Steve Morris — design/engineer. Steve has a Master's Degree in Mechanical Engineering from the University of Auckland, where he specialised in aerodynamics. Prior to joining Farr Yacht Design, he worked with Grant Dalton on *Fisher and Paykel* conducting performance analysis for the

1989–90 Whitbread Round the World Race. Steve joined Farr Yacht Design in 1989 as a designer with a specialty in research. He has developed skills in many areas of yacht design through work on multiple America's Cup and Whitbread programmes, as well as other racing and cruising yachts. He is equally at home performing structural engineering in high-performance composites or hydrodynamic design and analysis of hulls and appendages. In addition to responsibilities for rig design and performance estimation for Farr designs, Steve leads special research programmes and shares in administering the office computer systems.

Over the past two years, Steve has been spending time sailing on the Young America IACC yachts in Rhode Island and Auckland. He enjoys Wednesday night racing on the Chesapeake and is currently fixing up a Laser so he can teach his kids to sail this summer.

Patrick Shaughnessy — designer. Patrick was initially hired by Farr Yacht Design as a draughtsman on a part-time basis in 1990 after providing examples of his self-taught draughting skills. Prior to being hired, Patrick attended both the University of Maryland and Anne Arundel Community College where he studied Architecture. His draughting skills are exemplified by his 1st Place award in the 1988 Maryland State Draughting Competition. Since advancing into a design position with the company, Patrick's responsibilities have centered primarily on keel design, interior layout, and weight calculations. Additionally, his work includes rudder design, deck layout, and construction detailing.

Patrick spends his spare time yacht racing as a bowman and has in the past years begun to test the professional sailing market. He has raced recently with *Gaucho* (Farr 44), *Predator* (Farr 40 One Design), *Sayonara* (Farr ILC Maxi), *Seagoon* (Farr ILC 46), and *Woftam* (the Mumm 30 One Design).

Michael Price — designer. Mick graduated from the Tabor Academy in 1975, where he was a member of the New England Team racing championship team. He also attended the New York Institute of Technology where he studied mechanical engineering and from there began an Olympic Campaign in 470s. Mick joined the Farr Yacht Design staff late in 1990 to work as a designer with emphasis on keels, rudders, construction detailing, deck layouts and geometry, and sail plans. Prior to joining FYD, Mick designed and built a popular line of 30- and 40-foot fast cruising trimarans in Annapolis. His early experience includes working for a machine shop making yacht fittings as well as a sailmaker for Hard Sails and Ullman Sails. He was also an in-house design engineer for the Rhode Island boat-builder, C.E. Ryder, specializing in production boats as well as tooling and industrial fibreglass products.

Mick is an experienced offshore and dinghy sailor and participates actively in the racing scene.

Harry Dunning — designer. Harry studied both mechanical engineering and small craft design. He began his career as a designer in the offices of Gibbs & Cox and later took a position with Rosenblatt and Son. After several years of working with these two naval architecture firms, Harry branched out on his own as a New York-based yacht designer, working on large-scale cruising yacht projects. Since joining Farr in 1992, Harry's responsibilities have included preliminary design, detail design and construction detailing of both cruising and racing yachts.

Harry regularly races in major regattas, recently in the Farr 40 and the Mumm 36 classes where he has competed in the last few national and world championships. In the past, Harry regularly raced with *Full Cry* (Farr 50), *Gaucho* (Farr 44), *Rampant* (Farr 39), and other successful Farr-designed yachts. Before working for Farr, Harry raced internationally in the IOR Maxi Class circuit and has competed in such ocean races as the Fastnet, Buenos Aires–Rio de Janeiro, Sydney–Hobart, and every Newport–Bermuda race since 1984.

Britton R. Ward — designer. Britt graduated in 1996 from the Massachusetts Institute of Technology with a Masters Degree in Ocean Engineering. He received his Bachelors Degree in Naval Architecture and Marine Engineering from the Webb Institute of Naval Architecture in 1995. His emphasis was on experimental and computational hydrodynamics, with a concentration on yachts.

Britt is currently involved in research for the 2000 America's Cup, assisting with hull and rig development and other performance-prediction related issues. Additionally, he continues to assist with engineering and design-related tasks for other racing and cruising boat projects. He joined Farr Yacht Design in September 1996 to assist in the research of Whitbread 60s for the 97/98 Whitbread Round the World Race.

Britt began sailing at the age of 10 in Fremantle, Australia where he watched the 1987 America's Cup and 1988 Whitbread Round the World Race which stimulated his interest in the field of yacht design. Since moving to Annapolis, he has enjoyed racing aboard a variety of boats in both Wednesday night and weekend Chesapeake Bay races.

David Fornaro — designer. Dave holds a Master of Science degree in Mechanical Engineering from the University of Michigan and a Bachelor of Science degree in Mechanical Engineering from the University of Pennsylvania. Dave's responsibilities for Farr Yacht Design include keel and rudder design, structural calculations, construction detailing, and computer system administration. He joined Farr Yacht Design in 1998 after a nine-year career in the automotive industry, with five years at Ford Motor Company in the design and analysis of structural engine components and four years as a consultant in these areas.

Racing Tornado catamarans is Dave's primary sailing passion. He has competed twice in the Olympic Trials, was named in the 1992 US Sailing Team and has been ranked as high as fourth nationally. Dave also has a great deal of big boat experience trimming sails on the very successful 3/4 ton *Flyer* in the Great Lakes area, including multiple class wins in the Detroit NOOD regatta and Mackinac races and Detroit Regional Yachting Association Boat of the Year in 1996.

Bobbi Hobson — office manager. Bobbi, who has a Bachelor of Science degree from the University of Delaware, is responsible for management of the administrative section and is Bruce's personal assistant as well. Since joining the firm in 1983, Bobbi's role has expanded to include the generation of agreements for customers of Farr designs and account management. Prior to joining Farr Yacht Design, Bobbi worked for several years for an Annapolis-based sail brokerage firm where she acquired her initial knowledge of the sailing industry.

Bobbi previously owned and raced a Cal 25 and she now occasionally crews on a Farr 33, *Uh Oh*, and other Cal 25s on the Chesapeake Bay.

Jennifer Charnesky — administrative assistant. Jennifer joined Farr Yacht Design in 1990, bringing with her a strong background in office management and organisation. In addition to being Russell Bowler's personal assistant, Jennifer is responsible for providing administrative support to the design team for design and rating consultation work, preparing consulting and confidentiality agreements, project invoices, as well as database and file management.

Since 1982, Jennifer has worked in the Annapolis yachting community, having begun her career with Beneteau USA as an administrative assistant concentrating on public relations. Jennifer is presently restoring a 1977 O'Day DaySailer and looks forward to teaching her young son to sail.

Amy Fazekas — public relations. Amy graduated from Boston University in 1997, where she received a Bachelor of Science in Public Relations. Hired in 1998, Amy handles the public relations for Farr Yacht Design, including production of the newsletter and website, advertising, and acting as a press liaison. She is also responsible for providing administrative support to the design team. Amy's prior work experience includes public relations internships with a Broadway press office in New York, and with the Royal National Theatre's press office in London. She also acted as Public Relations Coordinator for a small full-service ad agency in New Hampshire.

Amy recently relocated to the Annapolis area and looks forward to becoming acclimated to the local sailing community.

Notes

Chapter 1

1. 'Anchor Me' by Don McGlashan. Reproduced with kind permission by Mana Music Publishing/Warner Chappell Music Australia Pty Ltd.
2. Paul Cayard, 'Paul Cayard Phone Interview No. 001' (unpublished, San Diego, 28 August 1995).
3. 'Mass Pakeha settlement burned off and logged as much or more forest in its first four decades than early Maori ... had done in their first four centuries.' James Belich, *Making Peoples* (Auckland: Allen Lane, The Penguin Press, 1996), p. 365.
4. Patricia Burns, edited by Henry Richardson, *Fatal Success: a history of the New Zealand Company* (Auckland: Heinemann Reed, 1989), p. 197.
5. John Owen Miller, *Early Victorian New Zealand: a study of racial tension and social attitudes, 1839–1852* (London; Wellington: Oxford University Press, 1958), p. 22.
6. By the end of the 1880s New Zealand's racial groups had more or less settled into the following ethnic mix of total population: English 45 percent, Scots 20 percent, Irish 18 percent, Maori 9 percent, Welsh 2 percent, German 2 percent, Scandinavia 2 percent, Chinese 2 percent. Refer James Belich, *Making Peoples*, p. 318.
7. John Owen Miller, *Early Victorian New Zealand: a study of racial tension and social attitudes, 1839–1852* (London; Wellington: Oxford University Press, 1958), p. 164.
8. Ibid., p. 165.
9. Legislation was passed on 20 September 1893 giving the vote to New Zealand women. It was another 40 years before Mrs E.R. McCombs, on 21 September 1933, became the first woman to be elected to New Zealand's parliament.
10. Ernest Rutherford was awarded the Nobel Prize for Chemistry in 1908. He discovered the atomic nucleus in 1909 and became the first person to disintegrate atoms on 26 March 1934. He died in 1937.
11. Although it is likely he had been working on his 'improved aerial or flying machine' as early as 1899 and may have flown in 1902, the majority of eye-witness accounts (nineteen) say they saw Richard Pearse fly his monoplane down the Main Waitohi Road on 31 March 1903. The question is not whether Pearse flew before the Wright brothers (17 December 1903) but whether he was able to control his craft once airborne. In his own sparse writing on the subject, Pearse was of the opinion that neither he nor the Wright brothers could claim a

properly controlled flight in 1903. For further reading on Pearse and his other inventions see Gordon Ogilvie, *The Riddle of Richard Pearse* (Auckland: Reed Publishing (NZ) Ltd, 1973).

12. Ileene Farr (nee Clayton), 'Jim & Ileene Farr Interview No. 001' (unpublished, Warkworth, 7 October 1995).

13. Sir Ian Hamilton as quoted by Alan Moorehead in *Gallipoli* (Sydney, NSW: Meed & Beckett Publishing, 1989).

14. Aubrey Herbert as quoted by John Strawson in *Gentlemen in Khaki* (London: Secker & Warburg Ltd, 1989), p. 134.

15. Of the 8556 New Zealanders who landed at Gallipoli, 2721 died and 4572 were wounded. Figures from Chris Pugsley and John Lockyer, *The Anzacs at Gallipoli: A Story for Anzac Day* (Auckland: Reed Publishing, 1999), p. 29.

16. Odile Farr (nee Jacquart), 'Odile Farr Interview No. 001' (unpublished, Auckland, 26 August 1996).

17. Ibid. Odile Jacquart was born on 21 April 1896. Her one-hundredth birthday was celebrated in Auckland, at Green Valley Lodge and Bob Farr's home. Bruce and his wife Gail had flown from the USA a week before to join in the celebrations. There was a message from the Queen, one from the Prime Minister, champagne and dancing (the guest of honour joining in) and the surprise arrival of a representative of the mayor's office in Wattrelos. Complete with a purse containing the medal minted in Odile's honour, the dapper Frenchman unfurled a scroll and delivered the mayor's citation in French. Sharp as a tack, the centenarian thanked him in her native tongue and insisted he stay for the rest of the festivities. Odile died peacefully on 28 July 1998.

Chapter 2

1. Geoff Stagg, 'Geoffrey Stagg Interview No. 001' (unpublished, Annapolis, 28 August 1995).

2. On 10 April 1968, with winds reaching 100 miles per hour, the car-passenger ferry *Wahine*, carrying 72 vehicles, sank in Wellington Harbour. Fifty-one people lost their lives.

3. Russell Bowler's major wins as skipper are:
   (a) 1962 *Tranquil* (P Class Tauranga Cup/NZ national title)
   (b) 1963 *Nieuffe* (NZ national OK Dinghy Freshwater title)
   (c) 1966 *Mecca* (NZ national Cherub title/Don McGlashan crew)
   (d) 1968 *Mecca* (NZ national Cherub title/Don McGlashan crew)
   (e) 1969 *Jennifer Julian* (Q Class/Bowler/Interdominion title/Don McGlashan crew)
   (f) 1969 *Jennifer Julian 2* (Australian national Cherub title/Peter Walker crew)
   (g) 1970 *Jennifer Julian 2* (World Cherub title/Peter Walker crew)
   (h) 1977 *Benson & Hedges* (Eighteen-footer/Bowler/NZ national Eighteen-footer trials/Simon Ellis and Graham Catley crew)
   (i) 1978 *Benson & Hedges* (Eighteen-footer/Bowler/NZ national Eighteen-footer title/Simon Ellis and Graham Catley crew).

   In addition, Russell won the P Class division of the Auckland Anniversary regatta (1962), the Auckland Provincial P Class Freshwater Championship (Lake

Rotorua (1962), the Auckland Secondary Schools Championship with Ron Blakey (1962), the OK Dinghy class in the Royal Regatta (1963), the Auckland trials for the NZ National Cherub Interdominion team with Don McGlashan crew (1966), the North Shore (1967) and Auckland Cherub Championships with Don McGlashan crew (1967 and 1968), the selection trials for the NZ Interdominion Twelve-footer team with Don McGlashan crew (1968), the Balokovic Cup as skipper of the Farr-designed *Anchor Challenge* (1980), and in two Laser match-racing series sailed on Chesapeake Bay, USA, defeated Grant Dalton 2:1 twice (1997 and 1998).

4. Cecil Bowler to his daughter Jeanette, 'Jeanette & Bob Holley Interview No. 001' (unpublished, Auckland, 15 November 1995).

5. Further information on Emma Beaufoy can be obtained from *The Book of New Zealand Women*, ed. Charlotte McDonald, Merimeri Penfold, Bridget Williams (Wellington: Bridget Williams Books Ltd, 1991) and Betty Beaufoy, *Emma of the High Country* (Wellington: Dorset Enterprises, 1997).

6. Ena Hutchinson, 'Ena Hutchinson Interview No. 001' (unpublished, Auckland, 22 December 1995).

7. Russ Bowler, 'Russell Bowler Interview No. 001' (unpublished, Annapolis, 3 June 1995).

8. The Don McGlashan referred to here is a civil engineer, not the singer-songwriter of the Mutton Birds.

9. Ena Hutchinson, 'Ena Hutchinson Interview No. 001' (unpublished, Auckland, 22 December 1995).

10. Ewart Townsend writing as 'Gaffer Gooseneck' in the 1961 Kawau Newsletter. From 'The Mark Paterson Collection' (unpublished, Auckland).

11. Ibid.

12. Peter Shaw, 'Peter Shaw Interview No. 001' (unpublished, Auckland, 21 February 1996).

13. Russ Bowler, 'Russell Bowler Interview No. 001' (unpublished, Annapolis, 3 June 1995).

14. The *8 O'Clock* was a weekly sports newspaper published by the Auckland Star. It was eagerly awaited on Saturday evenings by the sports-mad city it served. Its impact was arguably greatest in the early 1960s, in the years before television.

15. The Mark Paterson Collection (unpublished, Auckland).

Chapter 3

1. Excerpt from '"Good Morning Everybody" — Aunt Daisy', copyright Radio Zealand Ltd. Reprinted by permission.

2. Matheson family history based on a discussion with Ewan Matheson, Val Stern (nee Matheson) and Iris Chitty (nee Matheson) in 'Matheson Family Interview No. 001' (unpublished, Leigh, 18 February 1996).

3. Jim Farr, 'Jim & Ileene Farr Interview No. 001' (unpublished, Warkworth, 7 October 1995).

4. Bruce Farr, 'Bruce Farr Interview No. 001' (unpublished, Annapolis, 23 May 1995).

5. Ibid.

6. Lindsay Lovegrove, 'Lindsay Lovegrove Interview No. 001' (unpublished, Leigh, 13 December 1995).
7. Although related, the UK and European International Moth Class had slightly different rules (most notably in its sail-height restrictions, which produced a lower aspect ratio rig) from the Moth Class in Australia and its counterpart in New Zealand, the Restricted Moth. And all were different from the one-design Australian Moth Class, a scow type at that time, one design of which was also sailed in New Zealand as the one-design New Zealand Moth Class.
8. Hal Wagstaff, 'Hal Wagstaff Phone Interview No. 001' (unpublished, Auckland, 25 January 1996).
9. The Farr Mark I Restricted Moths weighed between 95 and 100 lb. The Mark II version, of which *Behemoth* was the prototype, weighed between 65 and 70 lb.
10. *Sea Spray* (Auckland: Review Publishing Co. Ltd., March 1966) p. 51.
11. At this stage of his boat-building career, Bruce Farr's charge-out rate was seven shillings an hour.
12. Rob Blackburn, 'Rob & Marilyn Blackburn/Peter & Denyse Hutchinson Interview No. 001' (unpublished, Hamilton, 27 April 1996).
13. Ibid.
14. Restricted Moth Class Reporter, *Sea Spray* (Auckland: Review Publishing Co. Ltd., 1968), p. 83.

Chapter 4

1. Marianne Gray, quoting Gerard Depardieu, in *Depardieu: A Biography* (London: Warner Books, 1992), p. 120.
2. In *Waitemata: Auckland's Harbour of Sails* (Auckland: Century Hutchinson, 1989), p. 101, Tessa Duder writes: '... it is believed (contrary to British claims of being first in 1937) that Jim Frankham's M3, *Manaia*, about 1935, was the first dinghy in the world to have a man hang out on a trapeze, using initially the spinnaker halliard and later a specially rigged masthead wire and a home-made canvas belt.'
3. Tom Clark, 'Sir Thomas Clark Interview No. 001' (unpublished, Auckland, 8 January 1996).
4. SOE is an acronym for 'State-Owned Enterprise', the name given to New Zealand government-owned organisations, of which some thirty at a total value of $10 billion were sold from the mid 1980s under the right-wing two-term Labour government. The practice was continued by the incoming National government who, like their predecessors, sought to simultaneously reduce their levels of debt and social responsibility. The Business Roundtable is an essentially androcentric ad hoc body composed of members from corporate New Zealand, dedicated to the corporatist social model and its corresponding diminution of the competitive market. Membership is by invitation.
5. Russ Bowler, 'Russell Bowler Interview No. 001' (unpublished, Annapolis, 3 June 1995).
6. Don McGlashan, 'Don & Katharine McGlashan Interview No. 001' (unpublished, Auckland, 2 November 1995).

Chapter 5

1. John Spencer (deceased) in a letter to the author (unpublished, Bay of Islands, 17 February 1997).
2. Jim Young, 'Jim Young Interview No. 001' (unpublished, Auckland, 22 November 1995).
3. Tim Shadbolt, *Bullshit and Jellybeans* (Wellington: A Taylor, 1971).
4. Although New Zealand gained full independence from Great Britain with the adoption of the Statute of Westminster in 1947, its ties have remained close.
5. From US$2 a barrel in the 1960s, the price of oil leapt to US$12 a barrel in 1973 and US$34 a barrel in 1979, having a profound effect on New Zealand's isolated economy.
6. Harry Bioletti, 'Harry Bioletti Interview No. 001' (unpublished, Warkworth, 13 December 1995).
7. C.T. Brooking, 'Mahurangi College letter of testimony concerning Bruce Kenneth Farr' (unpublished, Warkworth, 20 March 1967).
8. Bruce Farr, 'Bruce Farr Interview No. 001' (unpublished, Annapolis, 23 May 1995).
9. Bruce Farr, 'Bruce Farr Interview No. 002' (unpublished, Annapolis, 25 May 1995).
10. The first *Cool Leopard*, designed and built in 1967, was a four-handed eighteen-footer. *Guinness Lady*, which Bruce helmed to third in the 1969 Eighteen-footer Worlds, was an early three-hander. Thereafter the trend to three-handers increased, although Bruce was still designing four-handers for Australian clients in 1973.
11. Jim Young, 'Jim Young Interview No. 001' (unpublished, Auckland, 22 November 1995).
12. Pam Farr, 'Pamela Farr Interview No. 001' (unpublished, Auckland, 18 January 1996).
13. At this stage in his business career, circa 1970, Chris Bouzaid also owned Allspar. After the demise of Jim Young's business it took over the lease of Birkenhead wharf. It would later become Yachtspars.
14. Bruce Farr, 'Bruce Farr Interview No. 001' (unpublished, Annapolis, 23 May 1995) and 'Bruce Farr Interview No. 013' (unpublished, Auckland, 28 January 1998).

Chapter 6

1. *The Sun-Herald* (Sydney, NSW: John Fairfax, 15 January 1969).
2. Bruce Farr, as quoted by Michael Levitt in 'Farr Ahead', *Nautical Quarterly* Number 43 (Connecticut: Nautical Quarterly Co, 1988), p. 25.
3. Don Lidgard, 'Don Lidgard Interview No. 001' (unpublished, Auckland, 20 February 1996).
4. Ibid.
5. Russ Bowler, 'Russell Bowler Interview No. 001' (unpublished, Annapolis, 3 June 1995).
6. Ibid.
7. Don McGlashan, 'Don and Katharine McGlashan Interview No. 001' (unpublished, Auckland, 2 November 1995).

8. Ibid.

9. Sheila Patrick, *Australian Seacraft Power & Sail* (Sydney: Cavalcade Magazine Pty Ltd, April 1969), p. 26.

10. *Sydney Morning Herald* (Sydney, NSW: John Fairfax, 12 January 1969).

11. Harry Hayes, 'Spinnaker Tales', *The Sun-Herald* (Sydney, NSW: John Fairfax, 15 January 1969).

12. Jennifer Julian advertisement (Auckland, 1969).

Chapter 7

1. Lewis Francis Herreshoff, *The Common Sense of Yacht Design* (New York: Rudder Publishing Co, 1946-1948), p. 23. Francis Herreshoff is the son of Nathanael Herreshoff and father of Halsey Herreshoff.

2. Jim Young, quoted by Gary Baigent, *Light Brigade* (Auckland: Cox's Creek, 1997), p. 10.

3. The Aerodux range of resins was manufactured for Ciba Geigy, the trade-name holders, by ICI New Zealand Limited until the mid 1980s. These are the resorcinol formaldehyde resins (red colour) 'Aerodux 185' and the urea formaldehyde resins (clear) 'Aerolite 300'. 'Aerodux 185' is made by reacting resorcinol with sufficient formaldehyde to cause full cross linking. An acid catalyst is used to accelerate the polymerisation in the reactor, after which the resin is neutralised with caustic soda. The resin is cured using a paraformaldehyde-based hardener to complete polymerisation. It sets in two to three hours but needs seven days to reach full strength.

4. Froude's rule holds that a keel-boat in flat water is unable to travel faster than the product of 1.3 times the square root of its waterline length. For further information on R.E. Froude see Chapter 12, note 10, and Chapter 13.

5. 'Patiki' is the Maori word for flatfish or flounder. It was the name given to an original New Zealand centreboard-yacht-type racing at the turn of the century (few of which survived after the 1931 Napier earthquake and the Great Depression). It was the original name ('restricted eighteen-foot Patiki') given to the 1920s Arch Logan-designed M Class. And it was the actual name given to the 34-foot keeler designed by Bob Stewart for Peter Colemore-Williams. For further information on early New Zealand yacht design refer to Ronald Carter, *Little Ships* (Wellington: AH & AW Reed, 1947) and Gary Baigent, *The Light Brigade* (Auckland: Cox's Creek, 1997).

6. Gary Baigent, *The Light Brigade* (Auckland: Cox's Creek, 1997), p. 30.

7. Jim Young, 'Jim Young Interview No. 001' (unpublished, Auckland, 22 November 1995).

8. Rob Blackburn, 'Rob & Marilyn Blackburn and Peter & Denyse Hutchinson Interview No. 001' (unpublished, Hamilton, 27 May 1996).

9. Ibid.

10. Peter Hutchinson, 'Rob & Marilyn Blackburn and Peter & Denyse Hutchinson Interview No. 001' (unpublished, Hamilton, 27 May 1996).

11. The Winebox is the name given to the winebox of documents tabled in the New Zealand parliament in October 1992 and the commission of inquiry begun in October 1994 to look into them.

Although registered in the nearby Cook Islands, European Pacific (EP), the company to which the documents belonged, had as its principal shareholders three prominent New Zealand companies: Fay Richwhite and Company Ltd (later Capital Markets Ltd, Fay Richwhite's publicly listed company), Brierley Investments Ltd, and the Bank of New Zealand (wholly owned by the New Zealand government, then part-owned with minority shareholder Fay Richwhite and Company Ltd). Uncovered by the Winebox Commission was a massive defrauding of New Zealand's Inland Revenue Department by corporate New Zealand, spearheaded by EP. This was not the usual use of tax haven facilities. This was a premeditated system made possible by the arrangement between EP and the Cook Islands Government (CIG) — an arrangement that saw the government of a sovereign state robbing the revenue of another with one hand while accepting aid from it with the other.

In one transaction type ('Magnum'), EP would make a substantial tax payment to CIG on the bogus profit of its money go-round transaction one day and have that payment refunded the next — less a 'fee' for CIG's false tax credit certificate, the same certificate that would be presented to NZ's IRD. And having declared their tax payment (by way of the false CIG tax credit certificate) but not their tax refund, the participating company would be absolved from paying further tax on that transaction in NZ. Yet in spite of the Commission's knowledge of what lay behind these transactions, the overwhelming evidence of fraud to emerge during the hearings, the opinion of internationally recognised tax experts confirming that fraud, and even the public announcement of a former Cook Islands director of government auditing stating that CIG did indeed issue false tax credits to foreign companies, ex-Chief Justice Sir Ronald Davison chose to take the 'form-over-substance' route on his way to tell the Governor General, on August 15 1997, that he could find 'no evidence ... of ... fraud ... in ... the winebox'. Sir Ron then went fishing. It is not reported if the trout in Lake Taupo took his lure as well as the opposition politicians in Wellington, or if the trout ignored his lure as his findings seemed to have the evidence extracted from the winebox at a cost of more than NZ$17 million, including his NZ$1.5 million fee. These findings, however, would be resoundingly challenged by a strongly worded, unanimous five-judge Court of Appeal decision in November 1998. Not only were they described as being outside the Commissioner's terms of reference. Because he had based them on the 'form' and not the 'substance' of the transactions (when the law says that 'form over substance' does not apply to transactions constructed to defraud), the Appeal Court considered his conclusions to be based on 'apparent errors of law' and therefore 'erroneous'. See Chapter 27, Note 36.

12. In 1891, Nathanael Greene Herreshoff launched the first racing yacht with a fin keel and separate rudder. Called *Dilemma*, its lightweight hull construction was based on pioneering work he had done designing and building high-speed steam launches. *Dilemma*'s revolutionary fin keel was secured to the hull by way of attachment to deep floor timbers. It was soon followed by others from the inventor, and from copycat designers. With light construction above the waterline and the saved weight held down low in their new-style keel, the fin keelers were able to carry substantially more sail area without

increasing their waterline length. So successful was the concept that they immediately began winning races in all major classes in Great Britain and America. Soon they were labelled 'rule-cheaters' and 'unseaworthy', and banned from all racing.

Chapter 8

1. From the Russell Bowler collection (unpublished, Annapolis).
2. Russ Bowler, 'Russell Bowler Interview No. 02A' (unpublished, Annapolis, 4 June 1995); Jacqui Parks, 'Jacqui Parks Interview No. 001' (unpublished, Auckland, 5 March 1997); Basil Fuller, 'The Nullarbor Story,' (Adelaide: Rigby Limited, 1970).
3. Alan Bond was six years old when he arrived in Western Australia. He left school at fourteen to become a signwriter.
4. Jim Sharples, *Australian Seacraft Power & Sail* (Sydney: Cavalcade Magazine Pty Ltd, May 1970), pp. 24–25.

Chapter 9

1. Details taken from Taylor Marine Brokers, Notice of Sale (Auckland).
2. Rob Blackburn in a letter to the author (unpublished, Ujung Pandang, 25 August 1996).
3. When asked by Bert Woollacott why his designs continued to beat Charles Nicholson's, Nathanael Herreshoff is reported to have said: 'Nicholson studies the rules and designs a boat within it. I design the fastest boat I can for the overall length then make it fit the rule by altering shape as necessary at the measuring points.' *Sea Spray* (Auckland: Review Publishing Co. Ltd., February 1965), p. 51.
4. IMS stands for the International Measurement System which, from a boat's machine-measured shape, uses a computer to predict its actual speed which in turn determines its handicap. This method of handicapping is based on a Velocity Prediction Programme (VPP) originally developed by the Massachusetts Institute of Technology (MIT). A computer VPP predicts the speed a boat of a given set of dimensions and rating will travel at any given wind angle and strength over a given course. See Chapter 28 for further details.
5. The dialogue in this chapter is based on sections six and seven of 'Bruce Farr Interview No. 012' (unpublished, Hahei, 25 April 1996).

Chapter 10

1. Felipe Fernandez-Armesto, *Truth: A History and a Guide for the Perplexed* (London: Black Swan Books, 1998), p. 224.
2. Rob Blackburn, 'Rob & Marilyn Blackburn and Peter & Denyse Hutchinson Interview No. 001' (unpublished, Hamilton, 27 April 1996).
3. David Pardon, *Australian Sea Spray Weekly*, Vol 3 No. 138 (Sydney, NSW: April 1972), p. 1.
4. Bruce Farr, 'Bruce Farr Interview No. 012' (unpublished, Hahei, 25 April 1996).

5. This exchange is based on an article written by Alan J. Sefton in *The Auckland Star*, circa April 1972.

6. Laurie Davidson had designed *Blitzkrieg* for Tony Bouzaid and was its skipper when it won the New Zealand Half Ton Championship in 1971.

7. The first four place-getters in the Schweppes Half Ton Cup were: *Titus Canby* (Farr) with $118\frac{3}{8}$ points, *Blitzkrieg* (Davidson) with 114 points, *Conquero* (Salthouse) with $101\frac{1}{2}$ points and *Cavalier* (Salthouse) with $96\frac{1}{2}$ points.

Chapter 11

1. Jenny Farrell, *Sea Spray* (Auckland: Review Publishing Co. Ltd., July 1974), p. 33.

2. Tip-Over Ted, *NZ Boating World* (Auckland: Wilson and Horton Ltd, January 1974), p. 38.

3. 'Komutu' is the Maori word for surprise.

4. Olin Stephens' 1970 America's Cup defender *Intrepid* was the first Twelve Metre yacht to be designed with a knuckle bow, a future standard feature of Farr yachts. The almost plumb bow lengthened its waterline length when heeled. But because the knuckle was clear of the water when the boat was upright, it reduced *Intrepid*'s measured waterline, and so gave it a gain in rating.

5. Jim Lidgard, 'Jim Lidgard Interview No. 001' (unpublished, Auckland, 20 February 1996).

6. *TraveLodge NSW* was one of Bob Miller's less successful designs. He would later change his name to Ben Lexcen and funnel his erratic brilliance into the design of an inverted keel with wings. As Bob Miller, he had produced the first three-man eighteen-footer in 1960, the lightweight planing *Taipan*. The following year he won the eighteen-footer Worlds sailing *Venom*, another plywood shell of his own design. It was with *Venom*'s centreboard that he first experimented with 'wings'.

7. Helmer Pedersen, with Earle Wells as his forward hand, won the gold medal in the Flying Dutchman Class for New Zealand at the Tokyo Olympics in 1964.

8. Don Lidgard, 'Don Lidgard Interview No. 001' (unpublished, Auckland, 20 February 1996).

9. Robert David Muldoon, 1921–1992, was New Zealand's Finance Minister and its Prime Minister from 1975 to 1984.

10. *Patricia*, with its blunt bow and dish-like hull, was a sister ship to *Komutu*. It was among the first eighteen-footers to sport an extendable stacking-out board which slid from side to side across the hull and carried two men. *Sluefoot*, a sister ship to *Quandary*, was among the last eighteen-footers to be built by Jack Logan.

11. Kim McDell was seventeen when he ran 4:24 to break Murray Halberg's long-standing Inter-secondary mile record of 4:27. In his last year at school he broke Gary Philpott's New Zealand Junior Half Mile record with 1:54.1. Before his early retirement, he would run 1:50.1 for the 880 yards. He was the pacemaker, taking the field through the first half mile in 1 minute 54 seconds, when Peter Snell set his second world mile record at Western Springs, Auckland, in 1964.

12. The Farr 3.7 is Design No. 29, a single-handed twelve-foot (3.7 metre) trapeze centreboarder. Designed in 1970, it has a hard chine, rounded sections, full bow,

straight runs and a high-aspect rig (18.3 metres). There are approximately one hundred 3.7s still sailing around New Zealand. A national competition, for a cup donated by the designer, is held annually. Thirty-three competed for the national title in 1997, a title won by Bruce Farr in 1975. Twenty-seven years after he built *XL*, Bruce Farr's original wooden test boat is still competitive and has recently changed hands.

13. Kim McDell, 'Kim McDell Interview No. 001' (unpublished, Kawau Island, 11 January 1996).

14. Ibid.

15. Ibid.

16. Ian Hamlet, *NZ Power & Boating* (Wellington: IWL Press Limited, April 1978), p. 47.

17. Russ Bowler, 'Russell Bowler Interview No. 02B' (unpublished, Annapolis, 4 June 1995).

18. Ibid.

19. Keith Chapman (deceased), 'Keith Chapman Interview No. 001' (unpublished, Auckland, 17 February 1996).

20. Of the twenty competitors sailing for the Jas J. Giltinan Trophy and the world eighteen-footer title on Waterloo Bay in 1978, one each came from the United States and the United Kingdom, and six each from New South Wales, Queensland and New Zealand.

Chapter 12

1. Lewis Francis Herreshoff, *The Common Sense of Yacht Design* (New York: Rudder Publishing Co, 1946–1948), p. 45.

2. Bruce Farr in a letter to the editor of *Sea Spray* (Auckland: Review Publishing Co. Ltd., December 1973/January 1974), p. 71.

3. Timothy Jeffery, *The Champagne Mumm Admiral's Cup: The Official History* (London: Bloomsbury, 1994), p. 213.

4. The first recorded yacht race was between Charles II and the Duke of York in October 1661.

5. The original Royal Yacht Club measurement rule was based on the pre-1800 custom house rule for measuring 'tuns' of wine in order to levy duty. To try to bring some order to the growing sport, the Yacht Racing Association was formed in 1875. Its rule, expressed as $\frac{(L - B) \times (B \times \frac{1}{2}B)}{94}$, encouraged yachts of a narrow deep design.

6. The Sail Area or Scharenkreuzer Rule produced yachts of the Square Metre Classes, so-called because of the square metre content of their sails.

7. The formula of the Seawanhaka Corinthian Yacht Club read:
$$\frac{L + \sqrt{SA}}{2} = \text{Racing length.}$$
Its weakness was that it made no allowance for displacement. In general, the various American rating rules encouraged beamy, shallow-hulled yachts.

8. The New York Yacht Club was formed in 1844.

9. Adopted by the New York Yacht Club in 1902, the Universal Rule, using 18

percent of the product, expressed itself as: $R = 0.18 \dfrac{L \times \sqrt{SA}}{\sqrt[3]{D}}$

Arguably the most significant aspect of the Universal Rule was the use of the cube root of the displacement as the divisor. This meant the greater the vessel's displacement the lower would be its rating.

10. The International Rule, which produced the 6, 8, 10, 12 and 14.5 Metre Classes, was formulated thus: $\dfrac{L + \frac{1}{4}G + 2^d + \sqrt{s} - F}{2.5}$

This was based on the Yacht Racing Union's Rule which in turn was based on R.E. Froude's 1896 linear rule which added Length, Beam, Girth, Girth Difference and the square root of the Sail Area, the sum of which was divided by two to produce a linear rating.

11. Under the International Rule the scantlings of a yacht were to be built in accordance with the Lloyd's Register of Shipping scantling rule.

12. The CCA's first measurement rule for the Bermuda Race was Length Overall. This was modified to a length rule which took the mean of Overall Length and Water-Line Length. This was further modified when a sail area calculation was added.

13. Wells Lippincott originally developed his measurement rule for the Lake Michigan Yacht Racing Association.

14. The rating formula, based on Lippincott's 'base (or standard) boat' concept, which the CCA adopted in 1923, held that the rating in feet would be 95 percent of the Measured Length expressed as: Rating = 0.95 (Length ± Beam ± Draft ±Displacement ± Sail Area ± Freeboard − Iron Keel credit) x the Ballast Ratio x the Propeller factor. This meant, 'for example, [that] a sloop of 32-foot waterline in whose design speed takes precedence over comfort, may also be rated as a 37-footer. On the other hand, a 42-foot waterline ketch broad in beam and somewhat under-canvassed may also be rated as a 37-footer. In giving the two boats equal ratings the expected increase in speed due to the increase in water line length from 37 to 42 feet has been neutralized under the Rule by the increased beam, decreased sail area, less efficient rig, etc., of the larger boat.' Refer John Holden Illingworth, *Offshore* (London: Adlard Coles, 1958), Appendix XII, p. 287

15. The ORC received its Royal Warrant in November 1931 and was known thereafter as the Royal Ocean Racing Club (RORC). The RORC's rating formula of 1957, amended 1958, was expressed as: Measured rating = $0.15 \dfrac{L\sqrt{S}}{\sqrt{BD}} + 0.2 (L + \sqrt{S})$

16. John Holden Illingworth, *Offshore* (London: Adlard Coles, 1958), Appendix VII, p. 277.

17. In general, British measurement rules had favoured narrower, deeper hulls and American measurement rules hulls which were beamier, shallower and produced more displacement.

18. The members of the first ITC were: E.G. 'Ricus' van de Stadt (Holland), Gustav Plym (Sweden), Brigadier David Fayle and Major Robin Glover (UK), Dick Carter and Olin Stephens (USA).

19. Timothy Jeffery, *The Champagne Mumm Admiral's Cup: the official history* (London: Bloomsbury, 1994), p. 212.

20. Another significant fault of the IOR was its failure 'to weigh boats in order to

determine displacement. Instead, rather like the tonnage rules, point measurements were taken at predetermined locations on the hull to gauge its shape and hence how much of it was in the water. In time, displacement was scooped away between the measurement stations by designers as they sought to reduce wetted surface, and advances in glass-fibre — and then composite — construction meant that the hull shells themselves weighed less and less. In order to have the sort of displacement the IOR rule needed for an effective rating, more and more ballast was needed to replace what had been removed. By the mid-1980s, it was becoming difficult to get sufficient lead in the floor of boats, with great tombstones of lead often an integral part of the hull floor.' Timothy Jeffery, *The Champagne Mumm Admiral's Cup: The Official History* (London: Bloomsbury, 1994), p. 212.

21. Chris Bouzaid as quoted in *DB Yachting Annual*, edited by Alan J. Sefton (Auckland: Moa Publications, 1976), p. 101.

22. Roy Dickson as quoted in *DB Yachting Annual*, edited by Alan J. Sefton (Auckland: Moa Publications, 1977), p. 71.

23. The Leander Trophy, consisting of the original crest of HMNZS *Leander*, is competed for annually in the R Class. An open-design class with a crew of two, its main restrictions are a 12 ft 9 in round-bilged hull and a working sail area of 110 square feet with a mast height limit.

24. The IOR was published in updates as changes to it were made by the ITC. The IOR Mark II was the first version published. The IOR Mark III contained additional depth measurement points to better measure displacement. The Mark IIIa is discussed in Chapters 16 and 17.

25. Olin Stephens, for example, had modified the rig of *Prospect of Whitby* in 1973 by lowering the forestay marginally and increasing the height of the mast by the same amount. However, because the forestay had not been lowered to three-quarter height, he failed to achieve increased control over the narrow mainsail and large headsail, a configuration essentially the same as his masthead designs. With a loss of shooter and spinnaker area for no actual gain, *Prospect of Whitby* was converted back to a masthead rig.

26. *The Press* (GB: July 22, 1975).

27. New to the class when they bought *Cool Leopard* from Gary Banks, Murray Crockett and his father asked Bruce to join them as mainsheet hand for half a dozen races to help them 'learn the ropes'. This, the second *Cool Leopard*, designed by Bruce in 1969, was probably the first eighteen-footer to have wide built-in decks added. Prior to this, Auckland eighteen-footer helmsman John Lasher had experimented with tube extensions, with a tube across the end, in an attempt to get weight out wide. This, in the long run, would prove to be the right solution.

28. Murray Crockett, 'Murray Crockett Interview No. 001' (unpublished, Auckland, 14 December 1995).

29. Bill Endean, *Classic New Zealand Yachts: Four decades of successful yacht design — 1950–1990* (Wellington: GP Publications, 1992), p. 139.

30. The plug Noel Angus bought became the 727 called *Pinto*.

31. Bruce Farr, 'Bruce Farr Interview No. 002' (unpublished, Annapolis, 25 May 1995).

32. The crew of the One Ton-winning *Rainbow II* was: Chris Bouzaid, Dave Craig, Roy Dickson, Ward Schofield, Alan Warwick and John Woolley.

33. *Sea Spray* (Auckland: Review Publishing Co. Ltd., March 1975), p. 40.

34. Bruce Farr's national titles as skipper are:

   (a) 1966 *Mammoth* (Restricted Moth/Farr)

   (b) 1966 *Mammoth* (Junior National/Restricted Moth/Farr)

   (c) 1969 *Miss Beazley Homes* (Q Class/Farr)

   (d) 1969 *Guinness Lady* (18-footer/Frank Blackburn)

   (e) 1972 *Titus Canby* (Half Ton/Farr/co-helm with Rob Blackburn)

   (f) 1973 *Fantzipantz* (Quarter Ton/Farr/co-helm with Peter Walker)

   (g) 1975 *727* (Quarter Ton/Farr)

   (h) 1975 *3.7* (3.7 Class/Farr)

   In addition, in boats to his own design, Bruce won the Auckland Anniversary Regatta Pennant Class (1963), the Auckland Anniversary Restricted Moth Class (1966) and the Auckland Anniversary Regatta 3.7 Class (1972), two Auckland Restricted Moth Championships (1966 and 1967), four North Island Restricted Moth Championships (1965, 1966, 1967 and 1968), the Junior Restricted Moth title (1966) and the Interdominion Q Class title (1970).

35 Roy Dickson, 'Roy Dickson Interview No. 001' (unpublished, Auckland, 8 December 1995).

36. Ann Walker (née McGlashan), 'Ann Walker Interview No. 001' (unpublished, Auckland, 17 February 1996).

Chapter 13

1. Lewis Francis Herreshoff, *The Common Sense of Yacht Design* (New York: Rudder Publishing Co, 1946; 1948), p. 40.

2. Bruce Farr as quoted in *DB Yachting Annual*, edited by Alan J. Sefton (Auckland: Moa Publications, 1978), p. 83.

3. Third smallest of the ten-boat fleet, *Dorade* was two days clear of its nearest rival when it crossed the finish line. On handicap it won by almost four days.

4. The firm of Drake H. Sparkman was founded in 1928. Olin Stephens was invited to join it on a trial basis, providing designs for the company's clients. After his designs began winning races his position became permanent and the company's name was changed to Sparkman & Stephens, Inc. The year was 1930.

5. Among the vessels Sparkman & Stephens produced during World War II was the DUKW amphibian for which Rod Stephens, its designer, was awarded the Medal of Freedom.

6. Olin Stephens' Twelve Metre designs with which the New York Yacht Club successfully defended the America's Cup were: *Columbia* (1958), *Constellation* (1964), *Intrepid* (1967 and 1970), *Courageous* (1974 and 1977), and *Freedom* (1980).

7. Olin Stephens' winning design in the 1973–74 Whitbread was *Sayula II*, a 65-foot (19.8 m) ketch skippered by Ramon Carlin of Mexico.

8. Guy Cole, ed., *The Best of Uffa* (Lymington: Nautical Publishing Co. Limited, 1978), p. 32.

9. The Southern Cross Cup, a four-race series competed for by teams of three yachts each, is considered the Southern Hemisphere's equivalent of the Admiral's Cup. First sailed in 1967, it was won for New Zealand for the first time in 1971 when

Brin Wilson's *Pathfinder*, Chris Bouzaid's *Wai-Aniwa*, and John Lidgard's *Runaway* finished first, second and third in the Sydney–Hobart. The controversy referred to here took place prior to the 1977 Southern Cross Cup and is examined in Chapter 16.

10. Tank-testing of yachts in the United States was begun by Ken Davidson, an MIT graduate, who had set up an improvised towing tank in the swimming pool of the Stevens Institute of Technology. The first two models he tested were of the Stephens brothers' Six Metre yacht *Natka* and were made by Rod Stephens. The first towing tank to scientifically test models was set up by the English engineer William Froude in his Torquay garden in 1875. See *The Common Sense of Yacht Design* (New York: Rudder Publishing Co, 1946; 1948), p. 40.

'Froude's tests showed that while hull resistance at low speeds is almost entirely due to skin friction of the water passing across the hull's surface, above certain speeds another factor is introduced. This is wave-making resistance which grows steeper with increased speed. At the bow the water parts and slows, then it speeds up as the hull thickens out to its maximum beam, while there is a sudden decrease in pressure and slowing down as the hull narrows towards the stern. The water separates from the hull and a quarter wave forms, virtually a wall of water being dragged along by the boat. The earlier this wave is generated the greater its mass, further slowing the vessel down.' See Rik Dovey/Sally Samins, *12-Metre The New Breed: The Battle for The 26th America's Cup* (Sydney: Ellsyd Press Pty Ltd, 1986), p. 17.

11. While Bruce had in fact taken some notice of the shape of a dolphin's dorsal fin, his ideas were mostly derived from his observation of a wide variety of keels. It was only after he had a good grasp of the logic that lay behind them that he began designing what he thought would be the most efficient keels for his boats.

12. This conversation is based on 'Peter Beaumont Interview No. 001' (unpublished, Auckland, 5 December 1995).

13. Michael Levitt in 'Farr Ahead', *Nautical Quarterly*, Number 43 (Connecticut: Nautical Quarterly Co, 1988), p. 22.

14. *Sea Spray* (Auckland: Review Publishing Co. Ltd., December 1975/January 1976), p. 46.

15. *Afakazze* was sitting, unfinished, in John Wilson's garage when Murray Ross asked if he could finish it off in exchange for racing it in the 1973 Javelin Nationals. Sailing on Lyttelton Harbour during Easter, Andy Ball and Peter Newlands in the Farr-designed *Joshua* finished second to *Afakazze*. Peter Beaumont and Bob Eastmond were sixth in Murray Ross's old 1968 national Javelin champion *John Wesley Hardin*. Craig Gilberd and Denzil Ibbetson would sail *Afakazze* to another Javelin national title win in 1974.

16. Peter Beaumont, 'Peter Beaumont Interview No. 001' (unpublished, Auckland, 5 December 1995).

17. After setting up his board in the Burmester boat yard, Bruce spent four round-the-clock days drawing a full set of lines drawings, complete with construction details, for production of this cold-moulded vacuum-formed timber version of the 727.

18. See Chapter 14 for biographical notes on Paul Whiting.

19. Bruce Farr in 'Taped with the editor', *New Zealand Company Director*

(Christchurch: Mercantile Gazette of New Zealand, July 1976).

20. Keith Chapman (deceased), 'Keith Chapman Interview No. 001' (unpublished, Auckland, 17 February 1996).

21. Brin Wilson was widely regarded as New Zealand's foremost builder of wooden boats. He was also a gifted yachtsman and helmsman. The highlight of his racing career was winning the Sydney–Hobart in 1973 in *Pathfinder*. Brin died in 1973 from cancer. He is survived by his wife, Marie, a member of the well-known Lidgard family, and his two sons Robert and Richard, who continue to run Brin Wilson Boat Builders at Gulf Harbour, Whangaparaoa, Auckland.

22. Keith Chapman (deceased), 'Keith Chapman Interview No. 001' (unpublished, Auckland, 17 February 1996).

23. Following the disintegration of his business partnership with Craig Whitworth and the loss of his name for commercial purposes, Bob Miller changed his name to Ben Lexcen. The announcement of his new name was made public during the 1974 America's Cup series. For further reading on Ben Lexcen see Bruce Stannard, *Ben Lexcen: The Man, The Keel and The Cup* (London: Faber & Faber, 1984).

24. Of Dick Carter's innovations, Timothy Jeffery writes: '*Rabbit* had a small keel, spade rudder, mast with internal halyards, a pulpit mounted inside the jib, through-mast roller reefing and trim tab. All that in 1965. His 1969 Admiral's Cup offering, *Red Rooster*, had a swinging keel — extra deep for upwind bite and low resistance downwind.' From *The Champagne Mumm Admiral's Cup The Official History* (London: Bloomsbury, 1944), p. 60.

25. Ibid., p. 213.

26. Noel Angus was returning from the 1979 Auckland-Lautoka Race on the Bo Birdsall-designed *Ponsonby Express*, when he, co-owner Shirley Clay, Bill Kirk, Stephen Ball and Kevin Peers were lost without trace. They were last seen leaving Lautoka for Auckland on May 23.

27. Jenny Green in *Sea Spray* (Auckland: Review Publishing Co. Ltd., June 1976), p. 24.

28. Peter Spencer successfully protested the Ponsonby Cruising Club's sailing committee for not re-allocating points to the boats finishing behind the Davidson-designed *Tramp* when it received a 30 percent penalty for a port-starboard infringement in the intermediate race. This in spite of the IYRU's rules stating that a penalty (as distinct from a disqualification) 'shall not affect the score of other yachts.'

29. The lightweight Paul Whiting design *Magic Bus*, steered by Murray Ross, finished a convincing first with four wins and a second. Ten Farr 727s filled places two through four and six through twelve. Paul Whiting's career and competitive relationship with Bruce Farr and Laurie Davidson are examined more closely in Chapters 14 and 15.

30. Bruce Farr as quoted by Jenny Green in *Sea Spray* (Auckland: Review Publishing Co. Ltd., June 1976), p. 24.

31. The 1976–77 Sydney–Hobart, sailed in the worst conditions in eight years, proved the potency and seaworthiness of the light-displacement Farr designs over their heavy-displacement rivals. While Design 51 finished first and second overall, the 50-knot winds and big seas forced a then record fifteen yachts to

retire. Second placegetter *Rockie*, owned by Peter and Ray Kingston of Auckland, finished the 630 miles just 43 corrected minutes behind the Australian John Pickles. *Piccolo* had endured 49 hours of close-hauled work in heavy seas before making it into Hobart. Another Design 51, *Sweet Okole*, would go on to win the 1977 Southern Ocean Racing Conference (the USA's oldest offshore regatta) and the 1981 Los Angeles-to-Honolulu Transpac.

32. Steven Cross (deceased) in a letter to Nick Bailey (unpublished, Bonnington, Kent, 2 December 1992).

33. Cathie Cross (nee Cochran) in a letter to the author (unpublished, Bonnington, Kent, 30 April 1996).

34. *Burton Cutter*, the 80-foot British yacht designed by John Sharp, was co-skippered by Les Williams and Alan Smith. It finished sixteenth out of twenty in the 1973–74 Whitbread round-the-world race. Having suffered hull damage on the Cape Town-to-Sydney leg, it failed to complete Legs Two and Three of the four-leg race.

35. For the second Whitbread (1977–78), Peter Blake sailed on *Heath's Condor*. It lost its carbon fibre rig on the first leg and finished last in the fifteen-boat fleet. The race was won by Cornelis van Rietschoten's *Flyer*, an S&S design on which New Zealander Grant Dalton was sail coordinator.

36. Throughout his career Peter Blake has made a practice of imputing the ideas of others as his. Or, in a variation on this theme, he has accepted honours for ideas and efforts that should have rightfully gone to others. For example, in Chapter 2 of *Blake's Odyssey: The Round the World Race with Ceramco New Zealand* (Auckland: Hodder & Stoughton, 1982), the reader is given the impression that the design brief he gave Bruce Farr was based more on his ideas than those foundational to the Farr design philosophy developed almost a decade before *Ceramco*.

On page 16 of *Lion: The Round the World Race with Lion New Zealand* (Auckland: Hodder & Stoughton, 1986), the design requirements Blake says he gave Holland sound like a brief for the winning design Farr and Bowler produced for Fehlmann, even down to *UBS*'s displacement figure. (See Chapter 21, note 31.) In other words, the reader gains the impression that if Holland had met Blake's design brief which, from Blake's description, would have produced a boat like *UBS*, he, rather than Fehlmann, would have won the 1985–86 Whitbread.

On the video *The Steinlager Challenge: How New Zealand Won the 1989–90 Whitbread Round the Word Race* (Wellington: Television New Zealand Limited, 1990), Blake claims they won the race because they used sails they'd had made instead of those he says Farr had designed, which he knew from experience would be 'a waste of time'. Thus the viewer is led to believe that Blake and his crew won that Whitbread not because of the breakthrough maxi BFA had designed for them but because of the sails Blake had had made. Not only is this a simplistic nonsense, it is predicated on a lie: Bruce Farr had not designed a single sail for *Steinlager 2*; he had merely assumed some some sail measurements for his VPP studies and for rating purposes as he had for his other designs in the race.

The most spectacular example of Blake taking credit for the ideas and efforts of others was in his scooping of the major accolades for New Zealand's winning the America's Cup when, in fact, he contributed little in terms of time and talent compared with others in the challenge. But arguably his most shameful act in this regard was accepting an honorary Doctorate of Commerce from Massey University's College of Business, Albany Campus. Before Guest Orator Orams (a College of Business senior lecturer and Team New Zealand member) and 1500 people, who gave him a standing ovation, Blake allowed this honour to be bestowed on him for doing something he had never done: pioneering commercial sponsorship in sport.

Chapter 14

1. Francis S. Kinney, *You Are First* (London: Granada Publishing in Adlard Coles Limited, 1978), p 141.
2. Bruce Farr quoted by Michael Levitt in 'Farr Ahead' in *Nautical Quarterly* (Essex, Connecticut: Nautical Quarterly Co., 1988), p. 22.
3. 'Stan is whoever you like. And the family jewels come from Barry Crump's book. I had a pretty warped mind in my twenties!' Murray Ross in 'Murray Ross Interview No. 001' (unpublished, Auckland, 10 January 1996).
4. See Gary Baigent, *Light Brigade* (Auckland: Cox's Creek, 1997).
5. Ibid. p. 66.
6. Paul Whiting as quoted in *DB Yachting Annual*, edited by Alan J. Sefton (Auckland: Moa Publications, 1977), p. 34.
7. On IOR boats, the knuckle on the stem (which Bruce had first seen on Olin Stephens' Twelve Metre *Intrepid*) moved the forward inner girth measurement slightly further aft thus helping the forward overhang correction and producing a minor rating bonus — as well as producing an increase in waterline length when the yacht was heeled.
8. The hull of *Magic Bus* was foam sandwich (Airex core). At 2650 lb (1202 kg), it displaced 50 lb (23 kg) less than the Farr 727.
9. Bruce Farr, 'Bruce Farr Interview No. 013' (unpublished, Auckland, 28 January 1998).
10. John Mallitte in 'The Whiting Legacy', *Sea Spray* (Auckland: Review Publishing Ltd, August 1982), p. 19.
11. Gary Baigent, *Light Brigade* (Auckland: Cox's Creek, 1997), p. 77.
12. Murray Ross introduced a variety of rigging ideas to the Whiting-designed Quarter, Half and One Ton racers. These included an adaptation of the injection spinnaker pole system, wire-luffed genoas, an athwartship genoa track with barber-hauler attached, under-deck halyards and banks of cockpit-operated stoppers. He also 'had spinnakers launched through a specially designed pulpit with sails stored in sail cloth tubes below deck. This kept crew off the foredeck and allowed fast mark rounding, spinnaker hoists or last minute drops.' Ibid., p. 82.
13. See Chapter 13, note 28.
14. Locked together on 144 points at the start of the nearly 200-mile last race, *Magic Bus* beat the Ron Holland-designed *Business Machine* by just 71 seconds. The win gave it 214.0 points and the World Quarter Ton title. *Business Machine*, representing the USA, finished second overall with 212.0 points. Also sailing for the USA and finishing third and fourth respectively were the Gary Mull design

Expresso (205.5 points) and *Star Eyed Stella* (195.0 points), designed by Doug Peterson and helmed by Tom Blackaller. Next was Bonica's *Fun* on 177.5 points.

15. Laurie Davidson, 'Laurie Davidson Interview No. 001' (unpublished, Auckland 27 August 1996).

16. Gary Baigent, *Light Brigade* (Auckland: Cox's Creek, 1997), p. 71.

17. Based in Connecticut, Britton Chance Jnr was best known for his 5.5 Metre boats of the 1960s and his work on the Twelve Metre yachts *Nefertiti* (1962) and Olin Stephens' *Intrepid*, which he redesigned for the America's Cup in 1970.

18. Dick Carter's 1969 swing-keel *Red Rooster* was designed for the RORC rule. Although the Fastnet winner and top points scorer in that year's Admiral's Cup, it was considered to be 'unhealthy'. Ted Hood's *Robin*, a heavy-displacement centreboarder, finished fourth in the 1973 World One Ton Cup and won the North American One Ton Cup the following year. Bruce King's twin bilge-boarder *Terrorist* would have won the 1974 One Ton Cup had it not lost the top of its mast in the long ocean race.

19. Although used interchangeably with 'centreboard', 'daggerboard' is arguably the more correct term for this moving appendage.

20. Graham Woodroffe's *The Number* had been renamed *45° South II* in deference to its sponsor.

21. See Chapter 12.

22. Olin Stephens, 'Olin J. Stephens II Interview No. 001' (unpublished, Putney, Vermont, 1 September 1995).

23. Francis S. Kinney, *You Are First* (London: Granada Publishing in Adlard Coles Limited, 1978), p. 279.

24. Ibid., p. 272.

25. The principal reason for the failure of the theory was that the speed of a heavy-displacement Twelve Metre yacht was insufficient to make it work. As a result, tank-testing was discredited until the results of Ben Lexcen's winged keel once again proved the value of scientific research.

26. Roy Dickson in *Sea Spray* (Auckland: Review Publishing Co. Ltd., October 1976), p. 37.

27. Bruce Farr as quoted in *DB Yachting Annual*, edited by Alan J. Sefton (Auckland: Moa Publications, 1976), p. 19.

28. Ibid.

29. Ibid.

30. Doug Peterson as quoted in *DB Yachting Annual*, edited by Alan J. Sefton (Auckland: Moa Publications, 1976), p. 24.

31. Ibid.

32. Ibid.

33. Ron Holland as quoted in *DB Yachting Annual*, edited by Alan J. Sefton (Auckland: Moa Publications, 1976), p. 26.

34. Concerning this rule change, British designer Rob Humphreys reported in *Sea Spray* as follows: 'One of the major talking points over the past two seasons has been the success of the light-displacement, wide-stern racers designed by such as Bruce Farr and Paul Whiting, and as soon as the international technical committee of the ORC made known its intention to review the way the after over-hang component rates these boats, rumours started to fly as to how and by how

much ratings would change. It appeared that the ITC wanted to deal with extreme sloping transoms used in conjunction with a loophole surrounding the provision for measuring a skeg which fell on the 4 percent beam buttock. The new rule is more fundamental than just dealing with the length of the boat beyond the after girth station ... What it does is to assess (without requiring any measurements other than those already on the rating certificate) the slopes of the AOCG and AOCP lines (respectively the After Overhang Component Girth and the After Overhang Component Profile) and where the slope of the AOCG line is steeper than the slopes of the AOCP line, thus implying that the designer is deriving advantage from the wide skeg loophole, the rating of the stern is subject to a new set of rules. However, whether it intended to or not, the ITC appears to be penalising wide sterns per se, and some boats which have neither a wide skeg or sloping transom may be affected.' *Sea Spray* (Auckland: Review Publishing Co. Ltd., February 1977), p. 42.

35. From 'The Peter Beaumont Collection' November 1976 (unpublished, Auckland).

Chapter 15

1. Robert Perry, Technical Editor, *Sailing Magazine* (Wilmington, DE: Ports Publications Inc) as quoted in *Bruce Farr: A Catalog of his Boat Plans* (Annapolis, MD: Farr International Inc, 1982), p. 3.
2. Mike Spanhake in 'Boating '77 Supplement' in 'The Don Lidgard Collection' (unpublished, Auckland).
3. Originally called the Offshore Rating Council, the Offshore Racing Council is an international body set up to manage the rules by which offshore racing is run — not to organise the races. Originally composed of members of the RORC and CCA, its membership was extended to all participating members, though not proportionally. There was, for example, a large preponderance of delegates from the USA and England, and only one for both Australia and New Zealand which both had numerically strong IOR fleets.
4. Ann Walker (née McGlashan) is not related to Russell Bowler's brother-in-law Don McGlashan. Her brother is Don McGlashan, founder of the band The Mutton Birds. Her younger brother Sandy was tragically drowned with two companions in October 1974 as they were making their way at night to the McGlashans' island of Pakihi in the Hauraki Gulf.
5. Roger Hill owns and runs Roger Hill Yacht Design, Howick, Auckland, monohull and catamaran design specialists.
6. Roger Hill, 'Roger Hill Interview No. 001' (unpublished, Auckland, 15 February 1996).
7. Ibid.
8. *Bruce Farr: A Catalog of his Boat Plans* (Annapolis: Farr International Inc, 1982), p. 22.
9. The crew of *Joe Louis* when it won the World Three-quarter Ton Cup in 1977 was: Eric Simian (owner), Yves Pajot (skipper), Christophe Simian, Philippe Follenfant, Serge Milon and Christian Le Floch.
10. Eric Simian in a letter to the author (unpublished, Paris, 26 February 1996). Eric Simian is the former French heavy-weight boxing champion.

11. *Uin-na-mara III* was built of a high Kevlar content GRP by Tim Gurr's Ocean Racing Yachts for Hector Ross of Hong Kong. At 5594 kg (12334 lb) it displaced nearly 30 percent less than *Gerontius*. As the reserve boat for the Hong Kong Admiral's Cup team, it did not race in the HK team at the Admiral's Cup but became the 1977 Class 2 winner of the 'Round the Island' race at Cowes and the Morgan Cup. Its twin, *Monique*, would become the overall points winner in the 1978 Pan Am Clipper Cup when the all-Farr New Zealand A team of *Monique*, *Country Boy* and *Gerontius* finished second to Australia A. Sailing on *Monique* was Jock Sturrock, the America's Cup helmsman who was preparing to launch a twin to *Monique*.

12. Russ Bowler, 'Russ Bowler Interview No. 003' (unpublished, Annapolis, 11 June 1995).

13. Ibid.

14. Doug Bremner's first division keeler *Ta'Aroa*, designed by Sparkman & Stephens, had won the 1973 Auckland-to-Suva race in record time with Kim and Terry McDell as part of its crew.

15. Kim McDell, 'Kim McDell Interview No. 001' (unpublished, Kawau Island, 11 January 1996).

16. Ibid.

17. Ibid.

18. Laurie Davidson as quoted in *Light Brigade* (Auckland: Cox's Creek, 1997), p. 89.

19. The first four finishers in the 1977 Schweppes South Pacific Half Ton Cup were: 1. *Newspaper Taxi* (Whiting CB); 2. *Cotton Blossom* (Farr FK); 3. *Candu II* (Whiting FK); 4. *Spring Loaded* (Farr FK).

20. Murray Ross, 'Murray Ross Interview No. 001' (unpublished, Auckland, 10 January 1996).

21. Penny Whiting, 'Penny Whiting Interview No. 001' (unpublished, Auckland, 22 February 1996).

22. Murray Ross, 'Murray Ross Interview No. 001' (unpublished, Auckland, 10 January 1996).

23. Scantling requirements are rules to do with the structure and strength of a hull.

24. With 1000 lb (453 kg) in their retractable keels, Bruce Farr had calculated that Design 64 was as stable as his earlier fixed-keel One Tonners from Design 51.

Chapter 16

1. Jack Knights in *Sea Spray* (Auckland: Review Publishing Limited, July 1978), p. 59.

2. Robert Perry, Technical Editor, *Sailing Magazine* (Wilmington, DE: Ports Publications Inc) as quoted in *Bruce Farr: A Catalog of his Boat Plans* (Annapolis, MD: Farr International Inc, 1982), p. 4.

3. Ibid., p. 26.

4. See Chapter 11.

5. The original *Rangiriri*, a gunboat built for the New Zealand colonial government, was a steam-driven, stern paddle-wheeler, 140 feet in length. It served on the Waikato River during the 1860s, used mainly for towing troop and supply barges. It was armed with a twelve-pound gun and was capable of seven knots.

6. The yachts, with sails hoisted and two crew members in the cockpit, were to be hauled down by a line from the top of their mast to a block on a mooring at the bottom of Westhaven.

7. Category Two certificates are the minimum safety certificates required to be held by a yacht wishing to compete in any offshore event in New Zealand.

8. The New Zealand team to contest the 1977 One Ton world championship in Auckland, in order of trial placing, was: *Smackwater Jack, Mr JumpA, Smir-Noff-Agen, Jenny H, The Red Lion* and *Heatwave*.

9. Category One certificates are required for all yachts leaving New Zealand's territorial waters.

10. Bruce Farr as quoted by Alan J. Sefton in *The Auckland Star* (Auckland: New Zealand Newspapers, circa November 1977).

11. NZPA, Sydney, circa November 1977, in 'The Peter Beaumont Collection' (unpublished, Auckland) Volume 4.

12. *The Auckland Star* (Auckland: New Zealand Newspapers, 4 November 1977).

13. Peter Campbell in 'Aussie Yacht Boss is a Sore Loser' in the *Daily Telegraph* (Sydney, NSW: News Ltd, November 1977).

14. Bruce Farr as quoted by Alan J. Sefton in *The Auckland Star* (Auckland: New Zealand Newspapers, November 1977).

15. The 'I' point is where the forestay joins the mast.

16. Sandy Peacock in *Modern Boating* (Sydney, NSW: Federal Publishing, January 1978), p. 66.

17. Murray Ross, 'Murray Ross Interview' (unpublished, Auckland, 10 January 1996).

18. *The Red Lion's* winning crew in the 1977 World One Ton Cup was Stu Brentnall, Roy Dickson, Terry Gillespie, Dick Jones, Bruce Malcolm, Rob Martin and Carl Watson.

19. The top five boats and their placings in the 1977 World Half Ton Cup were: 1. *Gunboat Rangiriri* (Willcox NZ/Farr-lift-keel [LK]) 5/1/2/2/2; 2. *Silver Shamrock III* (Cudmore EEC/Holland-centerboard [CB]) 1/4/3/5/1; 3. *2269* (Oborn Aust/Farr-LK) 4/6/4/7/3; 4. *Waverider* (Bouzaid NZ/Davidson-CB) 2/3/1/3/9; 5. *Swuzzlebubble* (Gibbs NZ/Farr-LK) 3/2/5/1/7.

20. Ian Gibbs as quoted in *DB Yachting Annual*, edited by Alan J. Sefton (Auckland: Moa Publications, 1978), p. 63.

21. Don Lidgard, 'Don Lidgard Interview No. 001' (unpublished, Auckland, 20 February 1996).

22. *DB Yachting Annual*, edited by Alan J. Sefton (Auckland: Moa Publications, 1978), p. 71.

23. Ibid., p. 70.

24. Ibid., p. 72.

25. Ibid., p. 74.

26. Ibid., p. 78.

27. Ibid., p. 74.

28. Ibid., p. 66.

29. Ibid., p. 79.

Chapter 17

1. David Edwards as quoted by Jack Knights in 'Victory to the Conservatives', in *Sea Spray* (Auckland: Review Publishing Limited, July 1978), p. 59.

2. Ron Holland as quoted in *DB Yachting Annual*, edited by Alan J. Sefton (Auckland: Moa Publications, 1978), p. 87.

3. Jack Knights in 'Victory to the Conservatives', *Sea Spray* (Auckland: Review Publishing Ltd, July 1978), pp. 58–59.

4. 'The ratios came about like this: first the IOR equivalent of displacement (L x B x D) was related to L ... Next, the ratios (DSPL to L) were plotted and a line, which turned out to be a curve, was drawn through the mean to represent what was then regarded as the ratio of the wholesome majority. Finally a rule was mathematically devised to penalise those boats which were found to be furthest in the wrong (light) direction from the line. Ratios of sail area to displacement were plotted in much the same way, the mean line plotted and a tax for extra big rigs formulated.' Ibid., p. 59.

5. Bruce Farr, 'Bruce Farr Interview No. 014' (unpublished, Auckland, 31 January 1998).

6. Jack Knights in 'Victory to the Conservatives', *Sea Spray* (Auckland: Review Publishing Limited, July 1978), p. 61.

7. Ibid., p. 59.

8. Ibid.

9. Ibid., p. 61.

10. Bruce Farr, 'Bruce Farr Interview No. 005' (unpublished, Annapolis, 2 June 1995).

11. Ibid.

12. *Smackwater Jack* disappeared without trace in the Tasman Sea while competing in the Hobart–Auckland race. It was a victim of the 80-knot depression, Cyclone David, which produced conditions described as worse than those of the 1979 Fastnet. Alison Whiting was the last to be heard on the radio, speaking with the Royal Akarana Yacht Club watch while *Smackwater Jack* was hove to. Those lost were Scott Coombs, an American; Paul Whiting, Alison Whiting (nee Chambers) and John Sugden of Auckland.

13. The brilliantly performed *Waverider* created Ton Cup history in 1979 by winning the Half Ton Cup for the second year in a row. It once again confirmed Laurie Davidson's brilliance as a designer and co-skippers Tony Bouzaid and Helmer Pedersen's class as helmsmen. However, if it hadn't been for the 50 percent penalty collected by the Farr-designed *Swuzzlebubble* for a minor infringement, the fleet leader going into the fourth race, with its superior upwind speed, might have finally given Ian Gibbs that elusive Half Ton title.

14. Cathie Cross in a letter to the author (unpublished, Bonnington, Kent, UK, 30 April 1996.)

15. Robyn and her future husband, Air New Zealand pilot Ian Varcoe, had bought, together with friends, a restaurant in Ponsonby called Pabulum. In the years that lay ahead Robyn and Ian would remain close friends of Bruce and Cathie.

16. Today this disorder would be described as Bipolar Effective Disorder.

17. Cathie Cross (nee Cochran) in a letter to the author (unpublished, Bonnington, Kent, UK, 30 April 1996).

18. Russ Bowler, 'Russell Bowler Interview No. 001' (unpublished, Annapolis, 11 June 1995).
19. Marten Marine built approximately 140 Noelex 30s.
20. The trailer-sailer designs Sea Nymph ordered in addition to the Farr 6000 were: the Farr 5000, the Farr 7500 and the Farr 740 Sport.
21. Kim McDell, 'Kim McDell Interview No. 001' (unpublished, Kawau Island, 11 January 1996).

Chapter 18

1. Bill Endean, *Classic New Zealand Yachts: Four Decades of Successful Yacht Design— 1950–90* (Wellington: GP Publications, 1992), p. 188.
2. Keith Chapman (deceased), 'Keith Chapman Interview No. 001' (unpublished, Auckland, 17 February 1996).
3. The *Ceramco* challenge committee comprised Warwick White, Commodore of the Royal New Zealand Yacht Squadron, and the Devonport Yacht Club's Martin Foster and Peter Cornes. In the end, 418 individuals, partnerships, yacht clubs and businesses became the syndicated owners of *Ceramco*.
4. Tom Clark, 'Sir Thomas Clark Interview No. 001' (unpublished, Auckland, 8 January 1996).
5. For details, see Chapter 26, note 5 and Chapter 30, note 29.
6. Peter Blake and Alan J. Sefton, *Blake's Odyssey: The Round The World Race with Ceramco New Zealand* (Auckland: Hodder & Stoughton Ltd, 1982), p. 14.
7. Ibid., p. 12.
8. Ibid., p. 13.
9. Ibid., p. 15.
10. See Chapter 7, note 12 for the origin of light displacement racing yachts. For further reading on the origins of light displacement yacht design, see L. Francis Herreshoff, *The Common Sense of Yacht Design* (New York, Rudder Publishing Co., 1946–48).
11. The three previous yachts to win line and handicap honours in the Sydney–Hobart were John Illingworth's *Rani* in 1945, Ted Turner's *American Eagle* in 1972 and Jim Kilroy's *Kialoa* in 1977, the latter with a race record of 2 days 14 hours and 36 minutes. With 60 miles to go, *Ceramco* looked likely to take two hours off the *Kialoa*-held record. But on entering Storm Bay the wind dropped and it took *Ceramco* seven hours to cover the next nine miles.
12. Peter Blake and Alan J. Sefton, *Blake's Odyssey: The Round the World Race with Ceramco New Zealand* (Auckland: Hodder & Stoughton, 1982), p. 32.
13. Ibid.
14. Pierre Fehlmann was coming third in the OSTAR when his boat began breaking up. He was rescued by the container ship *The Atlantic Conveyor* in 50-foot seas. With his boat sinking beneath him, his was a split-second leap to the safety of the netting slung over the container ship's side.
15. Pierre Fehlmann, 'Pierre Fehlmann Interview No. 001' (unpublished, New York, 5 September 1995).
16. *Disque d'Or*, a Swan 65, was the same design as *Sayula II* which won the inaugural Whitbread round-the-world race. See Chapter 13, note 7.

17. Pierre Fehlmann, 'Pierre Fehlmann Interview No. 001' (unpublished, New York, 5 September 1995).

18. The Lievis' spar-making company, Licospars di L Lievi C.S.n. C., is based in Bogliaco, Italy.

19. The only restrictions for Libera Class yachts were that they had to be fixed-keel monohulls of less than 14.2 metres (46 ft) LOA. *Farneticante* and *Grifo* were both 13.45 metres (44 ft 2 inches) LOA and displaced 1875 kg (4125 lb), excluding crew but including keel weight of 640 kg (1410 lb).

20. The European Lake Boat circuit is sailed on the lakes of Switzerland (Geneva) and Austria (Boden See) as well as Italy. Raced on Lake Garda in September, the Centomiglia is the last race of the annual series. A history of the race can be found in: *Centomiglia del Largo di Garda* (Brescia: Puntografico spa, September 1986), La Mille Miglia Editrice.

21. Silvana Lievi, 'Silvana Lievi Interview No. 001' (unpublished, Carversville-Bogliaco, 29 August 1995).

22. Luciano Lievi would win the Centomiglia for a second time in 1996 sailing a popular one-design ASSO 99 called *Spitz*.

23. The Italian word is spelt 'farneticante'.

24. In total, Bruce Farr Yacht Designs would draw five boats for the European Lake Boat Circuit:

(a) The wooden *Grifo*, with America's Cup sailor Flavio Scala steering, won the Centomiglia in 1981 from 250 contestants. With Marco Holm on the helm it won again in 1982. It displaced 1875 kg (4125 lb) excluding crew and was 13.45 m (44 ft 2 in) LOA;

(b) the wooden *Farrneticante*, owned and helmed by Luciano Lievi, finished second in the Centomiglia in 1981 and 1984. In 1987 it finished fourth in the Centomiglia but won the European Libera Class Championship of that year. As *Pleasure* it finally won the Centomiglia Libera Class in 1991;

(c) the lightweight Nomex/Kevlar *OPNI* (*Object Planet Non-Identifie* or *Unidentified Flying Object*), displacing only 843 kg (1858 lb) and only 10.6 m (35 ft) LOA, won the Centomiglia in 1983. It was designed for Alain Golaz of Geneva and built by Mark Lindsay Boat Builders of Gloucester, Massachusetts, USA;

(d) *DF* (or *D'Marquis/Fehlmann*), designed for Claude Fehlmann of Geneva, finished third in the 1985 Centomiglia. Sold to Andrea Damiani and renamed *Lillo*, it won the Centomiglia in 1987;

(e) *Azzardissimo*, designed for C. Ribola, finished fourth in the 1985 Centomiglia.

25. Silvana Lievi, 'Silvana Lievi Interview No. 001' (unpublished, Carversville-Bogliaco, 29 August 1995).

26. The Laser 28 is Design 91. See Chapter 19 for further details.

27. Bruce Farr, 'Bruce Farr Interview No. 007' (unpublished, Annapolis, 9 June 1995).

28. *America* was housed at Trumpy's Yard on the edge of Spa Creek, Annapolis, when a heavy snowfall on 28 March 1942 caused the shed's roof to collapse, destroying the famous schooner's fragile remains.

29. Dinny White, 'Jean and Dinsmore White Interview No. 001' (unpublished, Annapolis, 26 August 1995).

30. Bruce Farr, 'Bruce Farr Interview No. 007' (unpublished, Annapolis, 9 June 1995).

Chapter 19

1. William Butler Yeats, 'Words for Music Perhaps: X. Her Anxiety', in *The Winding Stair and Other Poems* (1933) from *W.B. Yeats Collected Poems* (London: Picador, 1995), p. 297.
2. John Alden in 'Dreams of Glory' in *Baltimore Magazine* (Baltimore, MD: February 1991), p. 32.
3. These were the work boats of the area used for harvesting crabs and dredging for oysters in the Severn River and Chesapeake Bay. Like the turn-of-the-century mullet boats of Auckland, they carried vast amounts of canvas in the hope of beating their opposition to the docks to secure the best price for the day's catch. And as with the mullet boats, they became the boats in the area to be used for yacht racing.
4. John Alden in 'Dreams of Glory' in *Baltimore Magazine* (Baltimore, MD: February 1991), pp. 32 and 34.
5. Marmaduke's is now Ruth's Chris Steak House.
6. Before coming to Annapolis, George Hazen had been an undergraduate involved in MIT's Pratt Programme, helping write performance prediction programs for yachts as well as developing the Measurement Hull System (MHS) which would eventually become the basis of the empirical rating rule known as the IMS (see Chapter 9, note 4, and Chapter 28).
7. Roger Hill, 'Roger and Christine Hill Interview No. 001' (unpublished, Auckland, 14 November 1995).
8. It was Ian Bruce who took fellow-Canadian Bruce Kirby's lines and turned them into a piece of industrial design in 1969–70. Working in his garage where he produced the original tooling and the prototype, Ian Bruce pushed the boundaries of refinement to produce in the original Laser skiff what is commonplace today: simplicity of rig and deck form, simplicity of assembly of hull and deck, and simplicity of construction for mass production.
9. 'Kevlar' is DuPont's registered trade name for this arimid fibre. First spun into roving then woven into fabric, Kevlar was originally used in bullet-proof vests. It is not only much lighter than fibreglass. The same properties which make it ideal as a ballistic shield also made it suitable for use in a Whitbread hull. Its combination of high tensile and low compressive qualities means it is unlikely to collapse on impact should a hull strike pack ice in the Southern Ocean.
10. Although some 250 Laser 28s were eventually built by Ian Bruce, with insufficient funds to complete the research of high-speed production, the concept was thwarted and sales for the development company, Precis 99, fell far below those originally envisaged.
11. The problem wasn't with Marten Marine's immaculate hull construction. The problem was with the keel. In their wisdom the owners had gone to a third party to have it built but it twisted when the lead was poured into the steel fabrication. The only affordable way to fix the problem was to fill and fair the keel. This made the chord thicker than it should have been, most likely

producing more induced drag and adversely affecting the overall performance of the yacht.

12. Keith Chapman, 'Ceramco's Story', Sea Spray (Auckland, Review Publishing Limited, December 1981), p. 4.

13. The hounds or 'I' point is where the mast is braced fore and aft.

14. Gary Baigent, Light Brigade (Auckland: Cox's Creek, 1997), p. 117.

15. 'Grant Dalton Interview No. 001' (unpublished, Auckland, 21 December 1995).

16. There were six New Zealanders on Flyer's crew when it won the 1981–82 Whitbread. Four were on board for the whole race: Sail Coordinator Grant Dalton, Warwick Buckley, John Vitali and Earle Williams. Joe Allen and George Hendy joined the crew in Auckland for the remainder of the race.

17. Under the altered IOR, as a result of its wide stern, Disque d'Or III suffered a rating penalty of 3.6 feet.

18. Bruce Farr, 'Bruce Farr Interview No. 005' (unpublished, Annapolis, 2 June 1995).

19. The Farr and Lexcen design offices were the first race-boat designers to use George Hazen's VPP.

20. Duncan Spencer, 'Stagg, Farr and Bowler: A Team of Kiwi Magnificos', in Rags (Annapolis, MD: April 1995), p. 63.

21. Geoff Stagg, 'Geoffrey Stagg Interview No. 001' (unpublished, Annapolis, 28 August 1995).

22. Geoff Stagg and Keith Chapman were Ceramco's two watch captains in the 1981–82 Whitbread.

23. Bruce Farr, 'Bruce Farr Interview No. 007' (unpublished, Annapolis, 9 June 1995).

24. The shareholders of Farr International are Russ Bowler, Bruce Farr and Geoff Stagg, who is also its president.

25. Jean and Dinny White, 'Jean and Dinsmore White Interview 001' (unpublished, Annapolis, 26 August 1985).

26. Keen for the Farr office to link up with a builder in the area rather than seeing their IOR designs built in another state, the Marine Trades Association of Maryland, Inc and the Department of Economic Development, helped the Farr office to identify a production builder in the bay area. That was Dickerson Boat Builders of Trappe, on the Eastern Shore, a traditional builder in the process of modernizing their yard. They eventually built nineteen Farr 37s.

27. Cathie Cross (nee Cochran) in a letter to the author (unpublished, Bonnington, Kent, 30 April 1996).

28. Steven Cross (deceased) in a letter to Nick Bailey (unpublished, Bonnington, Kent, 2 December 1992).

29. Ibid.

30. Steven and Cathie remarried in 1984, nine and half years after their first marriage. They ran Scaramouche of Warwick for its owner until 1986 when they returned to England and settled into a small thatched cottage in Bonnington, Kent, where they had three children — Annabel, Isla and Jack — and ran a small book-binding business. Steven died on 29 May 1996 after a long battle with cancer.

31. Michael Ondaatje, Handwriting (London: Bloomsbury, 1998), p. 42.

Chapter 20

1. Roger Highfield and Paul Carter, *The Private Lives of Albert Einstein* (London: Faber and Faber, 1993) p. 67.

2. M. Scott Peck, *People of the Lie: The Hope for Healing Human Evil* (London: Arrow Books, 1990) p. 82.

3. Ibid., p. 141.

4. Tom Clark in a letter on Ceramco Limited letterhead to Bruce Farr dated 24 June 1982, from the correspondence files of Bruce Farr & Associates Inc (BFA), Annapolis, Maryland, USA.

5. Bruce Farr in a letter to Tom Clark dated 9 August 1982, from the correspondence files of Bruce Farr & Associates Inc, Annapolis, Maryland, USA.

6. A Maxi was 'so-called because it measures at the maximum of 70 ft under the IOR handicapping system.' Shane Kelly in 'Kelly's Eye', *Sea Spray* (Auckland: Review Publishing Co. Ltd., March 1985), p. 65.

7. Bruce Farr in a letter to Tom Clark dated 9 August 1982, from the correspondence files of Bruce Farr & Associates Inc, Annapolis, Maryland, USA.

8. Bruce Farr in letters to Tom Clark, and Tom Clark and Peter Blake, dated 5 January 1983, from the correspondence files of Bruce Farr & Associates Inc, Annapolis, Maryland, USA.

9. Peter Blake and Alan Sefton, *Lion: The Round the World Race With Lion New Zealand* (Auckland: Hodder & Stoughton Ltd, 1986), p. 16.

10. Telex from CERAMCO NZ2772, from the correspondence files of Bruce Farr & Associates Inc, Annapolis, Maryland, USA.

11. This figure includes an exclusivity fee as well as fees for research and development, design drawings and support services. Including the exclusivity factor, it was some two and a half times BFA's estimated one-off fee for such a design. Given that the Farr office eventually received three Maxi commissions for this race, and given the potential loss to both future income and reputation a single Whitbread entry may have represented, the figure appears reasonable-to-under-valued.

12. Telex to CERAMCO NZ2772, from the correspondence files of Bruce Farr & Associates Inc, Annapolis, Maryland, USA.

13. This Blake-Sefton statement is sophistic and untrue. Peter Blake received the negotiable figure for the exclusive use of the Farr office's services on February 3, 1983. It was not until 13 July 1983 that the Kuttel/*Atlantic Privateer* contract was signed with Bruce Farr & Associates, and 16 and 23 November respectively that the Fehlmann/*UBS Switzerland* and the Taylor/*NZI Enterprise* contracts were signed. A press release was made on 3 December 1983 announcing all three contracts.

14. By juxtaposing the false statement that BFA 'was already committed to designing Whitbread maxis for two other customers' with the 'cost to buy exclusivity', and by failing to declare that 'exclusivity' in this context meant the exclusive use of BFA's staff, resources, services and philosophy for an extended period and not an exclusive design, the reader is lead to believe that the prohibitive figure at which '[Clark and Blake] saw red' was an exorbitant one instead of the reasonable figure it was for all that Farr had offered. One can only surmise why Blake

and Sefton should have constructed this paragraph of mirrors. Was the experienced and normally lucid Sefton unable to express himself clearly? Or was it Blake? Was he concerned that if his readers (and sponsors) had realised that it was he who was entirely responsible for the design choice and associated decisions for the 1985–86 Whitbread Race and not those onto whom he had projected blame — Farr for asking a price which, by implication, prevented him from having the fastest boat, and Holland for providing a non-exclusive, overweight boat — he might not have gained both moral and sponsorship support for his next major project? Further, given BFA's original offer of an exclusive design and its well-known record of maintaining client confidentiality, and given the amount of design information that Farr had already sent to both Blake and Clark, for Blake to imply that BFA could not be trusted with 'the advantage of our thinking' was an uncalled-for besmirching of its name — an 'advantage', as it transpired, that turned out to be a huge liability for *Lion New Zealand*.

15. Peter Blake and Alan Sefton, *Lion: The Round the World Race With Lion New Zealand* (Auckland: Hodder & Stoughton, 1986), p. 16.

16. Ibid.

17. Telex from CERAMCO NZ2772, from the correspondence files of Bruce Farr & Associates Inc, Annapolis, Maryland, USA.

18. This Blake-Sefton comment is sophistic and untrue. Having been offered an exclusive design with full confidentiality as well as the exclusive use of BFA's resources and philosophy, Blake and Sefton tell their readers the opposite of that which in fact took place: '... that the Farr office would not even give us an assurance that they would work on no other Whitbread boat ...' Blake and Sefton then heap delusion upon falsity by suggesting that if they weren't protective of their ideas others might 'benefit from what ... [they] had to offer'. But not only was Blake bereft of any ideas that would have materially affected the design of a Whitbread Maxi. Those he did have — a bigger than normal crew living in relatively luxurious conditions in a boat 'built like sixteen brick shithouses' — contributed to making *Lion New Zealand* slower and heavier than it already was.

19. Peter Blake and Alan Sefton, *Lion: The Round the World Race With Lion New Zealand* (Auckland: Hodder & Stoughton Ltd, 1986), p. 18.

20. Telex to CERAMCO NZ2772, from the correspondence files of Bruce Farr & Associates Inc, Annapolis, Maryland, USA.

21. Bruce Farr, 'Bruce Farr Interview No. 006' (unpublished, Annapolis, 8 June 1995).

22. The proposed composition of the lightweight panels was Kevlar-reinforced sandwich skins laid over Nomex or foam core, Nomex and Kevlar being DuPont products.

23. This information is based on a file note dated June 2 1983 from the correspondence files of Bruce Farr & Associates Inc, Annapolis, Maryland, USA.

24. Telex from ROBNZ NZ2917, from the correspondence files of Bruce Farr & Associates Inc, Annapolis, Maryland, USA.

25. Before giving Peter Blake 'the green light to pursue design options' (*Lion*, p. 14), Tom Clark had already secured over half their budgeted project figure of $2.25 million. This included $1,000,000 from Lion Breweries for the naming rights,

sponsorship from six secondary corporate sponsors of $100,000 each (Alan Sefton, *The Inside Story of KZ-7: New Zealand's First America's Cup Challenge, Fremantle, 1986–87*, Auckland: Hodder & Stoughton Ltd, 1987, p. 59), and the support of shareholders in the *Ceramco New Zealand* syndicate who had agreed to roll their investment into another Blake boat. In the end the budget for the *Lion New Zealand* project would exceed $3,000,000, with the number of secondary corporate sponsors growing to twelve and the committed shareholders from the *Ceramco New Zealand* syndicate to 120.

26. This conversation is based on a file note dated June 14 1983, from the correspondence files of Bruce Farr & Associates Inc, Annapolis, Maryland, USA.

27. Details of these negotiations are taken from a file note dated June 23 1983, from the correspondence files of Bruce Farr & Associates Inc, Annapolis, Maryland, USA.

28. Telex from ROBNZ NZ2917, from the correspondence files of Bruce Farr & Associates Inc, Annapolis, Maryland, USA.

29. Graphnet Faxgram, from the correspondence files of Bruce Farr & Associates Inc, Annapolis, Maryland, USA.

30. Peter Blake and Alan Sefton, *Lion: The Round the World Race With Lion New Zealand* (Auckland: Hodder & Stoughton Ltd, 1986), p. 16.

31. Ibid.

32. Telex to CERAMCO NZ2772, from the correspondence files of Bruce Farr & Associates Inc, Annapolis, Maryland, USA.

33. Western Union Mailgram, from the correspondence files of Bruce Farr & Associates Inc, Annapolis, Maryland, USA.

34. Telex to CERAMCO NZ2772, from the correspondence files of Bruce Farr & Associates Inc, Annapolis, Maryland, USA.

35. Letter from Blake Offshore, Emsworth, Hampshire, UK, from the correspondence files of Bruce Farr & Associates Inc, Annapolis, Maryland, USA.

36. Letter from the files of Bruce Farr & Associates Inc, Annapolis, Maryland, USA.

37. See Chapter 30, note 29.

38. Bruce Farr & Associates were not the only designers maligned by Peter Blake and Alan Sefton in *Lion: The Round The World Race With Lion New Zealand*. On page 18 Blake says: 'The Holland office was confident it could do the boat and agreed to the exclusivity we required. We were later to discover that we in fact achieved neither of our goals. The Holland office had a major involvement in *Colt Cars* becoming *Drum England*, and even in helping to search out new owners in pop star Simon Le Bon and his business partners Paul and Mike Berrow.' When asked by the author for his version of events, in a facsimile dated 24 August 1998, Ron Holland replied as follows: 'I wasn't involved in "selling this project" to Simon Le Bon and our assistance with the project was extremely limited. A stunning example of this relates to our refusal to visit the project while she was being completed, and this lack of involvement had dire consequences. After *Drum's* keel broke off during the early stages of the Fastnet Race ... it was seen that the welding of the keel was to an extremely obvious substandard. I always regretted not seeing the keel being welded, as I would have insisted on better workmanship ...' As to the contentious issue of *Lion's* weight, on page 172 of *Lion: The Round the World Race With Lion New Zealand*

Peter Blake says this: 'We'd gone to [the Holland Design office] for the ultimate Whitbread Maxi, a boat which would be as light as the Farr boats turned out, but a lot stiffer with a lot more sail. *Lion* wasn't that boat, even though the Holland office assured us they could achieve the design features and the figures we speci-fied. Instead of being 31 tons actual weight, she was close to 38 tons. The keel was about right and so was the rig ... which meant that the hull and deck was about 50 to 60 percent overweight. The boat, I knew, had been built precisely to designer specifications ...'

In his facsimile of 24 August 1998 Ron Holland says: 'Peter's text tends to suggest I was responsible for *Lion* being overweight, which is not true. The con-struction for *Lion* involved SP Systems and was also influenced by her builder, Tim Gurr. It's important to note that we all were under Peter's and Tom Clark's insistence that the boat shouldn't break in any circumstances and I'm afraid in retrospect we all, as a team, took this too seriously and the result was that *Lion* came out overweight.'

Chapter 21

1. Peter Campbell in 'How the Kiwis Copped the Cup', *Sea Spray* (Auckland: Review Publishing Co. Ltd., February 1984), p. 6.

2. Woody Baskerville's *Migizi*, with Russ Bowler on board, won five wins of its six races to win Division F in the 1983 SORC.

3. Not only did the IOR's new cg rule give designers more scope to achieve greater stability. The rating penalty for designing a more stable boat was also reduced. While it might sound perverse to penalise a keel-boat's stability, the originators of the rule would have said that the IOR fairly rated stability. In other words, because the origins of the IOR's philosophy lay in Wells Lippincott's 'base boat' concept — which penalised those factors that made a yacht go fast and gave bonuses to those which slowed it down — stability, which increases a keel-boat's speed, was penalised under the IOR the further it moved away from that which the rule said was normative. Significantly, there was still no formula written into the IOR which assessed the performance advantages of removing weight from the ends of the hulls. This presented huge possibilities for those investigating the widespread use of exotic materials in race-boat construction.

4. Bruce Farr, 'Bruce Farr Interview No. 004' (unpublished, Annapolis, 7 June 1995).

5. Under the IOR, *Pacific Sundance* rated 30.5 feet and *Geronimo* and *Exador* 30.4 feet.

6. Tom Schnackenberg, 'Tom Schnackenberg Interview No. 001' (unpublished, Auckland, 18 December 1995).

7. John Mallitte in 'Freaked Out!' in *Sea Spray* (Auckland: Review Publishing Limited, October 1984), p. 12.

8. Dinny White, 'Jean and Dinsmore White Interview No. 001' (unpublished, Annapolis, 26 August 1995).

9. Pam Farr, 'Pamela Farr Interview No. 001' (unpublished, Auckland, 18 January 1996).

10. Ron Holland was not appointed to head the design programme because 'the steering committee [was] disenchanted with the lack of progress it was making

with the Farr office' as Alan Sefton claims in *The Inside Story of KZ 7*, p. 66. Ron Holland was appointed because he was considered the best man for the job. He had been in Auckland overseeing two major projects and had been able to meet in person with the committee. As well, Twelve-Metre yachts were very heavy-displacement and his thinking on this subject was far closer to Don Brooke's than it was to Bruce Farr's.

11. A.G. Malcolm, 'Close to the Wind' (unpublished manuscript, Auckland).

12. Alan Sefton, *The Inside Story of KZ 7: New Zealand's First America's Cup Challenge, Fremantle, 1986–87* (Auckland: Hodder & Stoughton Ltd, 1987), p. 78.

13. Belgian-born Sydney-based businessman Marcel Fachler, of his own initiative, had paid to the Royal Perth Yacht Club the A$12,000 deposit required to enter the RNZYS as a challenger for the 1987 America's Cup. Loosely connected to the RNZYS, the steering committee set up to investigate an America's Cup challenge was, in fact, an ad hoc body without legal status. Its members were: Cedric Allan, Donald Brooke (RNZYS vice commodore), Marcel Fachler, Rob Green (RNZYS commodore), Warwick Peacock (RNZYS accountant), sports administrator Sir Ronald Scott (chairman) and yachting writer Alan Sefton.

14. Aussie Malcolm was the Minister of Health and Immigration when he was approached by Donald Brooke, in May 1984, asking if he would meet with Marcel Fachler. Within 24 hours of his losing his seat in the July snap election, Cedric Allan and Marcel Fachler had asked the unemployed ex-Minister of the Crown if he would take on the tenuous job of running their steering committee.

15. *Enterprise* was the 1977 S&S-designed Twelve-Metre which Chris Dickson and crew had sailed with distinction in the 1984 World Championships in Sardinia.

16. A.G. Malcolm, 'Close to the Wind' (unpublished, Auckland). It is rumoured that *Australia II*'s lines plans had previously been offered to Englishman Peter de Savary for £50,000.

17. David McKellar Richwhite as quoted by Ian Wishart in *The Sunday Star-Times* (Auckland: Independent News Auckland Ltd, 17 August 1997), p. A5.

18. The New Zealand Court of Appeal takes a rather different view from David Richwhite's: '... where large tax fraud is concerned ... prosecution is the proper approach [and ...] imprisonment ... the normal punishment'. See Anthony Molloy, *Thirty Pieces of Silver* (Auckland: Howling at the Moon Productions Ltd, 1998), p. 271.

19. A.G. Malcolm, 'Close to the Wind' (unpublished manuscript, Auckland).

20. Ibid.

21. Ibid.

22. Ibid.

23. Ibid.

24. Ibid.

25. Skip Novak as quoted by Michael Levitt in 'Farr Ahead' in *Nautical Quarterly* Number 43 (Connecticut: Nautical Quarterly Co, 1988), p. 29, from Skip Novak, *One Watch at a Time* (Scranton, PA: W.W. Norton and Co., 1988).

26. *Sea Spray* (Auckland: Review Publishing Limited, June 1986), p. 15.

27. Ibid.

28. Ibid.

29. Tom Clark, 'Sir Thomas Clark Interview No. 001' (unpublished, Auckland, 8 January 1996).

30. Digby Taylor purchased only a basic design from the Farr office which included lines, keel geometry, rudder geometry and sail plan. But finding himself short of the requisite knowledge to do the hull structures and mast Digby would phone the Farr office requesting more design information. The Farr office also provided free of charge a mast section shape based on certain mast section calculations. The mast engineering, however, was carried out in Auckland under Digby Taylor's auspices.

31. *Lion New Zealand*'s certified DSPL (a rated displacement figure which is typically 5–10 percent lower than the boat's actual displacement) for the 1985–86 Whitbread was 35,024 kg (34.5 tons), down from its New Zealand certificate of 36,522 kg (35.9 tons) following modification to its keel in England before the race. By way of contrast, *Drum*'s DSPL for the Whitbread was 34,331 kg (33.8 tons), *UBS Switzerland*'s 28,544 kg (28.1 tons). In *Lion: The Round the World Race with Lion New Zealand* (Auckland: Hodder and Stoughton, 1986, [p. 172]), Blake and Sefton claim: 'Even though the Holland office assured us they could achieve the ... figures we specified, instead of being 31 tons actual weight, she [*Lion*] was close to 38 tons.' Given that this was close to *Lion*'s actual displacement before keel modification, and given the Holland-designed *Drum*'s displacement figures, it is hard to believe Blake could ever have targeted an actual displacement of 31 tons. It would seem more likely that from the moment he read the rating certificates he had been haunted by *UBS*'s real displacement figure: about 31 tons. This, of course, would prove to be one of the key differences between the two boats and a convenient figure with which to project blame onto Holland for producing a 'hull and deck [which] were approximately 50 to 60 percent overweight' (Ibid.) — yet another Blake statement difficult to believe. The reality was, Peter Blake had chosen to follow his own ill-judged philosophy which, among other things, saw various 'luxuries' such as two microwave ovens, a heater, a stove, an oven, hot and cold running water, two toilets and a shower adding to *Lion*'s weight.

32. Bob Fisher and Barry Pickthall, *Ocean Conquest: The Official Story Of The Whitbread Round The World Race Past, Present and Future* (London: Little, Brown, 1993), p. 87.

33. Chris O'Nial was a unique person, full of enthusiasm and energy for his trade. He died of cancer in 1998.

34. *Atlantic Privateer* had also been known as *Apple-Macintosh* and *Portatan* as its main sponsor changed.

35. Peter Blake as quoted by Bob Fisher, *The Greatest Race: The Official Story of the Whitbread Round-the-world Race 1985/1986* (London: Robertsbridge, 1986), p. 136.

36. In the interests of economy, *Atlantic Privateer* had used *Flyer*'s old mast.

37. The elapsed time results for the first three finishers were: *UBS Switzerland* 117 days, 14 hours; *Lion New Zealand* 122 days, 6 hours; *Drum* 122 days, 18 hours.

38. Michael Levitt in 'Farr Ahead', *Nautical Quarterly*, Number 43 (Connecticut: Nautical Quarterly Co., 1988), p. 29.

39. The Champagne Mumm World Ocean Racing Championship is a two-year event

in which the best three results from five events over that period are totalled.

40. Alan Sefton, *The Inside Story of KZ 7: New Zealand's First America's Cup Challenge, Fremantle, 1986–87* (Auckland: Hodder & Stoughton Ltd, 1987), p. 78.

Chapter 22

1. James Course as quoted by Chris Freer and Adrian Morgan in 'The Kiwi Dark Horses', Chapter 7 of *Sail of the Century* (London: Stamford Maritime, 1987), p. 88.

2. The first attempt by Australia to win the America's Cup was made with *Gretel* in 1962. There followed attempts with *Dame Pattie* and *Gretel II* before the first Bond-Lexcen campaign of 1974 with *Southern Cross*. It would take two more attempts by Bond and Lexcen — in 1977 with *Australia* and with a slightly modified *Australia* in 1980 — before their historic win with *Australia II* in 1983. New Zealand won the America's Cup at its second attempt — 'The challenging yacht *New Zealand* is declared to be the winner of the two races on September 7 and 9, 1988,' said New York Supreme Court Judge Carmen Ciparick on April 10, 1989 — before having that decision overturned by 'the shallow and flawed reasoning of the New York Court of Appeals'. [James Michael, 'Inside the America's Cup: The Kiwi Challenge, 1987–1990' (unpublished, USA, registered WGAW no 553855), p. 427] on 19 September, 1989 — and at its fourth in May 1995.

3. Peter Bateman, a Holland family friend, had won the World Fireball championship twice, had been the national yachting coach for the British Olympic team of 1980, and had been closely involved with the de Savary-Howlett programme that produced *Victory 83* — the loser to *Australia II* in the Louis Vuitton Cup (LVC) finals of 1983. — before he was seconded to the early New Zealand Challenge programme as a manager.

4. Paul Larsen and Russell Coutts, *Russell Coutts Course to Victory* (Auckland: Hodder Moa Beckett Publishers Limited, 1996), p. 62.

5. Chris McMullen, 'Interview No. 001' (unpublished, Auckland, 15 August 1998).

6. Prior to *Intrepid*, Twelve-Metre yachts had shallow long-run keels with the rudder hung from their trailing edge. When designing his 1967 defender, Olin Stephens borrowed ideas from ocean racing, replacing the long-run keel with a shorter, deeper one and separating the rudder from the keel by hanging it from a supporting aft skeg. To counteract the greater leeway (sideways movement) resulting from the new keel's reduced lateral area he attached a trim tab to its trailing edge. In much the same way as flaps work on aeroplane wings, the trim tab provided addition lift, reducing *Intrepid*'s leeway.

7. In order to overcome the Twelve-Metre Rule's minimum draught restriction, Ben Lexcen borrowed Uffa Fox's idea of an inverted keel, making his Twelve-Metre keel thicker and longer at the bottom than at the top. This lowered the keel's weight and made the boat more stable (and therefore more powerful). The unorthodox shape, though, produced a greater tip-vortex problem — the water flowing from the high pressure side to the low pressure, windward side — than had the more conventional tapered trapezoid shape. As Bob Miller, he had tried to lessen induced drag on his eighteen-footer *Venom* by putting an end plate on

its centreboard. From that rudimentary concept, and from known aeronautical research, he added winglets to his upside-down keel, thereby solving the tip-vortex problem and helping his countrymen to win the America's Cup in 1983.

8. Dennis Conner and Bruce Stannard, *Comeback: My Race for the America's Cup* (Sydney: Macmillan Company of Australia Pty Ltd, 1987), p. 180.

9. An architect by profession, Bill Ficker helmed *Intrepid* when it successfully defended the America's Cup in 1970 against the Royal Sydney Yacht Squadron's *Gretel II*. He also won the Congressional Cup and the International Star Class World Championships.

10. Bruce Farr as quoted by Charles Barthold in 'The Farr Vision' in *Yachting* (Greenwich, CT: Times Mirror Magazines, March 1988), p. 103.

11. The mathematical formula of the Twelve-Metre Rule, updated from its original version, reads: $\frac{L +2d - F + \sqrt{S}}{2.37} = 12$ metres

    L = length in metres, measured 180 mm above the static waterline;
    d = girth difference in metres, being the skin girth less the chain girth;
    F = freeboard in metres, being the sum of the freeboard at fore, amidships and aft hull measurement points;
    S = rated sail area in square metres of the mainsail and fore-triangle.

    As well, the Twelve-Metre Rule stipulated that the mast could be no more than 25 metres high, and the headsails attached at a point no higher than 75 percent of mast height. Maximum draught was set at 2.68 metres and minimum beam at 3.59 metres.

    Although the formula itself is simple, it takes some 25 pages of appended small print in the IYRU's Twelve-Metre Rule to interpret it.

12. Rik Dovey/Sally Samins, *12-Metre the New Breed: The Battle for the 26th America's Cup* (Sydney: Ellsyd Press Pty Ltd, 1986), p. 8.

13. Bill Endean, *Classic New Zealand Yachts: Four Decades of Successful Yacht Design — 1950–1990* (Wellington: GP Publications, 1992), p. 213.

14. The Twelve-Metre Class had been in existence since 1907, following the meeting in 1906 of representatives of yacht clubs from sixteen European and Scandinavian nations 'in London to form the International Yacht Racing Union ... At the same time they developed an International Rule which laid down metric formulas for boats that would measure 5, 6, 7, 8, 9, 12, 15, 19 and 23 metres ... In the interests of seaworthiness, the IYRU also stipulated that yachts built to the rule had to comply with [the] Lloyd's [Register of Shipping] scantling standards which define minimum construction strengths in the hull and rig ... Yachtsmen from the United States remained unimpressed by the European formula [and used Nathanael Herreshoff's 1902] Universal Rule ... which designated its classes by linear metre ratings ... [But] as it contained considerably fewer restrictions, the Americans were forced into continued amendments as the designers found loop-holes.' (See Rik Dovey/Sally Samins, *12-Metre the New Breed: The Battle for the 26th America's Cup* [Sydney: Ellsyd Press Pty Ltd, 1986], pp. 8–9).

15. America's Cup contests were suspended because of World War II. When the NYYC accepted challengers for a contest in 1958 it was decided to use the Twelve-Metre Rule. The decision was made primarily because Twelve-Metre yachts were smaller and therefore cheaper than the previously used J-Boats. It is

arguable that the New York Yacht Club also chose Twelve-Metres because *Vim*, designed by Olin Stephens, had proven markedly superior to any of the existing Twelves in Europe and England, thus giving the NYYC a distinct design advantage in that class.

16. In its initial protest under racing rule 27, the NYYC correctly claimed that when its hull was heeled *Australia II*'s wings gave it added depth. This, they calculated, increased its rating to 12.76 metres — 0.76 of a rated metre outside the Twelve-Metre Rule. It was not that the NYYC considered the winged keel illegal; the Twelve-Metre was a development class which permitted innovation. But there was valid argument for their requesting a penalty — such as a reduction in sail area — to offset *Australia II*'s unorthodox gains in rating. Heat was taken out of the NYYC's protest when, in August 1983, the International Measurement Committee reconfirmed that the Australian yacht was a legal Twelve-Metre. But it was only just before the finals, in the interests of tradition and to avoid an international backlash, that the America's Cup Committee decided not to cancel the contest. Two months later the IRYU approved winged keels for use on Twelve-Metre yachts. Overjoyed at the result, Ben Lexcen assigned all royalty payments on patent number 8200457, filed with the Netherlands Patent Office in February 1982, to the IYRU. Not so enamoured of the invention, perhaps because of the cost they thought it might add to construction, the ORC banned wings on the keels of all offshore yachts racing under its auspices.

17. Chris Freer and Adrian Morgan in 'The Kiwi Dark Horses', Chapter 7 of *Sail of the Century* (London: Stamford Maritime, 1987), p. 86, quoting from the IYRU keel-boat Committee 1982 Yearbook, p. 119.

18. Rik Dovey/Sally Samins, *12-Metre the New Breed: The Battle for the 26th America's Cup* (Sydney: Ellsyd Press Pty Ltd, 1986), p. 128.

19. The University of Southampton's Wolfson Unit for Marine Technology and Industrial Aerodynamics was the tank-testing facility chosen by the New Zealand Challenge Design Group for testing their hull shapes. Early work had been done by Ron Holland at the more expensive Netherlands Ship Model Basin in Holland. Keel testing would also be carried out in the wind tunnel facility at the Royal Swedish Institute of Technology in Stockholm, Sweden.

20. Bruce Farr in 'Bruce Farr and Russell Bowler Interview No. 001' (unpublished, Annapolis, 15 June 1995).

21. The hull and deck structure of a Twelve-Metre yacht accounts for approximately 10 to 15 percent of its total weight. Seventy to 80 percent of a Twelve's total weight is ballast, most of which is in the keel.

22. Steve Marten, 'Steve Marten Interview No. 001' (unpublished, Auckland, 11 and 12 December 1995).

23. An aluminium Twelve-Metre hull can shrink in excess of 50 mm when being welded.

24. Russ Bowler, 'Russell Bowler Interview No. 003' (unpublished, Annapolis, 11 June 1995).

25. High Modulus NZ Ltd was founded in 1979 by Bob Rimmer and was involved mainly in the buying and selling of materials used in the construction of reinforced plastic boats. These included Kevlar fabric, carbon fibre, structural non-

woven glass fabric, Zyron panels, fibreglass cloths, bonding tapes, Divinycell PVC core. From its humble beginnings in Warkworth, it branched out into engineering consultancy and the development of related new products and process. Richard Honey was its senior engineer.

26. These were the plans for *KZ-3* and *KZ-5*. The plans for *KZ-7* were approved by Lloyd's on 22 April 1986.

27. *KZ-5* finished second to the Lexcen-designed *Australia III* in the 1986 Twelve-Metre Worlds with a 1/3/5/3/3/3 result.

28. See Richard Becht, *Black Magic: Team New Zealand's Victorious Challenge* (Auckland: Hodder Moa Beckett Publishers Limited in association with TVNZ Enterprises, 1995), p. 10.

29. Ibid.

30. Andrew Johns, 'Andrew Johns Interview No. 001' (unpublished, Auckland, 21 November 1995).

31. Ron Holland, 'Ron Holland Interview No. 001' (unpublished, Auckland, 23 August 1996).

32. Michael Levitt in 'Farr Ahead' in *Nautical Quarterly* (Essex, Connecticut: Nautical Quarterly Co., 1988), p. 31.

33. Bruce Farr in 'Bruce Farr and Russell Bowler Interview No. 001' (unpublished, Annapolis, 15 June 1995).

34. Chris Freer and Adrian Morgan in 'The Kiwi Dark Horses', Chapter 7 of *Sail of the Century* (London: Stamford Maritime, 1987), p. 89.

35. Roy Dickson and Frenchman Laurent Esquier were the New Zealand Challenge's two sailing directors.

36. Steve Marten, 'Steve Marten Interview No. 001' (unpublished, Auckland, 11 and 12 December 1995).

37. Steve Marten, with Peter Nees as forward hand, won the New Zealand Cherub Nationals in 1970, and with Trevor Burgess as forward hand, the World Cherub Championship in 1972. With Con Lynton and Jack Scholes as crew, Steve Marten also represented New Zealand at the 1972 Munich Olympics in the Soling Class.

38. James Course as quoted by Chris Freer and Adrian Morgan in 'The Kiwi Dark Horses', Chapter 7 of *Sail of the Century* (London: Stamford Maritime, 1987), p. 88.

39. *The Weapon* was the moniker given by Michael Fay to the sleekest and most refined of the three fibreglass Twelve-Metres.

40. Born on March 2 1949, on a remote Hawke's Bay dairy farm, Naomi Powers had had no sailing experience when she met English sailor Rob James by chance in St Malo, France. Two years after they married, she had still not sailed solo when she set out on her round-the-world voyage. On 8 June 1978 she arrived back in Dartmouth Harbour on *Express Crusader*, to become the first woman to have sailed single-handed around the world. She did so in 272 days (including eight days in port), eclipsing Sir Francis Chichester's nine-year-old record by two days. For this outstanding achievement Naomi James was made Dame Commander of the Order of the British Empire on 1 January 1979. Her daughter was born ten days after her husband Rob was tragically drowned at sea in 1983.

Chapter 23

1. James Course as quoted by Chris Freer and Adrian Morgan in 'The Kiwi Dark Horses', Chapter 7 of *Sail of the Century* (London: Stamford Maritime, 1987), p. 82.

2. Anne Michaels, *Fugitive Pieces* (London: Bloomsbury, 1998), p. 162.

3. Dennis Conner's infamous cheating comment was made at the post-race press conference following *Stars & Stripes* 42-second defeat by Tom Blackaller's *USA* in the third race of Round Robin 3. But the first time Glassgate appeared at this regatta was after Race Five of Round Robin 1 during which *Stars & Stripes* had beaten *USA* by six seconds. It was then that Conner was asked why his syndicate had sent a letter to the Challenger of Record requesting that 'all composite construction yachts' be cored.

4. Dennis Conner quoting Dennis Conner, in Dennis Conner and Bruce Stannard, *Comeback: My Race for the America's Cup* (Sydney: Macmillan Company of Australia Pty Ltd, 1987), p. 100.

5. The first experimental structures in fibreglass (matted fine glass fibres) reinforced plastics were carried out in the USA, circa 1940. Its excellent strength-to-weight properties, its durability, its ability to minimise water absorption and withstand many of the elements which cause wood and metal to age, its ability to be moulded into compound curvatures and structures not requiring complex load-carrying joints, were characteristics quickly recognised as ideal for a wide variety of marine applications.

6. Bruce Farr in 'Bruce Farr and Russell Bowler Interview No. 001' (unpublished, Annapolis, 15 June 1995).

7. The Challenger of Record is the yacht club 'on record' which officially mediates between the challenging and defending parties. Before 1970, matches for the America's Cup had only been between two yacht clubs: the holder, the NYYC, and the challenging yacht club. But when *Gretel* and *France* presented themselves as challengers in 1970, an official trial regatta was held to determine which club would provide the challenging yacht. The term 'Challenger of Record' was coined by Douglas Philips-Burt and first used in his book *The Cumberland Fleet*. Since 1974, the Challenger of Record has been the yacht club which represents the challenging contestants, controls their regatta, and agrees with the defender to the conditions of the America's Cup regatta. For the 1987 contest, the new Cup holders, the Royal Perth Yacht Club, nominated Yacht Club Costa Smeralda as the Challenger of Record, the first yacht club from which it had received an official challenge.

8. Contents of the SAF letter as quoted by Alan Sefton in *The Inside Story of KZ 7: New Zealand's First America's Cup Challenge, Fremantle, 1986–87* (Auckland: Hodder & Stoughton, 1987), p. 191.

9. Contents of the SAF letter as quoted by Dennis Conner and Bruce Stannard in *Comeback: My Race for the America's Cup* (Sydney: Macmillan Company of Australia Pty Ltd, 1987), p. 100.

10. Russell Bowler in a facsimile to the author dated 5 February 1996.

11. Contents of Gianfranco Alberini's letter of 29 September 1996 as quoted by Alan Sefton in *The Inside Story of KZ 7: New Zealand's First America's Cup Challenge, Fremantle, 1986–87* (Auckland: Hodder & Stoughton, 1987), p. 192.

12. Tom Blackaller speaking in support of Dennis Conner in October 1986, as

quoted by Alan Sefton in *The Inside Story of KZ 7: New Zealand's First America's Cup Challenge, Fremantle, 1986–87* (Auckland: Hodder & Stoughton, 1987), pp. 190–191.

13. Tom Blackaller as quoted by Alan Sefton. Ibid., p. 238.

14. In his 'Off the Record' newsletter, July/August 1986, Gary Mull had this to say about the Saint Francis Golden Gate Challenge Twelve-Metre, *Revolutionary 1* (US 61): '... this boat, even if it only manages to get around the course without hurting anyone, will have yachtsmen and particularly yachting writers babbling even more incoherently than usual.'

15. Dennis Conner and Bruce Stannard, *Comeback: My Race for the America's Cup* (Sydney: Macmillan Company of Australia Pty Ltd, 1987), p. 180.

16. Bruce Farr as quoted in *SAIL* (Boston, MA: Primedia Inc. Publications, February 1987), p. 69.

17. Each Twelve-Metre yacht had triangles painted on both sides of its hull. The tips of the triangles had to be showing above the water for each boat to be declared by the measurers to be floating to their legal limit. Representatives from other syndicates could be present at these weigh-ins.

18. Clay Oliver, 'James Clayton Oliver III Interview No. 001' (unpublished, Annapolis, 18 August 1995).

19. 'When the challengers met during October to decide upon possible rule changes, *French Kiss* was one of the supporters of *KZ-7's* legality and never questioned the survey methods or the fibreglass construction at all throughout the entire contest until it came to the time of the semi-finals.' Andrew Johns (Auckland: notes dictated in response to a questionnaire from American James Michael, unpublished).

20. 'Fourteen holes [of very small diameter] were in fact drilled in random places along the hull. At the conclusion of the survey, Ken McAlpine commented on the remarkable degree of consistency which was apparent from the careful construction methods that had been employed in the lamination of the hull.' Andrew Johns (Auckland: notes dictated in response to a questionnaire from US lawyer James Michael, unpublished).

21. Steve Marten, 'Steve Marten Interview No. 001' (unpublished, Auckland, 11 and 12 December 1995).

22. Paul Larsen and Russell Coutts, *Russell Coutts Course to Victory* (Auckland: Hodder Moa Beckett, 1996), p. 73.

23. Ibid.

24. Ibid., p. 76.

25. Iain Morrison, Grant Cubis, Frank Haden, *Michael Fay on a Reach for the Ultimate* (Wellington: Freelance Biographies Ltd, 1990), p. 187.

26. Roy Dickson was an integral crew member in New Zealand's line honours win of the 1966 Sydney–Hobart (*Fidelis*), New Zealand's 1969 One Ton Cup (*Rainbow II*) and 1975 Quarter Ton Cup (*45° South*) wins. He was the winner of the inaugural Citizen match-racing series in 1979. Roy Dickson has also campaigned extensively for New Zealand in the Admiral's and Kenwood Cups and was the sailing coach to New Zealand's 1980 Olympic team.

27. Clay Oliver, 'James Clayton Oliver III Interview No. 001' (unpublished, Annapolis, 18 August 1995).

28. Riblets was the name given to the clear plastic sheets containing minute grooves which were applied to *Stars & Stripes'* hull. Originally designed for use on the outside of aircraft, it was estimated that riblets reduced *Stars & Stripes'* viscous resistance (the drag from water flowing across the surface of a yacht's hull and appendages) by two to four percent. From the thin layer of water around the hull which doesn't move at all to the successive layers away from the hull which move at different speeds, the riblets organise little bursts of turbulence which break down larger areas of disturbance and so reduce the overall drag.

29. Alan Sefton, *The Inside Story of KZ 7: New Zealand's First America's Cup Challenge, Fremantle, 1986–87* (Auckland: Hodder & Stoughton, 1987), p. 277.

30. Over the course of a ten-minute test, *KZ-7* would gain three to three and a half boat lengths over *KZ-5*. But in over eighteen knots of breeze, with its long wings attached, *KZ-7* would increase that to four and a half to five and a half boat lengths, representing a 30 percent increase in the advantage it already had over *KZ-5*.

31. Bruce Farr in 'Bruce Farr Interview No. 008' (unpublished, Annapolis, 10 June 1995) and Bruce Farr in *Bruce Farr and Russ Bowler Interview No. 001* (unpublished, Annapolis, 15 June 1995).

32. Alan Sefton in *The Inside Story of KZ 7: New Zealand's First America's Cup Challenge, Fremantle, 1986–87* (Auckland: Hodder & Stoughton, 1987), p. 271.

33. Audrey Young quoting from the Commission of Inquiry's winebox report in *The New Zealand Herald* (Auckland: Wilson & Horton Ltd, 15 August 1997), p. A4.

34. The Court of Appeal's Justice Thomas, as reported by Jenni McManus in 'Winebox re-opened', *The Independent Business Weekly* (Auckland: Pauanui Publishing, 18 November 1998), p. 6.

35 Jenni McManus, 'Winebox re-opened', *The Independent Business Weekly* (Auckland: Pauanui Publishing, 18 November 1998), p. 1.

36. Fay Weldon, *Splitting* (London: Flamingo, 1995), p 40.

Chapter 24

1. Einar Koefoed, 'Einar L. Koefoed Interview No. 001' (unpublished, Doylestown-Oslo, 31 August 1995).

2. Harunobu Takeda, 'Harunobu Takeda Interview No. 001' (unpublished, Tokyo, 9 September 1995).

3. Kaoru Ogimi, born of a Japanese father and British mother in San Sebastian, Spain, sailed the first Japanese entry in the Sydney–Hobart. That was in 1969, when he skippered his Japanese designed and built *Vago* into twenty-first place. In 1991, he and his associates bought a 180-ton Polish-built windjammer: *Kaisei* ('Ocean Planet'). As well as running his import-export business, Kaoru Ogimi is president of the Sail Training Association of Japan. Some 1000 young Japanese people sail on the *Kaisei* each year. The programme, he says, is of particular relevance to Japanese society as it teaches the participants that the wind and waves are not concerned with 'whatever titles or self-advertisements they may have on the shore, although these two factors determine basic Japanese human relationships.' (*Asahi Evening News*, 4 January 1995).

4. Kaoru Ogimi, 'Kaoru Ogimi Interview No. 001' (unpublished Tokyo, 9 September 1995).

5. Carbon fibre is a spun graphite, developed over the last twenty years by the aerospace industry. The end product is a woven black cloth with the tensile strength of steel at a fraction of its weight. These carbon fibre sheets, impregnated with resin and laid in laminations over featherweight core material such as Nomex honeycomb (hexagonal cells of Kevlar paper glued together with epoxy resin), are baked in computer-controlled ovens to form rigid sandwich hulls.

6. *Nakiri Daio* took 31 days, 19 hours and 6 minutes to complete the 5500-mile race.

7. Fram means 'forward' in Norwegian. It was the name of Fridtjof Nansen's ship, which took him and Roald Amundsen on their Arctic expeditions in the late 1890s and to Antarctica when the Norwegian beat Robert Falcon Scott to the South Pole in 1911. It had been the name of H.R.H.'s nine previous yachts.

8. H.R.H. Crown Prince Harald became H.M. King Harald V on the death of his father on 17 January 1991. In addition to winning a world title, H.M. King Harald V has won several national yachting titles as well as representing Norway in the 5.5-Metre Class at the 1964 and 1968 Olympics and in the Soling Class at the 1972 Olympics.

9. Text of this facsimile is courtesy of the *Fram X* files held by Einar L. Koefoed (unpublished, Oslo).

10. Ibid.

11. Ibid.

12. Mick Cookson, 'Mick Cookson Interview No. 001' (unpublished, Auckland, 18 January 1996).

13. The Beneteau family began business over one hundred years ago building fishing boats. In the early 1970s they built their first production yacht and over the next two decades became the biggest production yacht builder in the world. The Beneteau organisation has four assembly yards, three in France and one in the USA. It is run by a small management team that includes Mme Roux, a Beneteau, and Francois Chalain, a former Jeanneau employee. Beneteau's efficient, cost-effective building procedures produce fast, stylish cruising yachts noted for their quality and value. Beneteau's head office is located in St-Hilaire-de-Riez, France.

14. This was *Xeryus*, an IOR 43-footer, launched in the summer of 1988. In September 1988 it won the Nioulargue in St Tropez, winning all races ahead of *Corum*. In April 1989, it finished second overall in the French selection regatta for the Admiral's Cup team, representing France at Cowes that August. In 1990, *Xeryus* won in La Trinite, came second in Portofino and was leading a Round-the-Island race in July 1990 when it hit a submerged wreck off the Isle of Wight and sank.

15. An ex-lawyer from Paris, Olympic sailor and twice America's Cup skipper for France (1980 and 1983), Bruno Troublé represents the Louis Vuitton organisation in matters of public relations related to their yachting exploits.

16. Bruce Farr, 'Bruce Farr Interview No. 005' (unpublished, Annapolis, 2 June 1995).

17. Timothy Jeffery, *The Champagne Mumm Admiral's Cup: The Official History* (London: Bloomsbury, 1994), p. 169.

18. Ibid., p. 14.

19. The Fastnet Race was the idea of an amateur yachtsman, British-born/USA-based Weston Martyr. It was first raced in 1925, a time when ocean racing was in its infancy. Starting at Cowes and finishing at Plymouth, with turning marks at the Lizard, Fastnet and Bishop Rock, the course has remained unaltered since its inception.

20. Richard Gladwell in 'Firm Admiral's Trend to Bigger Yachts', in *Boating New Zealand* (Auckland: The Magazine Group, August 1987), p. 38.

21. With a long history of sailing in Zeddies, Idle Alongs, Arrows, Cherubs and Flying Dutchman, Bevan Woolley's name first came to prominence as *Wai-Aniwa*'s navigator when it won the One Ton Cup in 1972. Subsequently he was navigator on many boats contesting One Ton, Admiral's, Southern Cross and Clipper Cups and other international blue-water races. In his first season in *Acclaim* he became the national Farr 38 champion, a title he held for several seasons.

22. *Propaganda* (Design 182) was built by TP Cookson Boatbuilders Limited, Glenfield, Auckland.

23. Bevan Woolley, 'Bevan Woolley Interview No. 001' (unpublished, Auckland, 17 February 1996).

24. *Kiwi* (Design 181) was built by Ian Franklin Boatbuilders Ltd, Christchurch, New Zealand.

25. Bevan Woolley, 'Bevan Woolley Interview No. 001' (unpublished, Auckland, 17 February 1996).

26. Ibid.

27. Keith Chapman (deceased), 'Keith Chapman Interview No. 001' (unpublished, Auckland, 17 February 1996).

28. Timothy Jeffery, *The Champagne Mumm Admiral's Cup: The Official History* (London: Bloomsbury, 1994), p. 145.

29. Ibid., p. 152.

30. This *Jamarella* is Design 184, and was sailed for Britain in this regatta by Lawrie Smith and Rodney Pattisson. Although *Propaganda* had beaten *Jamarella* home in the Fastnet, their corrected time positions were reversed, with *Jamarella* finishing in third place, one minute and fifty-seven seconds ahead of *Propaganda*.

31. The winning New Zealand team was:
(a) *Goldcorp*
Designer: Laurie Davidson
Owner: Mal Canning
Crew: Dallas Bennett, David Brooke, Mal Canning, Simon Daubney, Richard Dodson, Aran Hansen, Terry McDell, Terry Smith, Peter Vitali
(b) *Kiwi*
Designer: Bruce Farr & Associates, Inc
Owner: Peter Walker, Admiral's Cup Challenge
Crew: Campbell Brooke, Keith Chapman, Tom Dodson, John Newton, Tony Rae, Owen Rutter, Rob Salthouse, Tom Schnackenberg, Alan Smith, Peter Spackman, Andrew Taylor, Peter Walker, Alan Walbridge
(c) *Propaganda*
Designer: Bruce Farr & Associates, Inc
Owners: Adrian Burr, Peter Tatham

Crew: Brad Butterworth, Ross Fields, Graeme Hanley, Mark Hauser, Peter Lester, Kim Rodgers, Jeremy Scantlebury, Kevin Shoebridge, Mike Spanhake, Bevan Woolley

(d) Management

Team Manager: Don Brooke

Team Captain: Graeme Dagg

Coach: Rod Davis

Sails: John Clinton

32. Erik von Krause in 'One Ton Cup', *Sailing Year: 1987–88 1st Edition: The International Racing Annual* (Richmond, Surrey, UK: Hazleton Publishing, 1987–88), p. 212.

33. The crew of the winning *Fram X* was:
Stein Foyen, Petter Hagelund, Kjell Arne Myrann, Einar Koefoed, Odd Roar Lofterod, H.R.H. Crown Prince Harald, Johan Foyen and Erik Otterstad.

34. Among the top ten finishers in the 1987 One Ton Cup, Farr boats finished first (*Fram X*), second (*Sirius IV*), sixth (*Jamarella*) and eighth (*Mean Machine*).

35. The allowable 'imports' among *Propaganda*'s crew were: John Bertrand (USA), Dennis Grudle (USA) and Kendall Law (UK)

36. Dennis Conner as quoted by Chris Chaswell in 'Acrimonious Cup' in *Sea Spray* (Auckland: Review Publishing Co. Ltd., October 1988), p. 18.

37. The top ten finishers (designer/country/points) in the 1988 OTC were:
1. *Propaganda*/Farr/NZ/142.25
2. *Bravura*/Farr/USA/121.50
3. *Fram X*/Farr/Norway/118.50
4. *Team Cirkeline*/Farr/Denmark/118.50
5. *Sagacious*/Farr/Australia/112.50
6. *Fair Share*/Farr/NZ/110.00
7. *Challenge 88*/Nelson/USA/103.00
8. *Pacific Sundance*/Farr/USA/ 98.00
9. *The Esanda Way*/Davidson/Aust/90.50
10. *Skedaddle*/Reichel-Pugh/USA/84.50
10. *Rush*/Farr/USA/84.50

38. As the oldest boat in the fleet, *Carat VII* would also win the last world 50-Footer championship before the association folded in 1993. By that stage Farr-Bowler designs accounted for half the fleet.

39. At the end of 1989 Bruce and Russell's office staff consisted of Bobbi Hobson and Suzanne Stout, their design staff of Jim Donovan, Ed Frank, recently joined New Zealander Steve Morris, Dave Ramos, Jim Schmicker and Graham Williams. Bryan Fishback had moved from design to Farr International in 1988 to work on project management.

40. The Eastport Maritime building was designed by Jim Miller of Miller and Associates, Alexandra, Virginia.

41. Duncan Spencer in 'Stagg, Farr and Bowler: A Team of Kiwi Magnificos', in *Rags* (Annapolis, MD: April 1995), p. 62.

42. Other awards won by BFA during this period were 'Architecte de L'Annee' awarded by *Bateaux* magazine in 1987 and 1988, and *Yachting Magazine*'s Designer of the Year awarded in 1988.

43. Bob Slaff, President of the Marine's Trade Association of Maryland, as quoted by Dan Guido 'Marine Traders cite Farr for Cup, industrial efforts', *The Capital* (Annapolis, MD: Capital Gazette Newspapers, circa 1988), p. B1.

Interlogue 4

1. See Chapter 25, pages 254 and 255.

Chapter 25

1. Clay Oliver, 'James Clayton Oliver III Interview No. 001' (unpublished, Annapolis, 18 August 1995).
2. Andrew Johns, 'Andrew Johns Interview No. 001' (unpublished, Auckland, 21 November 1995).
3. Malin Burnham as quoted in *America's Cup XXVII Stars & Stripes The Official Record 1988* (San Diego: Dennis Conner Sports, Inc., 1988), p. 22.
4. Bruce Farr, 'Bruce Farr Interview No. 009' (unpublished, Annapolis, 17 June 1995).
5. James Michael, 'Inside The America's Cup The Kiwi Challenge, 1987–1990' (unpublished, USA, 1992, registered WGAW No. 553855), p. 1.
6. Bruce Farr, 'Bruce Farr Interview No. 009' (unpublished, Annapolis, 17 June 1995).
7. *Ranger* was designed by Starling Burgess and Olin Stephens.
8. Steve Marten, 'Steve Marten Interview No. 001' (unpublished, Auckland, 11 and 12 December 1995).
9. Ibid.
10. Ibid.
11. Bruce Farr, 'Bruce Farr Interview No. 009' (unpublished, Annapolis, 17 June 1995).
12. Spreaders are the horizontal structures on a mast required to keep the stays out wide which in turn produce lateral stability for the structure.
13. Bruce Farr as quoted by Shane Kelly in *Sea Spray* (Auckland: Review Publishing Co. Ltd., May 1988), p. 24.
14. For a typical Whitbread Maxi sloop, the tube weight of the mast is approximately 760 kg, the whole rig 1600 kg.
15. Each sensor was a rosette of three gauges positioned in three directions to detect the principal strain.
16. Gennakers are large asymmetrical sails. A cross between a spinnaker and a genoa (or jib), they are used for downwind and cross-wind sailing.
17. Paul Larsen and Russell Coutts, *Russell Coutts Course to Victory* (Auckland: Hodder Moa Beckett Publishers Limited, 1996), p. 71.
18. Andrew Johns as quoted by Shane Kelly in *Sea Spray* (Auckland: Review Publishing Limited, May 1989), p. 26.
19. Dennis Conner as quoted in *America's Cup XXVII Stars & Stripes The Official Record 1988* (San Diego: Dennis Conner Sports, Inc., 1988), p. 44.
20. James Michael, 'Inside the America's Cup The Kiwi Challenge, 1987–1990' (unpublished, USA, registered WGAW no. 553855), p. 168.
21. These events occurred on 8 September 1987 at a press conference in New York

City, the day before Carmen Ciparick was due to hold her first hearings into the MBBC vs SDYC case. Participating in that press conference were Malin Burnham (SAF President), Robert Abrams (New York Attorney-General), Maureen O'Conner (San Diego's mayor) and Ed Koch (New York City's mayor).

22. Tom Ehman as reported on 20 November 1987 by the *San Diego Union*.
23. Dennis Conner as reported at a press conference in New York City on January 15, 1988.
24. Extract from Judge Ciparick's judgement of 28 March 1989, as quoted by James Michael in 'Inside the America's Cup The Kiwi Challenge, 1987–1990' (unpublished, USA, registered WGAW no. 553855), pp. 317–319.
25. Clay Oliver 'James Clayton Oliver III Interview No. 001' (unpublished, Annapolis, 18 August 1995).
26. For examples of Conner's low-life tactics, see pages 221, 222, 223 and 239 above.
26. James Michael, 'Inside the America's Cup The Kiwi Challenge, 1987–1990' (unpublished, USA, registered WGAW no. 553855), p. 336.
27. Ibid., p 335.
28. From 'The First Deed of Gift' — a letter dated July 8, 1857, from donors John C. Stevens (founder of the New York Yacht Club), Edwin A. Stevens, Hamilton Wilkes, J. Beekman Finley and George L. Schuyler to the Secretary of the New York Yacht Club.
29. George Tompkins Jr is a partner in the New York legal firm of Condon & Forsyth. He had worked in the defence of Air New Zealand in the Mount Erebus disaster case and had been retained by Andrew Johns during the 1986–87 America's Cup regatta.
30. James Michael, 'Inside the America's Cup The Kiwi Challenge, 1987–1990' (unpublished, USA, registered WGAW no. 553855), p. 86.
31. Ibid., p. 85.
32. On 31 August 1987, MBBC instituted legal proceedings with the New York Supreme Court, under whose jurisdiction the Deed of Gift fell, seeking that the SDYC accept its sole challenge or forfeit the Cup. The SDYC's petition to amend the Deed of Gift was filed with the New York Supreme Court on 4 September 1987. Both cases were assigned to Judge Ciparick.
33. James Michael, 'Inside the America's Cup The Kiwi Challenge, 1987–1990' (unpublished, USA, registered WGAW no. 553855), p. 87.
34. The Third Deed of Gift was adopted by the New York Yacht Club by way of resolution on October 24, 1887. The Second Deed of Gift had been adopted similarly on December 15, 1881. Both the Second and Third Deeds were drafted by the sole surviving donor, George L. Schuyler, in consultation with a committee appointed by the NYYC, in order to more adequately 'meet the intentions of the donors'.
35. James Michael, 'Inside the America's Cup The Kiwi Challenge, 1987–1990' (unpublished, USA, registered WGAW no. 553855), p. 357.
36. James Michael is a past Commodore of the CCA, past President of the Yacht Racing Association and was the NYYC's point man on the Deed of Gift, Twelve Metre measurement rule and conditions of the match from 1970–1983.
37. The seventeen amici curiae were: Robert N. Bavier Jr, John Bertrand, Alan Bond,

Courageous Sailing Centre of Boston Inc, William P. Ficker, Sir James Hardy, Frederick E. 'Ted' Hood, Gordon Ingate, Ischoda Yacht Club, Warren Jones, Arthur Knapp Jr, Graham Mann, Robert W. McCullough, Emil 'Bus' Mosbacher Jr, Noel Robbins, Jock Sturrock, and R.E. 'Ted' Turner.

38. Russ Bowler, 'Russell Bowler Interview No. 003' (unpublished, Annapolis, 11 June 1995).
39. Justice Thomas (a New Zealand High Court judge) as quoted by James Michael in 'Inside the America's Cup The Kiwi Challenge, 1987–1990' (unpublished, USA, registered WGAW no. 553855), p. 414.
40. James Michael, 'Inside the America's Cup The Kiwi Challenge, 1987–1990' (unpublished, USA, registered WGAW no. 553855), p. 393.
41. Ibid., p. 394.
42. Ibid., p. 396.
43. Ibid., p. 397.
44. Ibid., p. 419.
45. The mathematical formula of the IACC rule reads:
$$\frac{L + 1.25 \times \sqrt{(SAIL\ AREA - 9.8)} \times \sqrt[3]{DISPLACEMENT}}{0.388} = 42\ metres$$

The IACC rule stipulated a minimum/maximum displacement of 16816 kg (16.55 tons)/26000 kg (25.59 tons) and produced a yacht with the following minimum dimensions:
LOA = 22.86 metres (75 ft)
LWL = 17.37 metres (57 ft)
Beam = 5.49 metres (18 ft)
Draft = 3.96 metres (13 ft)
Mast height = 33.53 metres (110 ft)
Sail Area:
Mainsail and jib = 279 square metres (3003 square ft)
Spinnaker = 418 square metres (4499 square feet)
Crew = 16
This new class of boat was approximately twenty percent longer, had a draught 44 percent deeper, a mast 28 percent taller, carried 66 percent more sail area but was 33 percent lighter than yachts of the Twelve-Metre Rule they replaced for contesting the America's Cup.
46. Those on the committee that framed the IACC rule in 1988 were: Philippe Briand, Derek Clarke, John Marshall and Iain Murray. Bruce Farr was not a member.

Chapter 26

1. Peter Blake in *The Steinlager Challenge: How New Zealand Won the 1989–90 Whitbread Round the World Race* (Wellington: Television New Zealand Limited, 1990).
2. Steven O'Meagher, 'The Changing Face of the Whitbread Race' in *Sunday Magazine* (Auckland: Independent News Auckland Limited, 7 January 1990), p. 26.
3. Fiona Rotherham quoting Lion Nathan's chief executive, Kevin Roberts, in

'Sailing Into a Sponsorship Bonanza' in *National Business Review Weekly* (Wellington, Fourth Estate Newspapers, 29 June 1990), p. 18.

4. Ibid., p. 12.

5. In 1973–74 on *Burton Cutter*, Peter Blake scored a DNF, completing only two of four legs. In 1977–78 on *Heath's Condor* he finished fifth on elapsed time and fifteenth/last on corrected time. In 1981–82 on *Ceramco* he finished third on elapsed time and eleventh on corrected time. And in 1985–86 on *Lion New Zealand* he finished second on elapsed time and eighth on corrected time. By way of contrast, Grand Dalton has an eight-day four-hour faster monohull circumnavigation time than Blake and has won the Whitbread twice. In 1981–82 on *Flyer* he finished first on elapsed time and first on corrected time. In 1985–86 on *Lion New Zealand* he finished second on elapsed time and eighth on corrected time. In 1989-90 on *Fisher & Paykel* he finished second on elapsed time (there was no corrected time). In 1993–94 on *New Zealand Endeavour* he finished first on elapsed time (there was no corrected time, although there were two fleets, the Maxis and the W60s). In 1997–98 on *Merit Cup* he finished second on elapsed time (there was no corrected time). There are also other New Zealanders who have won the Whitbread more than once, including Ross Field (first W60 skipper in 1993–94). Mike Quilter, Tony Rae, Cole Sheehan, Kevin Shoebridge and Glen Sowry. Cornelis van Rietschoten of the Netherlands remains the only yachtsman in the world to have won the Whitbread twice as skipper.

6. Cover advertising on *The Steinlager Challenge: How New Zealand Won the 1989–90 Whitbread Round the World Race* (Wellington: Television New Zealand Limited, 1990).

7. Mike Quilter as quoted by Carroll du Chateau in 'Blakey from Bayswater' in *The New Zealand Herald* (Auckland: Wilson and Horton Ltd, 24–25 October 1998), H1.

8. Peter Blake in *The Steinlager Challenge: How New Zealand Won the 1989–90 Whitbread Round the World Race* (Wellington: Television New Zealand Limited, 1990).

9. Bruce Farr email (unpublished, Annapolis, 25 January 1999).

10. Grant Dalton, 'Grant Dalton Interview No. 001' (unpublished, Auckland, 21 December 1995).

11. See Chapter 20, note 14.

12. Grant Dalton, 'Ketch vs Sloop' in *Whitbread: The World's Toughest Race* (Auckland: New Zealand Yachting Magazine, 1989), p 31.

13. When his sponsors failed to produce, Novak's project was handed over to Swedish orthopaedic surgeon Roger Nilson and the ketch called *The Card*. Novak would eventually take up a commission as joint-skipper of the Russian entry *Fazisi*.

14. Bruce Farr, 'Bruce Farr Interview No. 015' (unpublished, Auckland, 22 November 1998).

15. Typically, when a commission is being discussed or let, Bruce and Russell and members of their design team will brainstorm conceptual ideas with the client. They will then further refine these ideas in later sessions with their client and/or among themselves. Bruce assumes responsibility for the lines drawings — none

of which leave the Farr office without his approval — the overall conceptual development, deck, interiors and sail plans, as well as research into other related areas such as tank-testing hull shapes and appendage development. Russell assumes responsibility for structural research and engineering detailing. Various aspects of their respective portfolios are then assigned to the appropriately qualified and experienced members of their design team for actioning/completing. Although Bruce and Russell retain overall responsibility for their respective areas, a large degree of autonomy is exercised within those by the other designers, with interaction between the various participants occurring frequently within their modern open-plan office.

16. Bruce Farr, 'Bruce Farr Interview No. 015' (unpublished, Auckland, 22 November 1998).

17. Carbon fibre is a spun graphite. It has been further developed over the last twenty years through extensive use in the aerospace industry.

18. *Fisher & Paykel's* crew for the 1989–90 Whitbread was: Grant Dalton (31), Keith Chapman (35), Shaun Connolly (28), Ken Davies (32), Eric Dewey (28), Ross Halcrow (22), John Jourdane (45), Lou Jones (31), Alec Rhys (22), Murray Ross (39), Jeff Scott (24), Matthew Smith (32), Grant Spanhake (29), Andrew Taylor (25), Steve Trevurza (35), Tom Warren (27).

19. Alan Sefton, 'Rating Rule Row Rears Up' in *New Zealand Yachting* (Auckland: Tasman Publishing, April 1989), p. 32.

20. See Chapter 28.

21. See Chapter 9, note 4. Also see Chapter 28.

22. Rear Admiral Charles Williams as reported by Alan Sefton in 'Rating Rule Rears Up' in *New Zealand Yachting* (Auckland: Tasman Publishing, April 1989, pp. 33 and 36.

23. In the early morning of 12 November 1989, the English Maxi *Creighton's Naturally* gybed then broached in 60-foot seas. Bart van den Dwey and watch-leader Tony Phillips were swept overboard. Both were recovered. Sadly, Phillips failed to respond to three hours of resuscitation and was buried at sea.

24. The entire *Martella OF* crew were rescued by Fehlmann's *Merit* and Couteau's *Charles Jourdan.*

25. The first three boats in the 1989–90 Whitbread were:
    1. *Steinlager 2* (Farr Maxi ketch) — 128 days, 9 hours, 40 minutes; average speed 10.69 knots;
    2. *Fisher & Paykel* (Farr Maxi ketch) — 129 days, 21 hours, 18 minutes; average speed 10.58 knots;
    3. *Merit* (Farr Maxi sloop) — 130 days, 10 hours, 10 minutes; average speed 10.54 knots.

26. From the video cover of *The Steinlager Challenge: How New Zealand Won the 1989–90 Whitbread Round the World Race* (Auckland: TVNZ Enterprises Limited, 1990).

27. Andrew Johns, 'Andrew Johns Interview No. 001' (unpublished, Auckland, 21 November 1995).

28. *Steinlager 2's* crew for the 1989–90 Whitbread was: Peter Blake (40), Brad Butterworth (30), Godfrey Cray (30), Ross Field (40), Graham Fleury (30), Barry McKay (22), Mark Orams (26), Dean Phipps (25), Mike Quilter (36), Tony Rae

(28), Cole Sheehan (27), Kevin Shoebridge (26), Glen Sowry (27), Craig Watson (23) and Don Wright (30).

29. Designs by Bruce Farr & Associates which finished in the top ten were: *Steinlager 2* (first), *Fisher & Paykel* (second), *Merit* (third), *The Card* (fifth), *Gatorade* (ex *NZI Enterprise*) (eighth), *Belmont Finland II* (ex *UBS Switzerland*) (tenth).

30. Jean White, 'Jean and Dinsmore White Interview No. 001' (unpublished, Annapolis, 26 August 1995).

31. Before they had arranged sponsorship, Kim McDell and John Street got together to form a joint-venture company, Foster McDell Limited, to build and manage a high-performance racing version of the successful Farr 10^{20}. New Zealand had been awarded the World Match-Racing championship for 1990 and it made sense to update the privately-owned Stewart 34 match-racing fleet with a fleet of Farr MRXs. These would be part-owned by a private owner (who would be allowed 200 sailing days a year) and the sponsor (who would be given three-year 'signage' rights), and would be rigorously managed by Foster McDell. Supported by the Royal New Zealand Yacht Squadron, the Farr MRX was also to be used as a training platform to develop the skills of New Zealand sailors for international competition.

32. To date Sea Nymph/McDell Marine has produced a total of 1035 Farr-designed yachts, 727 of which are trailer-sailers and 308 keel-boats. The totals by model are: 415 Farr 6000s; 97 Farr 5000s; 185 Farr 7500; 30 740 Sports; 149 Farr 10^{20}s; 17 Farr MRXs; 47 Farr 12^{20}s; 65 Farr Platu 25s; 5 Farr 41MXs; 10 Farr 940s (119 having previously been built by Marten Marine); 13 Mumm 30s; 2 Farr 11.6s.

Chapter 27

1. An amalgam of reported comments made by Peter Blake in San Diego both prior and subsequent to his appointment as General Manager of New Zealand Challenge Ltd in 1991.

2. Anne Michaels, *Fugitive Pieces* (London: Bloomsbury, 1998), p. 77.

3. Before it was lost in the Tasman Sea, *Smackwater Jack* had been competing in the Southern Cross Cup as part of the New Zealand North team, a team which also comprised the Farr One Tonner *Chick Chack* (Maurice Dykes) and the Lexcen-designed *Anticipation* (Don St Clair Brown). NZ North finished fifth out of eleven teams after *Smackwater Jack* produced a second placing in the Sydney–Hobart race. NZ South, which included the Farr One Tonner *Granny Apple* (Geoff Stagg), the Farr 1104 *Mardi Gras* (Del Hogg) and the S&S-designed *Koamaru* (Brian Millar), finished seventh, all boats returning safely to New Zealand.

4. Michael Fay, 'Sir Michael Fay Interview No. 001' (unpublished, Auckland, 26 August 1995).

5. Bob Fisher in 'Bob Fisher At Large' in *NZ Boating World* (Auckland: Sea Spray Publishing, August 1992), p. 32.

6. Russell Coutts and Paul Larsen *Russell Coutts Course to Victory* (Auckland: Hodder Moa Beckett, 1996), p. 154.

7. At no stage did Bruce Farr suggest that TNZ should use *NZL20* as their challenge

boat for AC95; only, in response to Coutts's questioning, and before the course was changed, that a revamped *NZL20* might do reasonably well. (This note is based on an email message from Bruce Farr to the author.)

8. A windward-leeward course has two legs: windward and running.
9. See Chapter 25, note 45.
10. Bruce Farr, 'Bruce Farr Interview No. 012' (unpublished, Hahei, 25 April 1996).
11. *NZL10* had been sold to the Bengal Bay Syndicate for delivery in May 1991. It remained undelivered when Masakazu Kobayashi's development company collapsed in the Japanese property crash.
12. The 'metre' bow was so-called because it resembled the shape of the bows on Twelve-Metre yachts. These overhanging bows were long and comparatively full in shape.
13. The 'destroyer' or 'IOR' bow is a more conventional bow, sharp, with an almost vertical stem.
14. A slam-dunk in yachting occurs when two boats are sailing upwind towards each other on opposite tacks and one boat is able to pass clear ahead of the other and execute an immediate tack to windward and in close proximity to the other. This effectively pins the second boat to leeward of the leading boat. In this position the leeward boat is unable to tack away without hitting the windward boat or, because of disturbed air, to sail forward of the windward boat through its wind shadow. Thus the leeward boat has to fall off further to leeward in order to create enough room to execute a safe and legal tack, losing a large amount of ground in the process. Typically, when the leeward boat has tacked away, the windward boat is able to also tack and take the leeward boat's wind again in a repeat of the slam-dunk manoeuvre.
15. David Barnes, 'David Barnes Interview No. 001' (unpublished, Auckland, 17 January 1996).
16. Bruce Farr, 'Bruce Farr Interview No. 014' (unpublished, Auckland, 31 January 1998).
17. David Barnes, 'David Barnes Interview No. 001' (unpublished, Auckland, 17 January 1996).
18. Ibid.
19. Barry Stevens and Bob Fisher, *Showdown in San Diego* (San Diego: Frontline U.S.A. Inc., 1992), p. 36.
20. Bruce Farr, 'Bruce Farr Interview No. 010' (unpublished, Annapolis, 17 June 1995).
21. Olin Stephens, 'Olin J. Stephens II Interview No. 001' (unpublished, Putney, Vermont, 1 September 1995).
22. The bowsprit is specifically mentioned in two areas of the IACC rule. The first is under 'spars' where construction limitations set the materials permissible when building one. The second is in its own section where the rule stipulates it must be fastened to the hull mechanically and be capable of being removed — this to stop those who would extend the hull and call it a bowsprit.
23. The afterguy (or preventer) is a rope used to control the spinnaker pole at the end farthest from the mast.
24. The foreguy applies downward pressure to the spinnaker pole.
25. Craig Davis, 'No Clear Favourites as Semifinals Begin' in *Boating New Zealand*

(Auckland: Boating News New Zealand Ltd, April 1992), pp. 34 and 36.

26. Ibid., p. 36.

27. Bob Fisher in 'Fisher At Large' in *NZ Boating World* (Auckland: Sea Spray Publishing, June 1992), p. 42.

28. Ibid.

29. Bruce Farr, Bruce Farr Interview No. 014' (unpublished, Auckland, 31 January 1998).

30. Bob Fisher, 'Fisher At Large' in *NZ Boating World* (Auckland: Sea Spray Publishing, June 1992), p. 42.

31. Bob Fisher, 'Fisher At Large' in *NZ Boating World* (Auckland: Sea Spray Publishing, August 1992), p. 32.

32. The sponsorship arrangement brokered by TVNZ worked as follows: the state broadcaster offered free advertising for the campaign's main sponsors — Steinlager, ENZA, Toyota and Lotto — in exchange for their logos on the boat but also with the rider that they would increase their advertising expenditure with TVNZ in the future. For providing 'free' advertising space on their network, TVNZ was also acknowledged as a sponsor. It would be the same arrangement Team New Zealand would target to fund its America's Cup campaign for 1995.

33. David Barnes won the Tanner and Tauranga Cups in 1973. He was New Zealand Junior OK Champion in 1975 and 76. With Hamish Willcox as forward hand he was New Zealand 470 Champion in 1982, 1983 and 1984 and World 470 Championship in 1981, 1983 and 1984. He helmed *KZ-7* to its World Twelve-Metre title in 1987 and *KZ-1* to its America's Cup victory in 1988. And he was tactician on *Beau Geste* when it won the 1997–98 Sydney–Hobart on corrected time.

34. During the financial year ended 31 March 1989 the BNZ made a NZ$600 million loss. In July 1989, through its publicly listed company Capital Markets Ltd (CML), the Fay Richwhite group (FRG) bought 30 percent (420 million 70-cent shares) of the BNZ from a panicking Labour government. Ensconced as new BNZ directors were Michael Fay and CML director Geoff Ricketts. In 1989, with shareholders' funds reduced to NZ$92 million, the bank's aggregate exposure to FRG was approximately five times that amount. In addition, through a raft of $100 companies, a further NZ$400 million of redeemable preference share deals was approved for CML by the BNZ. More controversy followed in 1990 when the new National government injected NZ$620 million to save the BNZ, purchased NZ$60 million of its shares from FR to help the latter meet their part of the bank's restructuring cost, then paid FR fees for their advice on that restructuring. Eyebrows met hairlines when the privately owned Fay Richwhite merged with CML in August 1990. Small wonder then that the National government would fail to fulfil its 1990 election promise to hold a fully fledged inquiry into the BNZ. Even more so when it was discovered during the Winebox Commission of Inquiry that the BNZ had been involved in the dodgy transactions.

35. See Charles Sturt's testimony to the Winebox Commission of Inquiry as quoted from the transcripts by Ian Wishart in *The Paradise Conspiracy* (Auckland: Howling at the Moon, 1995), p. 135.

36. Michael Fay was knighted for his services to merchant banking and yachting in the Queen's Birthday Honours List on June 16, 1990.

37. Following a challenge by the Hon. Winston Peters of the findings of the Winebox Commission of Inquiry, in November 1998 'the Court of Appeal ruled that winebox commissioner Sir Ron Davison had, arguably, made errors of tax and criminal law that lead to erroneous conclusions in his winebox report ... As Justice Thomas notes ... "the commissioner's apparent errors of law caused him to take a legally benign view of the Magnum transaction [the deal at the centre of the winebox inquiry] which, in turn, led to him being severely critical of Mr Peters ... [who] has a right to have that condemnation based upon a correct appreciation of the law" ... As Justice Thomas sees it, it is difficult to accept that Sir Ron would have concluded that the Magnum transaction was neither illegal nor fraudulent "if he had proceeded on a proper appreciation of the law".' [*The Independent Business Weekly*, (Auckland: Pauanui Publishing, 18 November 1998), pp. 1 and 6.]

38. Tom Clark, 'Sir Thomas Clark Interview No. 001' (unpublished, Auckland, 8 January 1996).

39. Ibid.

40. New Zealand Challenge Press Release, 18 May 1991, from the files of Bruce Farr and Associates, Annapolis, Maryland, USA.

41. Peter Blake, as quoted to the author by a New Zealand Challenge member of staff.

42. It was widely understood among the New Zealand Challenge shore crew that Blake was on a salary of $240,000 per annum, hence the moniker: the Million-Dollar Man.

43. Craig Davis, 'No clear favourites as semifinals begin' in *Boating New Zealand* (Auckland: Boating News New Zealand Ltd, April 1992), p. 36.

44. See page 290 above for an explanation of why the Italian sails were considered to be illegal.

45. Impressions of Peter Blake's management style from a NZC member of staff.

46. Peter Blake to Bruce Farr and Russell Bowler in a facsimile dated June 15 1992, from Blake Offshore Ltd, Emsworth, Hampshire, UK, from the correspondence files of Bruce Farr & Associates Inc, Annapolis, Maryland, USA.

47. Russell Coutts and Paul Larsen, *Russell Coutts Course to Victory* (Auckland: Hodder Moa Beckett Publishers Ltd, 1996), p. 120.

48. Richard Becht, *Champions Under Sail* (Auckland: Hodder Moa Beckett, 1995), p. 65.

49. Andrew Taylor as quoted in the *Sunday Star-Times* (Auckland: Independent News Auckland Ltd, 30 April 1995), p. B9. This statement is sophistic and untrue and typical of Team New Zealand members and other Farr detractors as they sought to lay blame for the entire challenge failure on Bruce Farr's shoulders. Neither Bruce Farr & Associates nor Bruce Farr ran the overall programme nor did all decisions, 'basically' or otherwise, 'go through his office'. Farr and Bowler were in charge of the design programme and, in that role, operated at a level below Fay, Barnes and Clinton in the first instance, and below Fay and Blake after Blake's appointment as GM.

50. Julie Christie, Producer/Director, *Born to Win: The Inside Story of Team New Zealand* (aired by TV One, Auckland, 1995).

51. Ibid.

52. Ibid.

53. Richard Becht, *Black Magic Team New Zealand's Victorious Challenge* (Auckland: Hodder Moa Beckett in association with TVNZ Enterprises, 1995), p. 10.

54. Russell Coutts and Paul Larsen, *Russell Coutts Course To Victory* (Auckland: Hodder Moa Beckett, 1996), p. 119.

55. Clay Oliver is a principal designer with Team New Zealand for America's Cup 2000.

56. Clay Oliver, 'James Clayton Oliver III Interview No. 001' (unpublished, Annapolis, 18 August 1995).

57. Bob Fisher, 'Fisher At Large' in *NZ Boating World* (Auckland: Sea Spray Publishing, August 1992), p. 32.

58. Rod Davis as quoted by Bruce Laybourn in 'Midnight Sacking' in *Boating New Zealand* (Auckland: Boating News New Zealand Ltd, June 1992), p. 18.

59. Ibid.

60. Barry Stevens and Bob Fisher, *Showdown in San Diego* (San Diego: Frontline U.S.A. Inc., 1992), p. 98.

61. See note 43 above.

62. Bob Fisher, 'Fisher at Large' in *NZ Boating World* (Auckland: Sea Spray Publishing, August 1992), p. 32.

63. David Barnes, 'David Barnes Interview No. 001' (unpublished, Auckland, 17 January 1996).

64. There was also the matter of *Il Moro di Venezia*'s spars — given New Zealander Chris Mitchell's apparent long-term absence from Italy while designing them, it seems unlikely that he was properly qualified as an Italian resident under the then America's Cup rules to legally design the Italian mast.

65. *NZL20*'s results for the 1992 LVC, excluding the annulment, were 28 wins, 10 losses; *Il Moro*'s were 26 wins, 12 losses.

66. Peter Blake, as quoted by a former New Zealand Challenge member of staff.

67. It was understood, from information purportedly taken from the NZC wages disk, that Blake's nanny was on a salary of $45,000 per annum. She also had the use of a NZC Toyota Previa wagon.

68. A former NZC member of staff, quoting another former NZC member of staff's public advice to Peter Blake regarding the suggestion that the afterguard be changed.

Chapter 28

1. Shane Kelly in 'Bruce Farr Tells All' in *Boating World* (Auckland: Sea Spray Publishing, April 1993), p. 23.

2. Bruce Farr, 'Bruce Farr Interview No. 004' (unpublished, Annapolis, 27 May 1995).

3. Patrick Shaughnessy, 'Patrick Shaughnessy Interview No. 001' (unpublished, Annapolis 6 June 1995); Mick Price and Patrick Shaughnessy, 'Marmaduke's Chat' (unpublished, Annapolis, 6 June 1995); Patrick Shaughnessy email correspondence March 1999.

4. Ian 'Tink' Chambers, NZCE, was taught to sail at the Wakatere Boating Club on Auckland's North Shore. He first went offshore sailing in the 1971 Sydney–Hobart on *Satanita II*. After working as an engineer on projects in Asia, he returned to New Zealand, helped organise the funding of New Zealand's first

Admiral's Cup team in 1975, was a member of the 1975 New Zealand Southern Cross team and, while in Sydney, met 'Jim' Kilroy. From 1976 to 1980 he sailed all over the world for Kilroy, the last two as skipper of *Kialoa 3*. He met and married Debbie Smith in 1980, and settled in Annapolis where he worked for ex-*Kialoa* mates in the Eastport Erection Company. In 1991 he joined BFA as a procurer of materials for New Zealand's America's Cup Challenge. When the campaign wrapped, 'Tink' joined FI as a broker. His younger sister was Alison Whiting.

5. See 'Meet the Crew of BFA', Appendix 3.
6. See Chapter 12, notes 13 and 14.
7. The six course types are: Windward-Leeward; Olympic; Circular Random (a theoretical one-mile race around a circular course with a stable wind from one direction); Linear Random (a theoretical one-mile race run in a straight line with the wind moving at a steady rate around the compass); Non-Spinnaker (which uses the Circular Random course without the yacht using its spinnaker); and Ocean Race (a mix of beating, reaching and running that changes with the wind speed). There is also a General Purpose handicap (the average of Circular Random's 8 and 12 time allowances).
8. The seven wind speeds, in knots, are: 6, 8, 10, 12, 14, 16, 20.
9. 'An introduction to IMS' in *US Sailing Measurer's Bulletin 93–6* (Rhode Island: US Sailing Association, 2 July 1993), P 7.
10. Bruce Farr as quoted in 'The IMS in 1993 — Coming Of Age', *Seahorse* (Lymington: Fairmead Communications Ltd, June 1993), p. 47.
11. Christian Schmiegelow was tragically killed in a motorcycle crash in November 1992.
12. Lisa Gosselin, editor of Yachting's Race Week News, as quoted in *The Farr Newsletter* (Annapolis: Bruce Farr & Associates, Inc, March 1992), p. 1.
13. Bruce Farr, 'Bruce Farr Interview No. 010' (unpublished, Annapolis, 17 June 1995).
14. Ibid.
15. See chapter 12, note 2.
16. Stuart Alexander, Sailing Correspondent, in *The Independent* (London: Newspaper Publishing PLC, circa 1994).
17. Carroll Marine, Bristol Rhode Island, was established by Barry Carroll in 1984. Using production boat-building methods, Carroll Marine produces quality high-performance racing yachts that have won many national and international regattas and Boat of the Year Awards from *Sailing World* magazine.
18. David Clarke is CEO of US Industries, Inc.
19. David Clarke, 'David H. Clarke Interview No. 001' (Annapolis, MD-Locust, NJ, 18 August 1995).
20. Ian Franklin began business in January 1971 with a $500 overdraft. Since 1975 he has been most closely associated with designers Laurie Davidson and Bruce Farr, moving into composite construction in the mid-eighties at Davidson's suggestion. Among the well-known boats built by Ian Franklin are *Kiwi*, the 43-footer and co-winner of the 1987 Admiral's Cup, and *Bodacious*, the Farr One Tonner SORC winner. Today Franklin Boat Builders Ltd exports Farr IMS 31s (Design 333) all around the world.

21. ILC stands for International Level Rating Class. The inaugural ILC 40 World Championship — the first level rating world championship held under IMS — was sailed in 1995, with Pasquale Landolfi's Farr design *Brava Q8*, helmed by Francesco de Angelis, the winner.

22. Timothy Jeffery, *The Champagne Mumm Admiral's Cup: The Official History* (London: Bloomsbury, 1994), p. 225.

23. Geoff Stagg, 'Geoffrey Stagg Interview No. 001' (unpublished, Annapolis, 28 August 1995).

24. The proposals came from: John Bertrand's US Racing Group and DDV Sailing offering a Reichel/Pugh design; a Scott Jutson design from Australia's Northshore Yachts; Jeanneau; X-Yachts; a Rob Humphreys design from Devonport Yachts; and the Farr IMS 36 from Farr International.

25. The committee was: John Dare (chairman), John Bourke (ORC chairman), Stuart Quarrie (RORC Rear Commodore), the RORC's Alan Green, Don Genitempo of the USA and David Kellet from Australia.

26. John Dare as quoted in a Farr International press release.

27. Dan Dickison in 'The Mumm Chronicles', *Sailing World* (Newport, RI: Cruising World Publications, July 1994), p 32.

28. The eight three-boat teams came from: Germany, Great Britain, Hong Kong, Ireland, Italy, Scandinavia, South Africa and the USA.

29. Bill Schanen, 'Lawbreaking Boats Push the Sailing Fun factor' in *Sailing Magazine* (Port Washington, WI: Port Publications Inc, October 1993), editorial.

30. Ibid.

Chapter 29

1. Richard Gladwell, 'Beyond the Laylines' in *Boating World* (Auckland: Sea Spray Publications, March 1994), p. 71.

2. Bob Fisher and Barry Pickthall, *Ocean Conquest: The Official Story of the Whitbread Round the World Yacht Race Past, Present and Future* (London: Little, Brown, 1993), p. 134.

3. See Chapter 12 on the CCA and RORC rules.

4. Stephen Pizzo (spp@quokka.com) reports that by the end of the 1989–90 Whitbread Race the Whitbread Website was receiving over 100,000 visitors per day. In total, for the duration of the race, it received one million unique visitors (those counted only once no matter how many times they frequent the site).

5. The rating formula for the IOR Maxi class was:

$$\frac{0.13 \times (L \times \sqrt{S})}{\sqrt{(B \times D)} + 0.25L + 0.2\sqrt{S} + DC + FC} = EPF \times CGF$$

L = Length (of a special sort)
S = Sail area
B = Beam
D = Draft
DC = Draft correction
FC = Freeboard correction
EPF = Engine propeller factor
CGF = Centre of gravity factor

6. Richard Gladwell quoting Bruce Farr in 'Beyond the Laylines' in *Boating World* (Auckland: Sea Spray Publications, March 1994), p. 70.

7. New Zealander Ross Field, an ex-detective sergeant, crewed on *NZI Enterprise* in the 1985–86 Whitbread and was a *Propaganda* crew member in the successful New Zealand Admiral's Cup team of 1987. He was a watch captain on *Steinlager 2* in the 1989–90 Whitbread. He won the Melbourne-to-Osaka two-handed race on *Nakiri Daio* in 1991. And with Kaoru Ogimi's help, he obtained sponsorship from Yamaha for the 1993–94 Whitbread.

8. Ivor Wilkins quoting Ross Field in 'Quick Deceiver' in *New Zealand Yachting* (Auckland: Tasman Publishing, June 1992), p. 35.

9. Dennis Conner quoted in 'New & Views' in *Boating World* (Auckland: Sea Spray Publications, August 1993), p. 10.

10. Chris Dickson — three times World Youth Champion and three times World Match-racing Champion — had gained the backing of Chuo Advertising. In his own version of a two-boat campaign, he ordered one design from West Australian John Swarbrick, designer of America's Cup defender *Kookaburra*, and one from BFA. After extensive two-boat testing, he selected the Farr design for the 93–94 Whitbread and named it *Tokio*.

11. *US Women's Challenge* would later change its name to *Heineken* when Dawn Riley took over as skipper from Nance Frank.

12. Steve Marten as quoted by Eugene Doyle in *In Magellan's Wake: The 1993–94 Whitbread Round the World Race* (NZ: Toyota New Zealand Limited, 1993), p. 5.

13. It was English sailor Lawrie Smith, *Fortuna's* skipper in this race, who called the ketches 'cheat boats' because of their clipper bows. In his opinion the extreme shape of their bows put them over the rated limit permitted by IOR. At issue was IOR Rule 311.1 which said that the two forward girth stations were not permitted to lie within a concave curve. BFA's interpretation was that this measurement clarification, issued in 1980, prevented bows with bulbs extending forward beneath the water in order to gain unrated waterline length. But it took until April 1993, just five months before the Whitbread was due to start, before the ORC's Chief Measurer cleared the proboscis bows from attracting rating penalty.

14. Handicap winner of this Sydney–Hobart was Syd Fischer's ninth *Ragamuffin*, the Farr 50-footer formerly called *Will*. Overall IMS race-winner was *Assassin*, a Farr 40-footer. Farr designs also filled first four places of the IOR section of the fleet.

15. The first three finishers in the '93 Fastnet were, in order: *Galicia 93 Pescanova*, *Winston* and *Intrum Justitia*. Fourth, fifth and sixth to cross the line were *New Zealand Endeavour*, *Merit Cup* and *La Poste*.

16. 'Hat Warning Prior to Whitbread '93' by Geoffrey Russell Bowler, first published in *Endeavour: Winning the Whitbread* by Grant Dalton and Glen Sowry (Auckland: Hodder & Stoughton Ltd, 1994), p. 64.

17. When *Fortuna's* winged mizzen fell over the side and Roger Nilson injured his knee on Leg One, Lawrie Smith found himself the skipper of *Intrum Justitia*.

18. Ivor Wilkins in 'What a Finish' in *Boating World* (Auckland: Sea Spray Publications, February 1994), p. 68.

19. First was *NZ Endeavour* (120 days, 5 hours, 9 minutes, 23 seconds, at an average speed for the race of 11.21 knots); second was *Yamaha* (120:14:55:00, at an average 11.17 knots); third was *Merit Cup* (121:02:50:47, at an average 11.12

knots); fourth was *Intrum Justitia* (121:05:26:26, at an average 11.11 knots); fifth was *Galicia 93 Pescanova* (122:06:12:23, at an average 11.02 knots); sixth was *Winston* (122:09:32:09, at an average 11.0 knots); seventh was *La Poste* (123:22:54:58, at an average 10.87 knots); eighth was *Tokio* (128:16:19:48, at an average 10.47 knots); ninth was the Bouvet-Petit design *Brooksfield* (130:04:29:27, at an average 10.35 knots) and tenth was *Hetman Sahaidachny* (135:23:17:52, at an average 9.9 knots). *Steinlager 2* had taken 128:09:40:30 to complete the same course in 1989–90.

20. Carroll du Chateau in 'Blakey from Bayswater' in *The New Zealand Herald* (Auckland: Wilson and Horton Ltd, 24–25 October 1998), H1.

Chapter 30

1. M. Scott Peck, *People of the Lie: The Hope for Healing Human Evil* (New York, NY: Touchstone Books, 1985), p. 241.

2. *The Strip* (Auckland: Local Publications Eastern Ltd, February 1994), Issue 6, p. 19.

3. John Matheson, 'Farr Warns Rivals' in *Sunday News* (Auckland: Independent News Ltd, 12 February 1995), p. 58.

4. Andrew Sanders, 'Conner Helped Kiwis Win the Cup' in the *Sunday Star-Times* (Auckland: Independent News Auckland Ltd, 13 October 1996), B11.

5. This statement is sophistic and untrue. Russell Coutts did not explain at this meeting (or his second meeting with the designers in December 1992) 'how the campaign would be managed'.

6. This statement is sophistic and untrue. Bruce Farr did not send this letter nor was it in the form of a report card. See Russell Bowler's fax to Peter Blake dated December 15, 1992.

7. Russell Coutts and Paul Larsen, *Russell Coutts Course to Victory* (Auckland: Hodder Moa Beckett Publishers Limited, 1996), p. 154.

8. Suzanne McFadden, 'Farr Tipped to Join Dickson Cup Challenge' in *New Zealand Herald* (Auckland: Wilson and Horton Ltd, circa February 1994).

9. Jan Corbett, 'Spar Wars' in *Metro* (Auckland: Australian Consolidated Press NZ Ltd, January 1995 No. 163, published 20 December 1994), p. 72.

10. This statement is sophistic. The reason BFA did not make a commitment to TNZ before the nationality deadline was because TNZ made no offer to BFA to which it could commit.

11. Russell Coutts and Paul Larsen, *Russell Coutts Course to Victory* (Auckland: Hodder Moa Beckett Publishers Limited, 1996), p. 154.

12. This statement is sophistic and untrue.

13. Andrew Sanders, 'Conner helped Kiwis Win the Cup' in the *Sunday Star-Times* (Auckland: (Auckland: Independent News Auckland Ltd, 13 October 1996,) B11.

14. Ibid.

15. Russell Coutts and Paul Larsen, *Russell Coutts Course to Victory* (Auckland: Hodder Moa Beckett Publishers Limited, 1996), p. 154.

16. Glen Sowry (script), *Black Magic: The Team New Zealand Story* (Wellington: Television New Zealand Limited, 1995).

17. This statement is sophistic and untrue.

18. This statement is a nonsense.

19. See the facsimile dated June 15 1992 from Blake Offshore Ltd, Emsworth, Hampshire, UK.

20. Russell Coutts and Paul Larsen, *Russell Coutts Course to Victory* (Auckland: Hodder Moa Beckett Publishers Limited, 1996), p. 154.

21. This statement is sophistic and untrue. Coutts did not outline 'our thinking [etc]' at either one of his two meetings with the designers in December 1992.

22. This statement is sophistic and untrue on several levels. (a) The letter referred to here was not written by Bruce Farr as the wider text implies. (b) It was not sent in response to Coutts' first meeting of December 9, 1992 with the designers. It was sent from Russell Bowler to Peter Blake in response to an accusatory facsimile from Peter Blake. (c) The letter was not in the form of a 'report card'. (d) 'The [Bowler] letter that gave us a D' (actually a 'D-') did not in fact give 'letter grades' plural. Nor did it suggest 'the grade might be downgraded further'. This occurred in a facsimile written by Russell Bowler to Russell Coutts dated December 28 1992, which was copied to Peter Blake. On the contrary, the so-called 'report card' letter described TNZ's early efforts as 'enthusiastic and commendable' while offering advice to make the campaign winnable, advice later used by TNZ. For the full text of the 'report card' letter see the Bowler-Blake fax dated December 15, 1992.

23. Russell Coutts and Paul Larsen, *Russell Coutts Course to Victory* (Auckland: Hodder Moa Beckett Publishers Limited, 1996), p. 154.

24. The meeting Coutts refers to in the third paragraph of p. 154 of *Course* took place on 9 December 1992 after he and Blake had decided not to use Farr and Bowler and some five months before 'the May 6 deadline was fast approaching'. The so-called 'report card' letter was sent neither following this meeting nor as a result of it. (See notes 21 and 22 above.) A further meeting between Coutts and the designers did take place on Christmas Eve 1992. But not even during this meeting did Coutts outline 'our thinking and objectives'. It was, however, following this meeting that Russell Bowler, unaware that he was being deceived by Coutts and Blake and that BFA had been discounted as TNZ's designers, criticised what he considered to be the then haphazard management approach of TNZ. (See Bowler fax to Coutts of 28 December 1992.)

25. Russell Coutts and Paul Larsen, *Russell Coutts Course to Victory* (Auckland: Hodder Moa Beckett Publishers Limited, 1996), p. 140.

26. See notes 30 and 32 below.

27. This statement is untrue. Coutts neither 'questioned some of the design aspects' nor 'explained what the consensus view was' to Farr (or Bowler) at either of the two December meetings he had with the designers.

28. Russell Coutts and Paul Larsen, *Russell Coutts Course to Victory* (Auckland: Hodder Moa Beckett Publishers Limited, 1996), p. 154.

29. The races Blake had won were: (a) a line honours win in the 1979 Fastnet on *Heath's Condor*; (b) a line and handicap win in the 1980 Sydney–Hobart on *Ceramco*; (c) a line honours win in the 1984 Sydney–Hobart on *Lion New Zealand*; (d) a line and handicap win in the 1989–90 Whitbread on *Steinlager 2*. Blake's Around Australia win with navigator Mike Quilter is discounted here as 'the main opposition was actually the weather, rather than the motley bunch of

craft that followed in the wake of his multihulled *Steinlager I.*' (Bob Fisher and Barry Pickthall, *Ocean Conquest: The Official Story of the Whitbread Round The World Race Past, Present and Future* [London: Little, Brown, 1993], p. 116). In the 1993 Jules Verne race, Blake was a co-skipper with the first man to sail non-stop around the world: Englishman Robin Knox-Johnston. They and their crew scored a DNF in *Enza*, losing to Bruno Peyron in *Commodore Explorer*. In 1994, Blake, Knox-Johnston and their five-man crew became the second (not the first) group to sail around the world in under 80 days, a voyage in which Blake spent fourteen percent (ten days) of the journey confined to his bunk with bruised ribs, unable to join his watch team. Their time of 74 days, 22 hours, 17 minutes and 22 seconds, for which they were awarded the Jules Verne Trophy in Paris, was beaten four years later by Olivier de Kersauson with a time of 71:14:18:08.

In his text on Peter Blake in '100 Champions of Sport', in the *Sunday Star-Times* (Auckland: Independent News Auckland Ltd, 12 April 1998, p. 3), Ron Palenski perpetuates the myth of Blake's racing success with his inanely foolish statement: 'If there is something in yachting that Peter Blake has not won, it's either not worth winning or he hasn't heard about it'. The reality is that even in his main event, the Whitbread Round the World Race, Peter Blake has a record inferior to Grant Dalton's, and that after sailing 250,000 more miles than his nine-year-younger rival. It is also difficult to understand how Palenski could credit Blake with organising and leading New Zealand's successful winning of the America's Cup, when Russell Coutts' *Course to Victory* makes it clear that Coutts and Schnackenberg were the key organisational figures for the 1995 event — a thesis given credence by Blake's facsimile of 15 December 1992 (see note 49 below), with his organisational failure in the 1992 America's Cup, and with his involvement in two Jules Verne attempts in 1993 and 1994. Further, in the *Sunday Star-Times* article 'Skipper Sinks Talk of Team NZ Rift' (30 May 1999, p. B11), Coutts appears to underscore this thesis by intimating that Blake's only significant contribution to TNZ was and is in liaising with the sponsors and having an iconic role.

The eagerness to bolster Blake's meagre record of wins has gone to absurd lengths. The New Zealand Challenge Press Release of 18 May 1991 listed his having had line honours wins on *Steinlager 2* in the 1989 Channel and Fastnet Races when *Steinlager 2* was not an official entrant and officially won neither. A television documentary aired in New Zealand after the 1995 America's Cup false-ly claimed he'd won the P Class Tanner Cup. New Zealand's National Maritime Museum in Auckland states that he won line honours on *Steinlager 2* in the 1989 Sydney–Hobart, a race won by Alan Bond's *Drumbeat* (line honours) and Lou Abrahams' *Ultimate Challenge* (handicap). In fact, it was a race neither Blake nor *Steinlager 2* could have competed in as both were sailing in the 1989–90 Whitbread, the line and handicap win the National Maritime Museum correctly states on the same board as its impossible claim. And the *North Shore Times Advertiser* (Auckland: Suburban Newspapers, 22 April 1999, p. 5), falsely report-ed that Blake 'skippered the crew which won the [America's] Cup in 1995' and 'was skipper of the America's Cup crew which won the cup in 1995'. (*North Shore Times Advertiser*, 4 May 1999, p. 1.) Perpetuating this myth might produce good copy, please Blake's sponsors and Roundtable members, and falsely bolster national self-esteem. But not only is it flawed. It does an enormous dis-service to the many New Zealanders who have won many more races and con-tributed far more than Peter Blake to making New Zealand yachting what it is today.

The myths being foisted on the public reached sublimely silly heights when Blake was awarded an Honorary Doctorate of Commerce by Massey University's Albany campus in April 1999 'in recognition of pioneering commer-cial sponsorship in sport'. (Auckland: North Shore Times Advertiser, Suburban Newspapers, 4 May 1999, p. 1.) Not only did the pioneering of commercial sponsorship in New Zealand keel-boat racing arguably begin with Chris Bouzaid's successful One Ton Cup (OTC) campaign in 1969, more than a decade before Blake's first commercially sponsored campaign, *Ceramco*. Blake, Martin Foster et al. failed to raise the necessary commercial sponsorship to fund Blake's proposed campaign for the 1977–78 Whitbread in a Bruce Farr-designed 65 ft ocean racer. In addition to Bouzaid's successful OTC campaign, there was also widespread commercial sponsorship of centreboarders from the mid-to-late sixties on both a national and interdominion level. Bruce Farr was one of the early yachtsmen to gain commercial sponsorship — this in 1967 from a builder for his first *Miss Beazley Homes*. There was Russ Bowler's Q Class *Jennifer Julian* in 1969 and his world champion *Jennifer Julian* Cherub in 1970, both sponsored by Evan Julian. The first widespread defiance of the IYRU's rule 26, which states that the 'hull, crew or equipment of a yacht owned or sponsored ... by a group or organisation shall not display any wording or emblem which specifically relates to such owner or sponsor', was by the eighteen-footer fleet on both sides of the Tasman. Among those New Zealanders who secured spon-sorship for their eighteen-footers during the 1960s and 70s were Gary Banks with his two *Cool Leopards*, the Fleet brothers with *Miss UEB*, Dave Kean with *Miss IGA*, Frank Blackburn with *Guinness Lady*, Wayne Innes with *Captain Morgan*, the McDell brothers with *TraveLodge*, the Lidgard brothers with *Smirnoff*, Russ Bowler with *Rank Xerox*, *Brownbuilt*, *Benson & Hedges* and *Stubbies*. There was the substantial sponsorship effort required to get New Zealand's first Admiral's Cup team to England for the 1975 regatta, and a grow-ing sponsorship of One Tonners in the latter part of the seventies: *Smir-Noff-Agen*, *Mr JumpA*, *The Red Lion*. Among the considerable number of pioneers of commercial sponsorship in New Zealand yachting, Peter Blake is nowhere to be seen. It was Tom Clark's Ceramco Group that initially underwrote *Ceramco* for the 1981–82 Whitbread. Clark is also credited by Blake with assembling the *Lion New Zealand* sponsorship package. (See *Lion: The Round the World Race with Lion New Zealand*, p. 13.)

Not only was Peter Blake not a pioneer of commercial sponsorship in sport in New Zealand. Massey University's award ignores the herculean efforts of the person who is arguably the real pioneer of major commercial sponsorship in New Zealand yachting: Trevor Geldard. His company, Healing Industries, was responsible for the 'sponsorship of countless club and class events, [youth train-ing, Olympic sailing], Admiral's Cups [and] Whitbread challenges ... [and] in 1985 announced NZ's biggest sporting sponsorship of $1 million for Admiral's Cup/America's Cup campaigns'. (*Sea Spray*, December 1990/January 1991,

p. 29.) Without the latter it is doubtful that the 1987 America's Cup campaign would have progressed through Aussie Malcolm's hands to Michael Fay's desk and become the catalyst for the flood of nationwide sponsorship for yachting that followed. It was Grant Dalton's *New Zealand Endeavour* Whitbread campaign that stitched together the innovative sponsorship deal that became the model for the New Zealand Challenge's 1992 America's Cup campaign, the self-same structure Team New Zealand used to fund its successful America's Cup bid in 1995. It was also one of the sponsors from that group that had previously funded the two Blake-Knox-Johnston Jules Verne attempts.

Over the last ten years, among the most prominent New Zealand yachtsmen to secure international sponsorship have been Ross Field, Chris Dickson, and Grant Dalton, all three of whom have brought their projects back to New Zealand, serving both their sponsors and New Zealand yachting well.

Is it not the measure of the man that, knowing he did not pioneer commercial sponsorship in sport, Peter Blake should accept an honorary doctorate for doing just that? The extent to which the New Zealand public has been duped is evidenced by 'the standing ovation [Blake received] from more than 1500 people at the [capping] ceremony'. (*North Shore Times Advertiser*, Tuesday May 4 1999, p.1.) But given the self-serving sellers of these myths — like 'Massey sports management lecturer and Team New Zealand member Dr Mark Orams [who] told graduates Sir Peter was revered by New Zealanders because he ... expresses through his actions what is best for all of us' (Ibid., p. 4.) — that is hardly surprising.

30. Russell Coutts and Paul Larsen, *Russell Coutts Course to Victory* (Auckland: Hodder Moa Beckett Publishers Limited, 1996), p. 153.

31. Apart from a brief involvement with the Mercury Bay Boating Cub's 1988 challenge for the America's Cup, and in spite of Bruce and Russell's attempts to get him involved in the 1992 America's Cup challenge, Tom Schnackenberg had worked only for non-New Zealand America's Cup syndicates prior to the RNZYS's Team New Zealand Challenge in 1995.

32. Russell Coutts and Paul Larsen, *Russell Coutts Course to Victory* (Hodder Moa Beckett Publishers Limited, 1996), p. 144.

33. The fraud became full-blown with (a) Richard Becht (text), *Black Magic Team New Zealand's Victorious Challenge: Official Team New Zealand Souvenir* (Auckland: Hodder Moa Beckett in association with TVNZ Enterprises, 1995); (b) Glen Sowry (script), *Black Magic: The Team New Zealand Story* (Wellington: Television New Zealand Limited, 1995); (c) Russell Coutts and Paul Larsen, *Russell Coutts Course to Victory* (Auckland: Hodder Moa Beckett, 1996); (d) Julie Christie, Producer/Director, *Born to Win* (Aired on TV One in 1995).

34. Glen Sowry (script), *Black Magic: The Team New Zealand Story* (Wellington: Television New Zealand Limited, 1995).

35. Douglas Myers as quoted by Carroll du Chateau in 'Blakey from Bayswater' in *New Zealand Herald* (Auckland: Wilson and Horton Ltd, October 24–25, 1998), p. H1.

36. Facsimile from Blake Offshore Ltd, Emsworth, Hampshire, UK, from the correspondence files of Bruce Farr & Associates Inc, Annapolis, Maryland, USA.

37. Facsimile from the correspondence files of Bruce Farr & Associates Inc, Annapolis, Maryland, USA.

38. Facsimile from Blake Offshore Ltd, Emsworth, Hampshire, UK, from the correspondence files of Bruce Farr & Associates Inc, Annapolis, Maryland, USA..

39. Facsimile from the correspondence files of Bruce Farr & Associates Inc, Annapolis, Maryland, USA.

40. Facsimile from Blake Offshore Ltd, Emsworth, Hampshire, UK, from the correspondence files of Bruce Farr Associates Inc, Annapolis, Maryland, USA.

41. In a 1995 issue, *The Independent Business Weekly* (Auckland: Pauanui Publishing) ran a story about the business structure and composition of TNZ. Part of that was reprinted in Anthony Molloy's *Thirty Pieces of Silver* (Auckland: Howling at the Moon, 1998) p. 2, as follows:

    'Team New Zealand Ltd, a $100 company, was incorporated in May 1993. One of these shares is held by Richard Green, a tax lawyer at Auckland solicitors firm Russell McVeagh. The other 99 are held by Team New Zealand Trustee Ltd. There are five directors: Green, fellow Russell McVeagh solicitor John Lusk; Fay Richwhite corporate financier Jim Hoare; Coopers & Lybrand partner George France; and former Ceramco supremo ... Sir Tom Clark. Team New Zealand's registered office is in the offices of Russell McVeagh and, as of the end of November ... [1994], it had $1.7 million of debt. Team New Zealand Trustee Ltd is a $10 company, also founded in May 1993. It owns Team New Zealand Ltd and has the same directors — with the exception of Sir Tom Clark — as Team New Zealand Ltd — two lawyers and two accountants'.

42. Facsimile from Blake Offshore Ltd, Emsworth, Hampshire, UK, from the correspondence files of Bruce Farr & Associates Inc, Annapolis, Maryland, USA.

43. Details from a Bruce Farr file note from his meeting with Russell Coutts on 9 December 1992, from the files of Bruce Farr & Associates Inc, Annapolis, Maryland, USA.

44. Russell Coutts and Paul Larsen, *Russell Coutts Course to Victory* (Auckland: Hodder Moa Beckett Publishers Limited, 1996), p. 154.

45. Ibid., p. 140.

46. Ibid., p. 154.

47. Blake and Robin Knox-Johnston were preparing for the Jules Verne race in *Enza*, departing on 31 January 1993. The race would be won by Bruno Peyron and his crew in *Commodore Explorer*, the first sail-boat to circumnavigate the world in under 80 days. Peyron's winning time was 79:06:15:56.

48. Facsimile from the correspondence files of Bruce Farr & Associates Inc, Annapolis, Maryland, USA.

49. Facsimile from Blake Offshore Ltd, Emsworth, Hampshire, UK, from the files of Bruce Farr & Associates Inc, Annapolis, Maryland, USA.

50. Facsimile from the files of Bruce Farr & Associates Inc, Annapolis, Maryland, USA.

51. Facsimile from the files of Bruce Farr & Associates Inc, Annapolis, Maryland, USA.

52. Facsimile from Blake Offshore Ltd, Emsworth, Hampshire, UK, from the correspondence files of Bruce Farr & Associates Inc, Annapolis, Maryland, USA.

53. Russell Coutts and Paul Larsen, *Russell Coutts Course to Victory* (Auckland: Hodder Moa Beckett Publishers Limited, 1996), p. 154.

54. Facsimile from the correspondence files of Bruce Farr & Associates Inc, Annapolis, Maryland, USA.

55. From a facsimile from Russell Coutts Yachting, San Diego, USA, from the correspondence files of Bruce Farr & Associates Inc, Annapolis, Maryland, USA.

56. Ibid.

57. Bruce Farr, 'Bruce Farr Interview No. 015' (unpublished, Auckland, 22 November 1998).

58. Facsimile from the correspondence files of Bruce Farr & Associates Inc, Annapolis, Maryland, USA.

59. From a facsimile from Alan Sefton to Bruce Farr, from the correspondence files of Bruce Farr & Associates Inc, Annapolis, Maryland, USA.

60. Russell Coutts and Paul Larsen, *Russell Coutts Course to Victory* (Auckland: Hodder Moa Beckett Publishers Limited, 1996), p. 162.

61. Facsimile from Team New Zealand, Auckland, New Zealand, from the correspondence files of Bruce Farr & Associates Inc, Annapolis, Maryland, USA.

62. Facsimile from the correspondence files of Bruce Farr & Associates Inc, Annapolis, Maryland, USA.

63. Facsimile from Blake Offshore Ltd, Emsworth, Hampshire, UK, from the files of Bruce Farr & Associates Inc, Annapolis, Maryland, USA.

64. Facsimile from Team New Zealand, Auckland, New Zealand, from the correspondence files of Bruce Farr & Associates Inc, Annapolis, Maryland, USA.

65. Peter Blake as quoted by Bob South in the *Sunday Star-Times* (Auckland: Independent News Auckland Ltd, 29 May 1994).

66. Facsimiles from the correspondence files of Bruce Farr & Associates Inc, Annapolis, Maryland, USA.

67. Facsimiles from the correspondence files of Bruce Farr & Associates Inc, Annapolis, Maryland, USA.

68. Facsimiles from the correspondence files of Bruce Farr & Associates Inc, Annapolis, Maryland, USA.

69. Russell Coutts and Paul Larsen, *Russell Coutts Course to Victory* (Auckland: Hodder Moa Beckett Publishers Limited, 1996), p. 154.

70. Ibid.

71. Refer note 16 above.

72. Russell Coutts and Paul Larsen, *Course to Victory* (Auckland, Hodder Moa Beckett, 1996), p. 155.

73. Design 336 was originally built by Mark Lindsay of Gloucester, Massachusetts, USA. Fifteen have been built in the original form, with a variation of the design also having been built by Cookson's of Auckland, NZ.

Chapter 31

1. David Barnes, 'David Barnes Interview No. 001' (unpublished, Auckland, 17 January 1996).

2. Philippe Briand as quoted in an interview with Marcus Hutchinson in 'The Expert's Expert: Racing Yacht Designers' in *Observer magazine* (GB: 17 February 1991), p. 10.

3. Olin Stephens in a letter to the author (unpublished, New Hampshire, 12 December 1995).

4. The crew of *TAG Heuer* averaged 27 years of age. They were: Chris Dickson

(skipper/helm), Peter Lester (tactician), Jon Bilger (navigator), Greg Flynn (runner/floater), Mike Sanderson (runner/floater), Ian Stewart (runner/floater), Steven Cotton (runner/floater), Gavin Brady (runner), Kelvin Harrap (traveller), Jim Close (trimmer), Grant Loretz (trimmer), Sean Clarkson (grinder), Graham Fleury (grinder), Denis Kendall (pit), David Brooke (mast), Rodney Ardern (mid-bow), Kevin Batten (bowman), Brad Webb (reserve bow), Brad Jackson (reserve trimmer), Chris Salthouse (reserve trimmer).

Details taken from Sarah Ell, 'Dickson Watches Clock as Time Draws Near' in *Boating New Zealand* (Auckland: The Magazine Group, January 1995), pp. 70–71.

5. Peter Blake as quoted by Dave Gendell, *Rags* (Annapolis, MD: June 1995), p. 83.
6. Andrew Sanders, 'US spy boat stalks rival Black Magic', *Sunday Star-Times* (Auckland: Independent News Ltd, 27 June 1999), pp. A1 and A2.
7. Team New Zealand's acts of intimidation continued with their ramming of 'an America's Cup challenger craft ... [that] was playing by the rules ... Chris Main from Japan's Nippon Challenge syndicate told *The New Zealand Herald* that he feared for his safety when a Team NZ chase boat rammed his craft in the Hauraki Gulf last week. Main, a 24-year-old Aucklander, said he had been abiding by cup protocol, which stipulated that craft should not venture within 200 m of competitors' yachts. The incident followed jibes about his sailing for the Japanese ... [Russell] Coutts said on radio yesterday that Main's rubber boat "wasn't within the 200 m rule at the time" of the incident. However, both the Nippon Challenge and the New York Yacht Club syndicates [the latter involving Halcrow] had been doing "what we would deem as shadowing".' [*The New Zealand Herald* (Auckland: Wilson and Horton Ltd, 10 February 1999), A3. Said Coutts of the incident when speaking to radio sports-talk host Murray Deaker on the ZB Network on Sunday afternoon February 14, 1999: 'I cannot understand how a Kiwi could do this to us', naming Halcrow and Main as two New Zealanders working for foreign challenge syndicates allegedly spying.
8. See Douglas Myers on Peter Blake in Carroll du Chateau's 'Blakey from Bayswater' in *The New Zealand Herald* (Auckland: Wilson and Horton Ltd, October 24–25, 1998), p. H3.
9. From a 'Steinlager' advertisement in *Boating New Zealand* (Auckland: The Magazine Group, January 1995), p. 71.
10. For their defence of the America's Cup Team New Zealand has employed two principal designers: Laurie Davidson and American Clay Oliver who, since 1988, had worked extensively for BFA. Doug Peterson, ex *America³* and principal designer of *NZL32* with Laurie Davidson, has been hired by the Italian Prada syndicate.
11. Ross Halcrow was a Team New Zealand trimmer and member of the winning team in 1995. He subsequently moved to Canada, married a Canadian, and joined Young America. Soon after this announcement was made a campaign was begun by Russell Coutts and Peter Blake to discredit him. Said Blake to the Takapuna Rotary Club, projecting fault and blame on his target victim: 'People are allowed to do what they want ... Halcrow was very important to our team in 1995. I'm disappointed he made a promise to Russell [Coutts, the skipper], said he was on board and that he's left after Russell outlined our whole design and

test programme.' (Felicity Anderson quoting Peter Blake in 'Call for Unity in Cup Challenge' in *North Shore Times Advertiser* (Auckland: Suburban Newspapers, 14 February 1997, p. 2.)

12. PACT2000 is an alliance between the NYYC, the challenger *Young America* and four partner clubs: Portland Yacht Club (Falmouth, ME), Bayview Yacht Club (Detroit, MI), Annapolis Yacht Club (Annapolis, MD) where Farr and Bowler are members, and St Petersburg Yacht Club (St Petersburg, FL).

13. Press release from the New York Yacht Club America's Cup Challenge, from the correspondence files of Bruce Farr & Associates Inc, Annapolis, Maryland, USA.

13. Russell Coutts and Paul Larsen, *Russell Coutts Course to Victory* (Auckland: Hodder Moa Beckett Publishers Limited, 1996), p. 154.

15. Ibid., p. 155.

16. The NYYC engages syndicates to represent it in America's Cup defences or challenges which in turn engage the designers, sailors, builders, sail- and spar-makers etc. The first non-USA-born yacht designer used by a syndicate engaged by the NYYC was Johan Valentijn. The Dutch-born Valentijn designed *Liberty* for the Freedom Campaign Syndicate of the State University of New York Maritime College at Ft Schuyler. *Liberty*, skippered by Dennis Conner, lost the America's Cup 4–3 to *Australia II* in 1983. Valentijn had previously worked with Ben Lexcen for Alan Bond.

17. The principal members of the PACT2000 design team are:
John Marshall, President and CEO — founded the Partnership for America's Cup Technology (PACT), a cooperative research organisation, in 1992, and is in his ninth America's Cup;
Duncan MacLane, Project Manager — five-times winner of the Little America's Cup and lead designer of the 1988 catamaran *Stars & Stripes*;
Dr 'Jerry' Milgram — MIT professor specialising in rig and sail aerodynamics;
Bruce Rosen — a Northrop Grumman Aviation colleague of Joe Laiosa's and developer of the SPLASH code, the world's leading computational fluid dynamic (CFD) software that predicts wave-length drag;
Joe Laiosa — an aero-hydro design and analysis consultant who, with Bruce Rosen, was part of the winning *Stars & Stripes* design team of 1987;
Dr Mark Drela — an associate professor in aeronautics at MIT specialising in aerodynamics and computational fluid dynamics;
Professor Horst Richter — professor, Dartmouth College Thayer School of Engineering, specialising in investigating all aerodynamic aspects of rig, sails and hull;
Olin Stephens — eight times winner-as-designer of the America's Cup;
James Teeters — ex Sparkman & Stephens, specialising in tank test analysis and VPPs;
Roberto Biscontini — aerospace engineer specialising in code developments to refine the VPPs;
Dirk Kramers — structural engineer specialising in composite construction;
Paul Bogataj — a former aerospace engineer at Boeing specialising in keel design as well as mast and sail design work;
Dr Robert Ranzenbach — developer of the first US-based off-wind sail testing capability, specialising in wind tunnel testing;

Rob Pallard — a leading naval architectural experimentalist, responsible for the tank-test programme;

Dr Alexis Mantzaris — with a PhD in marine hydrodynamics, he will be supervising the tank test programme;

Henry Elliot — ex *America³*, he is the builder of the one-third scale models for tank testing;

Ed Baird — ex Team New Zealand and 1995 World Match Champion of Match Racing is skipper of the NYYC/'Young America' Challenge;

Dave Hulse — manages the on-water data collection programme providing data analysis for the design, sailing and boat-building teams;

Ross Halcrow — ex Team New Zealand, is sail development manager and trimmer;

Steve Calder — a Soling Bronze Medallist, is principal sail designer;

Christopher Bedford — is a meteorologist who specialises in meso scale marine meteorology and forecasting and has provided forecasting services to the last four America's Cups;

Dick McCurdy — is a software writer and was a member of the winning *Stars & Stripes* team in 1987;

Dr William Unkel — a former MIT professor and mechanical engineer, his computer information display technology helped *America³* win the America's Cup in 1992. His Sail Vision technology will interface with sail design;

Hall Spars — from Bristol, R.I., are the makers of PACT2000's spars, rigging and sail handling equipment.

18. Larry Ellison is the founder and CEO of US software powerhouse Oracle Computer Company.

19. The *Big Boat* of the USA team was the Farr IMS 49 *Flash Gordon 3*, owned by Helmut Jahn of Chicago and steered by Ken Read. The second Farr boat in the USA team was the Farr Mumm 36 *Jameson*, leased off Tom Roche of Ireland and led by Chris Larson and Dee Smith.

20. The overall finishing positions of the Farr W60s in the 1998–99 Whitbread were: 1. *EF Language* (Farr Design 378) Paul Cayard (Sweden); 2. *Merit Cup* (Farr Design 390) Grant Dalton (Monaco); 3. *Swedish Match* (Farr Design 394) Gunnar Krantz; (Sweden) 4. *Innovation Kvaerner* (Farr Design 392) Knut Frostad (Norway); 5. *Silk Cut* (Farr Design 396) Lawrie Smith (UK); 6. *Chessie Racing* (Farr Design 386) George Collins and John Kostecki (USA); 7. *Toshiba* (Farr Design 382) Dennis Conner and Paul Standbridge (USA); 9. *EF Education* (Farr Design 378) Christine Gillou (Sweden). The 2002 round-the-world will be sponsored by Volvo.

21. Ivor Wilkins in 'The Winning Kiwi Combination' in *Sea Spray* (Auckland: Review Publishing Limited, February 1998), p. 25.

22. The two articles, both written by Suzanne McFadden, are: 'Champers on Ice as Crew Ready to Celebrate', which appeared in *The New Zealand Herald* on August 12 1998, Section C, and 'Offshore Drought Ends as NZ Clinch Kenwood Cup', which appeared in *The New Zealand Herald* on August 15–16, 1998, Section C. The three boats used by the winning New Zealand team were: Hideo Matsuda's Farr IMS 45 *Big Apple III*, Brett Neill's Farr-designed Cookson 12MT *WHITE CLOUD Stackerlee* and Shizue Kanbe's Farr ILC 40 *G'NET*.

23. New Zealand won the Kenwood Cup in 1986 with Michael Clark's Farr 40 *Exador*, Del Hogg's Farr 43 *Equity* and Don St Clair Brown's Farr 43 *Thunderbird*.

24. Of the estimated worldwide figure of finished Farr-Bowler product, approximately 35 percent is produced in Europe, 30 percent in the USA, 20 percent in New Zealand and the SW Pacific, and the balance in countries such as South Africa and Argentina.

25. Mick Cookson, 'Mick Cookson Interview No. 001' (unpublished, Auckland, 18 January 1996).

26. Michael Fay, 'Sir Michael Fay Interview No. 001' (unpublished, Auckland, 12 September 1995).

27. Ian Franklin in a letter to the author (unpublished, Christchurch, 8 August 1996).

28. Kim McDell, 'Kim McDell Interview No. 001' (unpublished, Kawau Island, 11 January 1996). The Thailand Sailing Academy commissioned McDell Marine Ltd to produce a small racing keel-boat after seeing the benefits of the Auckland Farr MRX fleet as trainers and one-design racers. Currently there are 50 Farr Platu 25s sailing in the Gulf of Thailand.

29. Steve Marten, 'Steve Marten Interview No. 001' (unpublished, Auckland, 11 and 12 December 1995).

30. From an interview with Marcus Hutchinson in 'The Experts' Expert: Racing Yacht Designers' in *Observer magazine* (GB: 17 February 1991), p 10.

31. Ibid.

32. Ibid.

33. Olin Stephens, 'Olin J. Stephens III Interview No. 001' (unpublished, 1 September 1995).

34. From an interview with Marcus Hutchinson in 'The Experts' Expert: Racing Yacht Designers' in *Observer magazine* (GB: 17 February 1991), p. 10.

35. The Royal Society of New Zealand is responsible for promoting scientific and technological achievements, and awards the Silver Medal in recognition of exceptional career contributions to science and technology.

36. Excerpts from the speech of New Zealand's Ambassador to the USA, John Wood, given at the New Zealand Embassy, Washington, D.C. on 8 February 1996.

Acknowledgements

A project of this kind is a cooperative effort, impossible to complete without the support and contribution of a great many people.

First, I would first like to express my grateful thanks to Reed Publishing (NZ) Ltd and staff – especially Peter Janssen for his encouragement when it mattered most, and Alison Southby for the many gifts she brought to bear on the project, but most of all for her acuity and patience over the last few months. I would also like to thank Ian Watt who picked up the idea and ran with it.

I owe a special word of thanks to my mother Joan, sister Sue and brother-in-law Rich who have been as steadfast in their love as in their practical support. As has my old friend, Gjoko Ruzio-Saban. Thanks, mate, for encouraging me with this book and bearing with me when I discovered the writing process and nearly burnt down your Bucks County farm in the process (twice)!

My thanks to two very special people — Jodi and Nathan — for your love and encouragement and for forgiving me my too-long absence from your lives.

A thank you to those who made possible my peripatetic life during the writing of this book by providing me with a roof over my head at various stages along the way: Jenny Baris-Wheeler, Russ and Lynda Bowler, Jeanette Brawn, Pene Brawn-Douglas, Tink and Debbie Chambers, John and Elizabeth Evans, Sue and Richard Gillard, Gjoko Ruzio-Saban, Audrey Smith, and especially Graham and Jenny Lawry and Rob Towner for the extended use of their seaside holiday home. I would also like to thank the Hahei community for making me feel at home in their magical part of the world.

I am indebted to those who read for me during the writing, who entered into debate, offered words of advice and encouragement and generally helped me distinguish the waves from the water: Gary ('Chainsaw') Baigent, Pene Brawn-Douglas, Graham Lawry, Russ McLachlan, Gjoko and Sinda Ruzio-Saban, Brian Stephenson.

A thank you too to Aussie Malcolm for a wide-ranging discussion that set the tenor for the early part of *Shape*. Thanks also to Jenny McLachlan for helping me with the transcribing load, Mike Miron and the Eastport Historical Society, Nicole Smale for translation, Brian Stephenson for research, and Trikitiki for handling my email correspondence. And a special thank you to Bobbi Hobson and Jennifer Charnesky of Farr Yacht Design for their professional and cheerful assistance over the last four years.

Perhaps more than any other sport, yachting incites a passion that once it has taken a grip rarely ever lets go. I think that is true for most of my interviewees. To you my sincere thanks for your contributions:

David Barnes, Peter Beaumont, Harry Bioletti, Rob and Marilyn Blackburn, Chris Bouzaid, Russ and Lynda Bowler, Paul Cayard, Tink Chambers, Keith Chapman, Iris Chitty, David Clarke, Sir Tom Clark, Mick Cookson, Don Cowie, Murray Crockett, Cathie Cross, Grant Dalton, Laurie Davidson, Roy Dickson, Rick Dodson, Bruce and Gail Farr, Jim and Ileene Farr, Odile Farr, Pam Farr, Sir Michael Fay, Pierre Fehlmann, Ian Franklin, Ian Gibbs, Bobbi Hobson, Carol Horvath, Roger and Christine Hill, Ron Holland, Jeanette and Bob Holley, Ena Hutchinson, Peter and Denyse Hutchinson, Gary Jobson, Andrew Johns, Evan Julian, Denis Kendall, Einar Koefoed, Peter Lester, Silvana Lievi, Don Lidgard, Jim Lidgard, Lindsay Lovegrove, Kim McDell, Ann McGlashan, Don and Katharine McGlashan, Barbara McMahon, Chris McMullen, Aussie Malcolm, Steve Marten, Ewan Matheson, Andie Ogilvie, Kaoru Ogimi, Clay Oliver, Mark Paterson, Mick Price, Jacqui Parks, Murray Ross, Tom Schnackenberg, Tim Shadbolt, Pat Shaughnessy, Peter Shaw, Eric Simian, John Spencer, Geoff and Mary Stagg, Olin Stephens, Val Stern, Harunobu Takeda, Robyn Varcoe, Hal Wagstaff, Tom Whidden, Jean and Dinny White, Penny Whiting, Chris Wilkins, Graham Williams, Bevan Woolley and Jim Young.

A special thank you to Peter Beaumont for giving me the words for the title. (And a 'sorry' to Jeremy Lawry for just missing out.)

I would also like to thank those who have kindly allowed me access to their bookshelves, albums, videos and scrapbooks — Peter Beaumont, Russ Bowler, Keith Chapman, Bruce Farr, Jim and Ileene Farr, Don Lidgard, Don and Katharine McGlashan, Mark Paterson, Geoff Stagg — and the staff at Takapuna Library. And for the colour and excitement you have added to *Shape* by allowing the use of your pics, a thank you to Farr Yacht Design, Jim Lidgard, Steve Marten, Franco Pace, PPL Oxbow Ltd and Ivor Wilkins.

Shape would have been an impossible task without the support of its subjects. In the midst of their very busy lives and the unrelenting pressure of their business, Bruce and Russell gave often and generously of their most precious commodity: time. A huge thank you to you both. It was through your help and encouragement that the challenge became an experience, a privilege more than a project.

Lastly, I want to thank Debbie. If you hadn't entered my life, this book would not have been written.

John Bevan-Smith
July 1999